The Complex Connection between Cannabis and Schizophrenia

The Complex Connection between Cannabis and Schizophrenia

Edited by

Michael T. Compton

Marc W. Manseau

Academic Press is an imprint of Elsevier
125 London Wall, London EC2Y 5AS, United Kingdom
525 B Street, Suite 1800, San Diego, CA 92101-4495, United States
50 Hampshire Street, 5th Floor, Cambridge, MA 02139, United States
The Boulevard, Langford Lane, Kidlington, Oxford OX5 1GB, United Kingdom

Library of Congress Cataloging-in-Publication Data
A catalog record for this book is available from the Library of Congress

British Library Cataloguing-in-Publication Data
A catalogue record for this book is available from the British Library

ISBN: 978-0-12-804791-0

Publisher: Nikki Levy
Acquisition Editor: Joslyn Chiaprasert-Paguio
Editorial Project Manager: Timothy Bennett
Production Project Manager: Anusha Sambamoorthy
Cover Designer: Miles Hitchen

Typeset by SPi Global, India

Contents

Contributors

Numbers in parentheses indicate the pages on which the authors' contributions begin.

Jeff K. Abbott (37), Northwestern University, Chicago, IL, United States

David L. Atkinson (37), University of Texas-Southwestern, Dallas, TX, United States

Peter Bosanac (289), St Vincent's Hospital, Melbourne, VIC, Australia

Beth Broussard (247), Lenox Hill Hospital, New York, NY, United States

John D. Cahill (75), Yale School of Medicine; Connecticut Mental Health Center, New Haven, CT, United States

David J. Castle (289), St Vincent's Hospital, Melbourne, VIC, Australia

Angelo B. Cedeño (157), Lenox Hill Hospital, New York, NY, United States

Michael T. Compton (1), Columbia University College of Physicians and Surgeons, New York, NY, United States

Wilson M. Compton (9), National Institute on Drug Abuse, National Institutes of Health, Bethesda, MD, United States

Jose Cortes-Briones (75), Yale School of Medicine, New Haven; VA Connecticut, West Haven, CT, United States

Serena Deiana (309), Boehringer Ingelheim Pharma GmbH & Co. KG, Biberach an der Riss, Germany

Deepak C. D'Souza (75), Yale School of Medicine, New Haven; Connecticut Mental Health Center, New Haven; VA Connecticut, West Haven, CT, United States

Papanti Duccio (199), Drug Addiction Centre, Latisana, Italy; University of Hertfordshire, Hatfield, United Kingdom

Anne Eden Evins (221), Massachusetts General Hospital; Harvard Medical School, Boston, MA, United States

Schifano Fabrizio (199), University of Hertfordshire, Hatfield, United Kingdom

Ana Fresán (271), Instituto Nacional de Psiquiatría Ramón de la Fuente Muñíz, Mexico City, Mexico

Jodi M. Gilman (221), Massachusetts General Hospital; Harvard Medical School, Boston, MA, United States

Swapnil Gupta (75), Yale School of Medicine; Connecticut Mental Health Center, New Haven, CT, United States

Beth Han (9), Substance Abuse and Mental Health Services Administration, Rockville, MD, United States

Orsolini Laura (221), University of Hertfordshire, Hatfield, United Kingdom; Polyedra Research Group, Teramo, Italy

Ana Lusicic (289), Orygen Youth Health, Melbourne, VIC, Australia

Marc W. Manseau (1), New York University School of Medicine, New York, NY, United States

Patrick McGorry (167), Orygen, The National Centre of Excellence in Youth Mental Health; University of Melbourne, Melbourne, VIC, Australia

Meredith McHugh (167), Orygen, The National Centre of Excellence in Youth Mental Health; University of Melbourne, Melbourne, VIC, Australia

Barnaby Nelson (167), Orygen, The National Centre of Excellence in Youth Mental Health; University of Melbourne, Melbourne, VIC, Australia

Brian O'Donoghue (167), Orygen, The National Centre of Excellence in Youth Mental Health; University of Melbourne, Melbourne, VIC, Australia

Luca Pauselli (183), University of Perugia, Perugia, Italy; Columbia University Medical Center; New York State Psychiatric Institute, New York, NY, United States

Rajiv Radhakrishnan (75), Yale School of Medicine, New Haven; Connecticut Mental Health Center, New Haven; VA Connecticut, West Haven, CT, United States

Claire Ramsay Wan (247), Cambridge Health Alliance, Somerville, MA, United States

Rebeca Robles-García (271), Instituto Nacional de Psiquiatría Ramón de la Fuente Muñíz, Mexico City, Mexico

Mohamed Sherif (75), Yale School of Medicine, New Haven; VA Connecticut, West Haven, CT, United States

Sara M. Sobolewski (221), Massachusetts General Hospital; Harvard Medical School, Boston, MA, United States

Nadia Solowij (129), University of Wollongong, Wollongong, NSW, Australia

Carlos-Alfonso Tovilla-Zarate (271), Universidad Juárez Autónoma de Tabasco, Comalcalco, Mexico

Erica Zamberletti (309), University of Insubria, Busto Arsizio, Italy

Chapter 1

An Introduction to the Complex Connection Between Cannabis and Schizophrenia

Marc W. Manseau*, Michael T. Compton†
*New York University School of Medicine, New York, NY, United States
†Columbia University College of Physicians and Surgeons, New York, NY, United States

BACKGROUND AND HISTORY

The *Cannabis sativa* plant has two main subspecies, *sativa* and *indica*, and contains more than 400 compounds, with approximately 80 of these classified as *cannabinoids* (Borgelt, Franson, Nussbaum, & Wang, 2013). People most often smoke parts of the plant to obtain psychoactive effects of the drug, but it can also be vaporized and inhaled or ingested orally in multiple forms (oil infusions, edible products, etc.). Having been cultivated for thousands of years, there are hundreds of different strains of *Cannabis sativa* (hereafter referred to as "cannabis"), making for a wide breadth of cannabinoid blends with corresponding subtly varying psychoactive effects when ingested. However, the principal cannabinoids thought to be responsible for psychoactive effects are Δ^9-tetrahydrocannabinol (THC) and cannabidiol (CBD). THC is responsible for most psychoactive effects from cannabis use, whereas CBD does not produce intoxication but may counteract some effects of THC. Many differences between strains of cannabis are likely explained by variation in THC and CBD content, as well as the ratio between the two cannabinoids, and there is evidence that the THC-to-CBD ratio has been increasing in cannabis strains over the past few decades (ElSohly et al., 2016).

Cannabis is the most widely used illicit drug in the world, with about 13.1 million people (prevalence rate = 0.19%) globally meeting criteria for cannabis dependence in 2010, and with prevalence rates being highest among young adult males in higher income countries (Degenhardt et al., 2013). Furthermore, cannabis use rates have recently been stable on average worldwide, with about 3.8% of people estimated to have used cannabis in 2014, but use has generally been rising in the Americas over the past decade, following a long period of decline since the 1970s (United Nations Office on Drugs and Crime, 2016).

The Complex Connection between Cannabis and Schizophrenia. http://dx.doi.org/10.1016/B978-0-12-804791-0.00001-X

While it is not entirely clear what has contributed to recent increasing cannabis use in some global regions, changing social norms and legal status around cannabis use may be at least partially implicated. Since the 1961 United Nations' Single Convention on Narcotic Drugs (United Nations, 1962), cannabis cultivation, sale, and use has been illegal in most countries. In fact, suppressing cannabis production and distribution has been a major priority for the Drug Enforcement Administration's (DEA) efforts in the "War on Drugs" within the United States (US) (US Drug Enforcement Administration, 2017). However, more recently, many countries and states of the US have moved to decriminalize (reduce or remove criminal penalties) or legalize (remove criminal penalties and establish a system for taxation and regulation of production and distribution) recreational cannabis, and/or approve its legal use for medical purposes. Specifically, as of 2017, at least 40 countries and eight states of the US as well as the District of Columbia (DC) have removed some legal penalties for recreational cannabis use (covering more than 20% of the US population), while 29 states of the US and DC have allowed some form of medical cannabis use (Carnevale, Kagan, Murphy, & Esrick, 2017). However, many still strongly oppose legal cannabis access and use, and governments continue to restrict cannabis production and distribution, creating enormous controversy and ambivalence. For instance, cannabis remains classified as a schedule I substance (high potential for abuse; no currently accepted medical treatment use) by the Food and Drug Administration (FDA) and DEA on the federal level within the US. In 2013, the US Department of Justice under the Obama Administration decided not to pursue legal action against physicians "prescribing" medical cannabis in states that allow it, a decision that could be reversed under a new Presidential Administration. Presumably, recent shifts in legal regulation have been driven at least in part by changing public opinion toward cannabis use. Indeed, there is evidence within the US that attitudes are increasingly favoring arguments about the potential societal benefits of cannabis legalization over messages about public health risks (McGinty, Niederdeppe, Heley, & Barry, 2017).

One reason for shifting attitudes about cannabis use could be related to changing perceptions about the availability of cannabis or risk of its use. For instance, there is evidence that state-level medical cannabis laws have led to perceptions of increased availability, and in turn to higher use, of cannabis among adults in the US (Martins et al., 2016). It is likely that increasing legalization efforts affect public perceptions and vice versa, creating political and social dynamics that support removing restrictions to cannabis access and increase use rates within many societies. However, there is strong reason for concern that cannabis use is not without substantial risks. Putting aside the physical health risks of inhaling cannabis smoke, there is evidence that cannabis exposure carries mental health risks, especially when used by youth or used heavily. For instance, when used regularly and/or in adolescence, cannabis use has been convincingly linked to cognitive and motivational problems, lower educational attainment, unemployment, use of more dangerous substances, and a range of psychiatric symptoms. In addition, for those who have already developed psychiatric disorders (e.g., mood, anxiety, or psychotic disorders), cannabis use may be associated with poorer engagement in care, worse symptom control, and lower psychosocial functioning (Agosti, Nunes, & Levin, 2002; Fergusson, Boden, & Horwood, 2015; Wilkinson, Stefanovics, & Rosenheck, 2015). With this said, it must be recognized that many adults use cannabis sporadically or even regularly without significant detriment to their mental health.

CANNABIS AND PSYCHOSIS

What about the topic of this book, the connection between cannabis use and psychosis? The relationship is indeed complex, and with a long and storied history. The cult classic film from the late 1930s, "Reefer Madness," dramatized (and spurred on) longstanding public fear about the connection between cannabis use and psychosis. While controversial, such fear around this perceived connection may have contributed to increasing efforts to restrict legal access to cannabis (Barcott, 2015). Partially driven by a recognition that THC administration can transiently cause many symptoms of schizophrenia (i.e., positive, negative, and cognitive symptoms), the scientific literature over the past half-century has paid increasing attention to the role of cannabis use in the development of psychotic disorders.

Chapters 10 and 11 address, respectively, two core aspects of the relationship between cannabis use and psychosis: whether cannabis use *causes* psychosis-spectrum disorders and whether it leads to an earlier age at onset of psychosis (AOP). Specifically, Chapter 10 discusses the literature on cannabis use as an independent, causal risk factor in the development of schizophrenia and related psychotic disorders. While many studies have found a consistent and robust association between cannabis use and the development of psychotic disorders, the exact nature of this relationship has been controversial, with competing arguments for direct causation, reverse causation, and confounding (Manseau & Goff, 2015). Gilman and colleagues carefully and systematically review the literature on this topic, making a strong argument that cannabis use is an important component cause in the development of psychotic disorders, particularly among those who are biologically vulnerable to psychosis. While it is impossible to definitively demonstrate a causal relationship between cannabis use and schizophrenia in a controlled experiment, the authors demonstrate that the following causal criteria have been established in the literature: a strong and consistent association between cannabis use and psychosis; cannabis use commonly precedes the development of psychosis; there appears to be a dose-response relationship; confounding factors can be largely ruled out as entirely explaining the relationship; and there is a plausible biological mechanism for the relationship between cannabis use and the development of psychotic disorders.

Chapter 11 covers the relationship between cannabis use and AOP. Similar to the discussion on cannabis use as an independent risk factor in the development of psychotic disorders, the authors review the substantial literature showing a replicated, significant association between cannabis use and a younger AOP. The authors argue that cannabis use likely causes an earlier AOP by demonstrating that cannabis use often precedes the onset of psychotic symptoms (and indeed often prodromal symptoms), that there is a dose effect of cannabis use on AOP, and that while certain vulnerability factors do exist that predispose individuals to an earlier AOP in the setting of cannabis use, none of them likely completely confound the relationship between cannabis use and AOP.

Aside from the discussion of the causation of schizophrenia and related disorders, cannabis is the most commonly used illicit substance among individuals with schizophrenia, and there is substantial evidence that its use in this population is associated with worsening psychotic symptoms, illness relapse, and decreased functioning over time (Clausen et al., 2014; Radhakrishnan, Wilkinson, & D'Souza, 2014). Chapter 12 addresses this very matter. That chapter describes the prevalence of cannabis use and cannabis use disorders among those with psychotic disorders, and explores the effects of cannabis use on the course and outcomes of psychosis-spectrum illnesses.

BRIEF OVERVIEW OF THE BOOK

The remainder of the chapters in this book thoroughly explore various aspects of the relationship between cannabis and psychosis, from epidemiology to biological mechanisms to treatment, thereby providing important context to the core issues of causation and effects on the course of these devastating illnesses. Compton and Han address the epidemiology of cannabis use in the US in Chapter 2. This chapter is unique in that it reports the results of an original analysis of data from the 2002–2014 National Surveys on Drug Use and Health. It shows that cannabis use and use disorder rates have actually been decreasing recently among youth in the US, but increasing among adults. The trends, which will be interesting to continue to follow, have important public health implications related to the effects of tobacco use control efforts and perceived risk of cannabis on cannabis use rates.

Chapter 3 reviews the biological effects of cannabinoids on the brain. The chapter includes an in-depth discussion of the natural endocannabinoid system, which has fascinating and important implications for psychosis, both in the context of cannabis use and entirely on its own (Muller-Vahl & Emrich, 2008; Manseau & Goff, 2015). The chapter also covers the biology of addiction as it applies to cannabis use, and the effects of cannabinoids on a wide array of brain functions. Chapter 4 provides a comprehensive overview of the effects of THC in the laboratory setting, by systematically reviewing the large literature on the psychoactive and cognitive effects of THC and related cannabinoids in human laboratory studies.

Chapter 5 reviews the psychotomimetic and cognitive effects of cannabis use in the general (i.e., nonclinical) population. While cannabis use is an important risk factor for the development of schizophrenia, the fact remains that a small proportion of people who use cannabis develop a full-threshold psychotic disorder. Solowij compiles evidence supporting the notion that cannabis use may be linked to transient and/or subclinical psychotic-like symptoms, cognitive deficits, and related functional impairment in a portion of the population. Chapter 6 provides an overview of the literature connecting cannabis use to schizotypy, thereby providing further evidence for the relationship between cannabis use and psychotic-like outcomes in the wider population of people not diagnosed with psychotic disorders. The discussion of the impact of cannabis use within populations at mounting risk for psychosis progresses further with Chapter 7, which addresses the effects of cannabis use in those at ultra-high risk (UHR) for psychosis. In this chapter, the authors review the effects of cannabis use on the illness course and outcomes of those at UHR for psychosis, including research on whether and how cannabis use influences the likelihood of progression to a full-threshold psychotic disorder.

Chapter 8 describes the scientific literature on cannabis-induced psychotic disorders, providing further evidence that cannabis use can cause full-threshold psychotic syndromes, even if only transiently in some users. In this chapter, the author provides a broad overview of this topic, from epidemiology to diagnosis to management approaches. Chapter 9 addresses a new class of cannabinoid compounds, synthetic cannabinoids (SCs). In this chapter, the authors detail the emergence of SC use, which some have called an epidemic, and describe how and why SC use may lead to more immediate and serious psychotic symptoms than natural cannabis use.

Chapters 10–12—on cannabis use as a component cause, a risk factor for earlier onset, and a poor prognostic indicator in the context of schizophrenia and related psychotic disorders—have already been introduced above. Chapter 13 addresses treatment approaches to cannabis use disorder among individuals with psychotic disorders. In this chapter, the authors discuss principles of pharmacological and psychosocial management of patients with comorbid psychotic and cannabis use disorders. However, recognizing the paucity of effective, evidence-based approaches to this all-too-common and impairing combination of disorders, they also discuss the critical research necessary for progress in the field. Finally, Chapter 14 discusses the exciting potential of CBD as a novel treatment for schizophrenia and related disorders. While recognizing that conclusions in the field remain preliminary, they delineate both animal and human studies that support the possible use of CBD as an effective, safe, and well-tolerated antipsychotic compound. They also discuss legal barriers to further research on cannabinoids in general and CBD specifically.

CONCLUSIONS

Cannabis sativa is the most commonly used illicit drug globally. In recent years, its possession is being increasingly decriminalized and even legalized for both medical and recreational use in the US and worldwide. Among the mental health concerns related to the drug, there is mounting evidence of an intricate link between cannabis use and schizophrenia and related psychotic disorders. Cannabis use is more prevalent among people with schizophrenia than in the general population; young people who use cannabis are twice as likely to develop schizophrenia; premorbid cannabis use appears to hasten the onset of psychosis among those who develop a psychotic disorder; and cannabis use can induce schizophrenia-like symptoms in otherwise healthy individuals. At the same time, there is promising evidence to suggest that CBD, one of the many cannabinoid compounds found in the *Cannabis sativa* plant, could prove to be an effective antipsychotic to treat schizophrenia.

This book provides an in-depth overview of the current state of knowledge on the role that cannabis plays in psychotic symptoms, psychotic disorders, and schizophrenia, covering both pathophysiological and pharmacological implications. It addresses wide-ranging topics including the epidemiology of cannabis use in the US, the risks associated with its use, the biological aspects of the drug and its effects on the brain, management of cannabis use disorder, and the pharmacological possibilities of using CBD to treat psychotic disorders.

Key Chapter Points

- The *Cannabis satvia* plant contains many cannabinoids, including the principal psychoactive component, THC, as well as CBD.
- Cannabis is the most widely used illicit drug in the world, with over 13 million people globally meeting the criteria for cannabis dependence recently, and with prevalence rates being the highest among young adult males in higher income countries. Cannabis use rates have recently been stable on average worldwide, but use has generally been rising in the Americas over the past decade.
- There is mounting evidence of an intricate link between cannabis use and schizophrenia and related psychotic disorders. Cannabis use is more prevalent among people with schizophrenia than in the general population; young people who use cannabis are twice as likely to develop schizophrenia; premorbid cannabis use appears to hasten the onset of psychosis among those who develop a psychotic disorder; and cannabis use can induce schizophrenia-like symptoms in otherwise healthy individuals.

REFERENCES

Agosti, V., Nunes, E., & Levin, F. (2002). Rates of psychiatric comorbidity among U.S. residents with lifetime cannabis dependence. *American Journal of Drug and Alcohol Abuse*, *28*, 643–652.

Barcott, B. (2015). *Weed the people: The future of legal Marijuana in America*. New York, NY: Time.

Borgelt, L. M., Franson, K. L., Nussbaum, A. M., & Wang, G. S. (2013). The pharmacologic and clinical effects of medical cannabis. *Pharmacotherapy*, *33*, 195–209.

Carnevale, J. T., Kagan, R., Murphy, P. J., & Esrick, J. (2017). A practical framework for regulating for-profit recreational marijuana in US States: Lessons from Colorado and Washington. *The International Journal on Drug Policy*, 1–15. EPub.

Clausen, L., Hjorthoj, C. R., Thorup, A., Jeppesen, P., Petersen, L., Bertelsen, M., et al. (2014). Change in cannabis use, clinical symptoms and social functioning among patients with first episode psychosis: a 5-year follow-up study of patients in the OPUS trial. *Psychological Medicine*, *44*, 117–126.

Degenhardt, L., Ferrari, A. J., Calabria, B., Hall, W. D., Norman, R. E., McGrath, J., et al. (2013). The global epidemiology and contribution of cannabis use and dependence to the global burden of disease: results from the GBD 2010 study. *PLoS ONE*, *8*, e76635.

ElSohly, M. A., Mehmedic, Z., Foster, S., Gon, C., Chandra, S., & Church, J. C. (2016). Changes in cannabis potency over the last 2 decades (1995-2014): Analysis of current data in the United States. *Biological Psychiatry*, *79*, 613–619.

Fergusson, D. M., Boden, J. M., & Horwood, L. J. (2015). Psychosocial sequelae of cannabis use and implications for policy: Findings from the Christchurch Health and Development Study. *Social Psychiatry and Psychiatric Epidemiology*, *50*, 1317–1326.

Manseau, M. W., & Goff, D. C. (2015). Cannabinoids and schizophrenia: Risks and therapeutic potential. *Neurotherapeutics*, *12*, 816–824.

Martins, S. S., Mauro, C. M., Santaella-Tenorio, J., Kim, J. H., Cerda, M., Keyes, K. M., et al. (2016). State-level medical marijuana laws, marijuana use and perceived availability of marijuana among the general U.S. population. *Drug and Alcohol Dependence*, *169*, 26–32.

McGinty, E. E., Niederdeppe, J., Heley, K., & Barry, C. L. (2017). Public perceptions of arguments supporting and opposing recreational marijuana legalization. *Preventive Medicine*, *99*, 80–86.

Muller-Vahl, K. R., & Emrich, H. M. (2008). Cannabis and schizophrenia: towards a cannabinoid hypothesis of schizophrenia. *Expert Review of Neurotherapeutics*, *8*, 1037–1048.

Radhakrishnan, R., Wilkinson, S. T., & D'Souza, D. C. (2014). Gone to pot—a review of the association between cannabis and psychosis. *Frontiers in Psychiatry*, *5*, 54.

United Nations. (1962). *Single convention on narcotic drugs, 1961*. New York, NY: United Nations.

United Nations Office on Drugs and Crime. (2016). *World Dug Report 2016, Chapter 1: Cannabis*. Vienna: United Nations.

US Drug Enforcement Administration. (2017). *Domestic cannabis eradication/suppression program*. Washington, DC: US Drug Enforcement Administration. Available at: https://www.dea.gov/ops/cannabis.shtml Accessed 18.04.17.

Wilkinson, S. T., Stefanovics, E., & Rosenheck, R. (2015). Marijuana use is associated with worse outcomes in symptom severity and violent behavior in patients with posttraumatic stress disorder. *Journal of Clinical Psychiatry*, *76*, 1174–1180.

Chapter 2

The Epidemiology of Cannabis Use in the United States*

Wilson M. Compton*, Beth Han†
**National Institute on Drug Abuse, National Institutes of Health, Bethesda, MD, United States*
†Substance Abuse and Mental Health Services Administration, Rockville, MD, United States

INTRODUCTION

Cannabis is the most commonly used federally illicit substance in the United States (US) and has become even more common than cigarette use among youth (SAMHSA, 2015; Lopez, Compton, & Volkow, 2009). Perhaps a consequence of or a contributor to the changes in prevalence, laws and policies related to cannabis have been shifting markedly in the US over the past 20 years. Legalization for medical purposes has been adopted by 24 states and the District of Columbia (Pacula & Sevigny, 2014; Roy-Byrne et al., 2015; Hasin, Wall, et al., 2015, Hasin, Saha, et al., 2015), and several jurisdictions have legalized nonmedical cannabis (Roy-Byrne et al., 2015). Simultaneously, the potency of cannabis has been increasing markedly over the past 20 years in the US, with the average Δ^9-tetrahydrocannabinol (THC) content in cannabis from police seizures across the country increasing from 6.35% in 2002 to 13.8% in 2014 (ElSohly, 2015). Given these changes and the adverse effects of cannabis use (Silins et al., 2014; Volkow, Baler, Compton, & Weiss, 2014), understanding the detailed patterns of cannabis use and use disorders among both youth (including school dropouts) and adults in the US, and how they have changed over time, is essential.

To provide a framework for understanding the epidemiology of cannabis use in the US, this chapter first summarizes data on national trends in cannabis use and use disorders from 2002 through 2014 among youth aged 12–17 and among adults aged 18 or older in the US, using data from the 2002–14 National Surveys on Drug Use and Health (NSDUHs). The NSDUHs are large, nationally representative surveys conducted by the Substance Abuse and Mental

*Disclaimers: The findings and conclusions of this study are those of the authors and do not necessarily reflect the views of the National Institute on Drug Abuse of the National Institutes of Health, the Substance Abuse and Mental Health Services Administration, or the US Department of Health and Human Services.

The Complex Connection between Cannabis and Schizophrenia. http://dx.doi.org/10.1016/B978-0-12-804791-0.00002-1

Health Services Administration (SAMHSA) annually. In addition to describing trends, this chapter examines sociodemographic characteristics and other factors (e.g., risk perceptions of cannabis use, tobacco use, and alcohol use) that may drive these trends among youth and adults. Finally, we discuss the implications of the trends and correlates of cannabis use and use disorders among youth and adults. Unlike other chapters, this chapter presents new, primary data analyses and interpretations, rather than an exclusive review of the existing research reported in the literature.

DATA SOURCES AND METHODS

Data Sources

We examined data from people aged 12 or older who participated in the 2002–14 NSDUHs. Each NSDUH provides nationally representative data on cannabis use and use disorders (in addition to other drugs, alcohol, and tobacco) among the US civilian, noninstitutionalized population aged 12 or older. Excluded from the survey are persons without a household address (e.g., homeless persons not living in shelters), active-duty military, and institutional residents. The NSDUH data collection protocol was approved by the Institutional Review Board at RTI International.

The key advantages of using NSDUH data include the consistent survey design, methodology, and questionnaire content, and large sample sizes, allowing the sensitive detection of changes in cannabis use trends across every year during 2002–14 (Han, Compton, Jones, & Cai, 2015; Grucza, Agrawal, Krauss, Cavazos-Rehg, & Bierut, 2016). The annual mean weighted response rate of the 2002–14 NSDUHs was 66.0% (SAMHSA, 2016) according to the definition of Response Rate 2 by the American Association for Public Opinion Research (AAPOR, 2015). Details regarding NSDUH methods are provided elsewhere (SAMHSA, 2016).

Measures

NSDUH collected 12-month use of tobacco, alcohol, cannabis, cocaine, hallucinogens, heroin, and inhalants; 12-month nonmedical use of prescription opioids, sedatives, and stimulants; and past-month heavy alcohol use (drinking 5 or more drinks on the same occasion on each of 5 or more days in the past 30 days) from all respondents. Past-year cannabis users were asked to state the number of days they used cannabis. "Daily or near daily users" were those reporting on average using ≥5 days per week, ≥20 days per month, or ≥240 days in the past year (SAMHSA, 2016). For persons reporting cannabis use, NSDUH collected the source of last used cannabis.

NSDUH also collected perceived risk of smoking cannabis once or twice a week (hereafter called "perceived risk of smoking cannabis"); perceived parental strong disapproval of using cannabis once a month or more (for youth respondents only); perceived peer's strong disapproval of using cannabis once a month or more (for youth respondents only); having talked to parents about the dangers of tobacco, alcohol, and drugs (yes/no, for youth respondents only); perceived state legalization of medical cannabis use (whether respondents think that medical cannabis use is legalized in their residing state); perceived ease of cannabis availability (fairly or very easy to obtain cannabis); and age at first cannabis use. Based on state and year information, we also created a variable indexing state legalization of commercial sales or personal possession. Note that the actual state legal policy may be relevant for some analyses. However, our results show that for our examined outcomes (personal consumption patterns), perceived state legalization of medical cannabis use is conceptually more relevant and statistically stronger than the actual state medical cannabis legalization.

NSDUH estimated past 12-month major depressive episode (MDE) and each specific substance use disorder (dependence on or abuse of alcohol, cannabis, cocaine, hallucinogens, heroin, inhalants, or nonmedical use of prescription opioids, sedatives, or stimulants) based on the assessments of individual diagnostic criteria from the *Diagnostic and Statistical Manual of Mental Disorders, Fourth Edition* (DSM-IV) (American Psychiatric Association, 1994). Nicotine dependence among cigarette smokers was assessed using the Nicotine Dependence Syndrome Scale (Shiffman, Waters, & Hickcox, 2004). These measures have good validity and reliability (Grucza, Abbacchi, Przybeck, & Gfroerer, 2007; Jordan, Karg, Batts, Epstein, & Wiesen, 2008; SAMHSA, 2010). Sociodemographic characteristics included age, gender, race/ethnicity, educational attainment, employment status, marital status (adults only), health insurance, metropolitan statistical area, census region, and year.

Statistical Analyses

For each examined year, we estimated the 12-month prevalence of cannabis use and use disorders, daily or near-daily use, prevalence of risk perceptions of smoking cannabis, perceived parental and peer's strong disapproval of using cannabis, perceived state legalization of medical cannabis use, and the mean number of days of cannabis use. For percentage estimates, bivariable logistic regression models were applied to: (1) estimate prevalence, (2) test for differences (Bieler, Brown, Williams, & Brogan, 2010; RTI, 2015) between estimates for 2002 and each year during 2003–14 (two-sided t-test with a significance level of 0.05), and (3) test p values of beta coefficients of the year variable. For mean numbers of days of cannabis use, linear regression models were applied to examine differences between estimates for 2002 and each year during 2003–14 (two-sided t-test with a significance level of 0.05) and to test p values of beta coefficients of the year variable.

Bivariable and multivariable logistic regression modeling were applied to assess unadjusted and adjusted relative risks (Bieler et al., 2010; RTI, 2015) for cannabis use among youth and for cannabis use disorders among cannabis users. Because MDE was unavailable in the 2002–03 NSDUH (SAMHSA, 2012), separate multivariable models were conducted for 2004–14 with this additional variable included and for 2002–14 without it. This study used SUDAAN software (RTI, 2015) to account for the complex sample design and sampling weights of NSDUH. Multicollinearity and potential interaction effects between examined factors were assessed and were not identified in final multivariable models.

EPIDEMIOLOGY OF CANNABIS USE AND USE DISORDERS AMONG YOUTH

Trends in Cannabis, Tobacco, and Alcohol Use Among Youth

Based on 288,300 sampled people aged 12–17 from the 2002–14 NSDUHs, the prevalence of cannabis use among youth decreased from 15.8% in 2002 to 13.1% in 2014 (absolute difference −2.7%, $p < 0.0001$) (Fig. 1). The prevalence of cannabis use disorders among youth decreased from 4.3% in 2002 to 2.7% in 2014 (absolute difference −1.6%, $p < 0.0001$). The prevalence of tobacco use decreased from 23.6% in 2002 to 12.7% in 2014 (absolute difference −10.9%, $p < 0.0001$). The prevalence of alcohol use decreased from 34.6% in 2002 to 24.0% in 2014 (absolute difference −10.6%, $p < 0.0001$).

Trends in Risk Perceptions of Smoking Cannabis and Perceived Cannabis Availability Among Youth

The prevalence of perceiving no risk of smoking cannabis increased from 5.0% in 2002 to 12.8% in 2014 (absolute difference 7.8%, $p < 0.0001$) (Fig. 1). The prevalence of perceiving great risk of smoking cannabis decreased from 51.5% in 2002 to 37.4% in 2014 (absolute difference −14.1%, $p < 0.0001$; data not shown: see Han, Compton, Jones, & Blanco, 2017). Perceived parental strong disapproval of using cannabis decreased from 92.0% in 2002 to 90.0% in 2014 (absolute difference −2.0%, $p < 0.0001$). The prevalence of perceived availability decreased from 55.0% in 2002 to 47.8% in 2014 (absolute difference −7.2%, $p < 0.0001$), suggesting that cannabis was perceived to be less available by youth in 2014 than in 2002.

Trends in Cannabis Use Disorders and Perceived Risk of Smoking Cannabis Among Youth Cannabis Users

Among cannabis users, the prevalence of cannabis use disorders decreased from 27.0% in 2002 to 20.4% in 2014 (absolute difference −6.6%, $p < 0.0001$). The prevalence of perceiving no risk of smoking cannabis increased from 17.4% in 2002 to 47.4% in 2014 (absolute difference 30.0%, $p < 0.0001$). The prevalence of perceiving great risk of smoking cannabis decreased from 15.8% in 2002 to 5.9% in 2014 (absolute difference −9.9%, $p < 0.0001$; data not shown: see Han et al., 2017). The prevalence of perceiving parental strong disapproval of using cannabis decreased from 74.4% in 2002 to 63.2% in 2014 (absolute difference −11.2%, $p < 0.0001$).

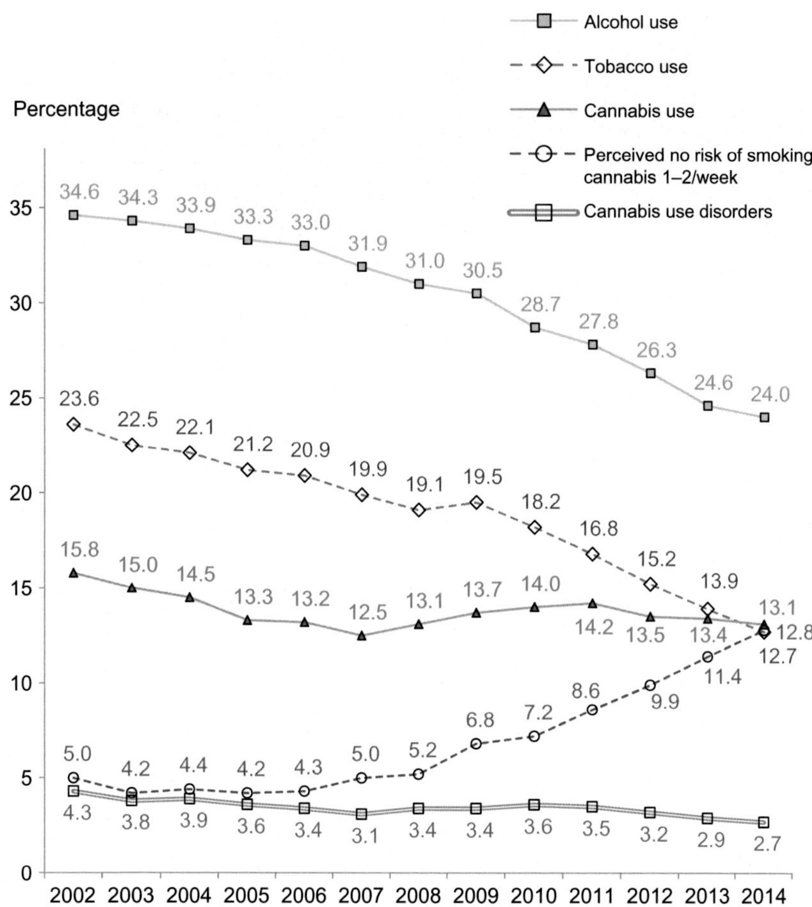

FIG. 1 Twelve-month prevalence of cannabis use, cannabis use disorders, tobacco use, alcohol use, and perceived no risk of smoking cannabis once or twice a week among youth in the US: 2002–14.

Trends in and Associations of Cannabis Use Among Youth Tobacco and Alcohol Users

Among tobacco users, the prevalence of cannabis use increased from 51.9% in 2002 to 57.1% in 2014 (absolute difference 5.2%, $p<0.0001$; data not shown: see Han et al., 2017). Among alcohol users, the prevalence of cannabis use increased from 40.5% in 2002 to 43.0% in 2014 (absolute difference 2.5%, $p<0.0001$).

Bivariable logistic regression results showed that youth were less likely to use cannabis during 2004–14 than in 2002 (unadjusted relative risks (URRs) = 0.8–0.9) (Table 1). After controlling for other covariates (See Table 1 footnotes and Table 2, entering the risk perceptions of cannabis use and perceived cannabis availability separately did not impact the trend), but without adjusting for alcohol and tobacco use, youth were still less likely to use cannabis during 2005–14

TABLE 1 Twelve-Month Unadjusted and Adjusted Relative Risks of Cannabis Use Among Youth as well as Cannabis Use Disorders Among Youth Cannabis Users in the US

Factors Associated With Cannabis Use Among Youth	Cannabis Use Among Youth: 2002–14 NSDUH, $n=288{,}300$[a]				
	Bivariable Model Without Adjusting for Any Covariates	Multivariable Model Controlled for Other Covariates, but Not Alcohol or Tobacco Use[b]	Multivariable Model Controlled for Alcohol Use and Other Covariates, but Not Tobacco Use[b]	Multivariable Model Controlled for Tobacco Use and Other Covariates, but Not Alcohol Use[b]	Multivariable Model Controlled for Other Covariates and Alcohol and Tobacco Use[b]
	URR (95% CI)	ARR (95% CI)	ARR (95% CI)	ARR (95% CI)	ARR (95% CI)
Year					
2002+	1.0	1.0	1.0	1.0	1.0
2003	1.0 (0.90–1.01)	1.0 (0.95–1.04)	1.0 (0.95–1.03)	1.0 (0.96–1.04)	1.0 (0.95–1.03)
2004	**0.9 (0.87–0.98)**	1.0 (0.95–1.04)	1.0 (0.95–1.03)	1.0 (0.96–1.04)	1.0 (0.95–1.03)
2005	**0.8 (0.80–0.90)**	**0.9 (0.89–0.98)**	**0.9 (0.89–0.97)**	**0.9 (0.90–0.98)**	**0.9 (0.90–0.97)**
2006	**0.8 (0.79–0.89)**	**0.9 (0.90–0.99)**	**0.9 (0.90–0.98)**	**0.9 (0.91–0.99)**	**0.9 (0.90–0.98)**
2007	**0.8 (0.75–0.84)**	**0.9 (0.87–0.95)**	**0.9 (0.87–0.95)**	**0.9 (0.89–0.97)**	**0.9 (0.89–0.96)**
2008	**0.8 (0.78–0.88)**	**0.9 (0.89–0.97)**	**0.9 (0.90–0.99)**	1.0 (0.93–1.01)	1.0 (0.93–1.01)
2009	**0.9 (0.82–0.92)**	**0.9 (0.86–0.94)**	**0.9 (0.88–0.96)**	**0.9 (0.91–0.99)**	**0.9 (0.92–0.99)**
2010	**0.9 (0.84–0.95)**	**0.9 (0.86–0.95)**	**0.9 (0.91–0.99)**	1.0 (0.93–1.02)	1.0 (0.95–1.03)
2011	**0.9 (0.85–0.96)**	**0.9 (0.85–0.93)**	**0.9 (0.91–0.99)**	1.0 (0.96–1.04)	1.0 (0.98–1.06)
2012	**0.9 (0.81–0.91)**	**0.8 (0.80–0.88)**	**0.9 (0.87–0.96)**	1.0 (0.94–1.02)	1.0 (0.96–1.05)
2013	**0.9 (0.80–0.91)**	**0.8 (0.77–0.84)**	**0.9 (0.86–0.94)**	1.0 (0.93–1.01)	1.0 (0.97–1.06)
2014	**0.8 (0.78–0.89)**	**0.8 (0.73–0.80)**	**0.9 (0.82–0.90)**	1.0 (0.91–1.00)	1.0 (0.95–1.04)

Alcohol use					
Yes			**2.9 (2.77–2.95)**		**2.1 (2.01–2.13)**
No+			1.0		1.0
Tobacco use					
Yes				**3.0 (2.92–3.09)**	**2.3 (2.28–2.40)**
No+				1.0	1.0
Factors Associated With Cannabis Use Disorders Among Youth Cannabis Users	**Cannabis Use Disorders Among Youth Cannabis Users: 2002–14 NSDUH, $n=41{,}100$[a]**				
	Bivariable Model Without Adjusting for Any Covariates	**Multivariable Model Controlled for Other Covariates, but Not Alcohol or Tobacco Use[c]**	**Multivariable Model Controlled for Alcohol Use and Other Covariates, but Not Tobacco Use[c]**	**Multivariable Model Controlled for Tobacco Use and Other Covariates, but Not Alcohol Use[c]**	**Multivariable Model Controlled for Other Covariates and Alcohol and Tobacco Use[c]**
	URR (95% CI)	**ARR (95% CI)**	**ARR (95% CI)**	**ARR (95% CI)**	**ARR (95% CI)**
Year					
2002+	1.0	1.0	1.0	1.0	1.0
2003	0.9 (0.85–1.04)	1.0 (0.88–1.07)	1.0 (0.88–1.07)	1.0 (0.89–1.07)	1.0 (0.89–1.07)
2004	1.0 (0.90–1.09)	1.0 (0.92–1.10)	1.0 (0.92–1.10)	1.0 (0.92–1.09)	1.0 (0.92–1.10)
2005	1.0 (0.90–1.10)	1.0 (0.94–1.13)	1.0 (0.94–1.13)	1.0 (0.93–1.13)	1.0 (0.93–1.13)
2006	1.0 (0.86–1.07)	1.0 (0.89–1.09)	1.0 (0.90–1.09)	1.0 (0.88–1.08)	1.0 (0.89–1.08)
2007	0.9 (0.83–1.02)	1.0 (0.87–1.06)	1.0 (0.87–1.06)	1.0 (0.87–1.05)	0.9 (0.87–1.05)
2008	1.0 (0.87–1.07)	1.0 (0.93–1.13)	1.0 (0.93–1.13)	1.0 (0.93–1.13)	1.0 (0.93–1.13)
2009	0.9 (0.83–1.02)	0.9 (0.86–1.04)	0.9 (0.86–1.05)	0.9 (0.86–1.05)	0.9 (0.86–1.05)
2010	0.9 (0.85–1.04)	1.0 (0.89–1.08)	1.0 (0.89–1.08)	1.0 (0.90–1.09)	1.0 (0.90–1.09)
2011	0.9 (0.83–1.02)	0.9 (0.85–1.03)	0.9 (0.85–1.04)	0.9 (0.87–1.06)	1.0 (0.87–1.06)

Continued

TABLE 1 Twelve-Month Unadjusted and Adjusted Relative Risks of Cannabis Use Among Youth as well as Cannabis Use Disorders Among Youth Cannabis Users in the US—cont'd

2012	**0.9 (0.79–0.99)**	0.9 (0.83–1.01)	0.9 (0.83–1.02)	1.0 (0.86–1.06)	1.0 (0.86–1.06)
2013	**0.8 (0.71–0.89)**	**0.9 (0.76–0.95)**	**0.9 (0.77–0.96)**	**0.9 (0.80–0.99)**	**0.9 (0.80–0.99)**
2014	**0.8 (0.67–0.86)**	**0.8 (0.70–0.89)**	**0.8 (0.70–0.89)**	**0.8 (0.75–0.95)**	**0.8 (0.75–0.96)**
Alcohol use					
Yes			**1.1 (1.07–1.23)**		**1.1 (1.01–1.16)**
No+			1.0		1.0
Tobacco use					
Yes				**1.5 (1.43–1.63)**	**1.5 (1.42–1.61)**
No+				1.0	1.0

Notes: CI, confidence interval; URR, unadjusted relative risk; ARR, adjusted relative risk; +Reference group. Significant relative risks are in bold.

[a]*SAMHSA requires that any description of overall sample sizes based on the restricted-use data files have to be rounded to the nearest 100 to minimize potential disclosure risk.*

[b]*Each multivariable model also adjusted for the following variables not showing in the table above: age; gender; race/ethnicity; health insurance; metropolitan statistical area; region; use of heroin, cocaine, hallucinogens, or inhalants; nonmedical use of prescription pain relievers, sedatives, and stimulants; perceived state legalization of medical cannabis use; state legalization of commercial sales or personal possession; perceived risk of smoking cannabis once or twice a week; perceived parent disapproval of using cannabis once a month or more; perceived peer's disapproval of using cannabis once a month or more; perceived cannabis availability; talked to parents about dangers of tobacco, alcohol, and drugs; and major depressive episode (see Table 2). Entering risk perceptions of cannabis use and perceived cannabis availability separately did not impact the trend.*

[c]*Each multivariable model also adjusted for the following variables not shown in the table above: age; gender; race/ethnicity; health insurance; metropolitan statistical area; region; use of heroin, cocaine, hallucinogens, or inhalants; nonmedical use of prescription pain relievers, sedatives, and stimulants; age at first cannabis use; perceived state legalization of medical cannabis use; state legalization of commercial sales or personal possession; perceived cannabis availability; source of cannabis; perceived risk of smoking cannabis once or twice a week; perceived parent disapproval of using cannabis once a month or more; perceived peer's disapproval of using cannabis once a month or more; talked to parents about dangers of tobacco, alcohol, and drugs; and major depressive episode (see Table 2). Entering risk perceptions of cannabis use and perceived cannabis availability separately did not impact the trend.*

TABLE 2 Other Correlates of Cannabis Use Among Youth as well as Other Correlates of Cannabis Use Disorders Among Youth Users in the US

Factors	Cannabis Use among Youth 2002–14 NSDUH, $n = 288{,}300$[a] Adjusted Relative Risk (95% CI)	Cannabis Use Disorders Among Youth Users 2002–14 NSDUH, $n = 41{,}100$[a] Adjusted Relative Risk (95% CI)
Age		
12–13	**0.7 (0.70–0.75)**	**0.8 (0.72–0.88)**
14–15	**0.9 (0.92–0.95)**	**0.9 (0.89–0.98)**
16–17+	1.0	1.0
Gender		
Male	1.0 (0.98–1.01)	1.0 (0.98–1.07)
Female+	1.0	1.0
Race/ethnicity		
NH white+	1.0	1.0
NH black	**1.3 (1.26–1.32)**	**1.2 (1.17–1.33)**
NH Native American/Alaska Native	**1.4 (1.31–1.53)**	**1.2 (1.04–1.43)**
NH Hawaiian/Other Pacific Islander	1.1 (0.97–1.30)	**1.3 (1.03–1.69)**
NH Asian	1.0 (0.92–1.06)	**1.2 (1.02–1.48)**
NH more than one race	**1.2 (1.11–1.22)**	**1.2 (1.03–1.31)**
Hispanic	**1.1 (1.05–1.10)**	**1.2 (1.16–1.31)**
Health insurance		
Private only+	1.0	1.0
No insurance coverage	1.0 (1.00–1.07)	1.0 (0.95–1.12)
Medicaid	**1.1 (1.09–1.14)**	**1.1 (1.01–1.12)**
Other	1.0 (0.96–1.05)	1.0 (0.89–1.12)
Metropolitan statistical area		
Large+	1.0	1.0
Small	1.0 (0.97–1.01)	1.0 (0.97–1.07)
Nonmetropolitan	**0.9 (0.90–0.94)**	1.0 (0.93–1.04)
Region		
Northeast	**1.0 (0.94–0.99)**	1.0 (0.93–1.07)
Midwest	**0.9 (0.92–0.97)**	1.0 (0.97–1.10)
South	**0.9 (0.88–0.92)**	1.1 (0.94–1.07)
West+	1.0	1.0
Cocaine use		
Yes	**1.9 (1.71–2.00)**	**1.2 (1.13–1.29)**
No+	1.0	1.0

Continued

TABLE 2 Other Correlates of Cannabis Use Among Youth as well as Other Correlates of Cannabis Use Disorders Among Youth Users in the US—cont'd

Factors	Cannabis Use among Youth	Cannabis Use Disorders Among Youth Users
	2002–14 NSDUH, $n = 288{,}300$[a]	2002–14 NSDUH, $n = 41{,}100$[a]
	Adjusted Relative Risk (95% CI)	Adjusted Relative Risk (95% CI)
Hallucinogen use		
Yes	**1.8 (1.71–1.86)**	**1.2 (1.18–1.32)**
No+	1.0	1.0
Heroin use		
Yes	1.2 (0.96–1.44)	1.0 (0.83–1.22)
No+	1.0	1.0
Inhalant use		
Yes	**1.0 (1.01–1.08)**	**1.2 (1.14–1.30)**
No+	1.0	1.0
Pain reliever nonmedical use		
Yes	**1.2 (1.20–1.27)**	**1.3 (1.21–1.34)**
No+	1.0	1.0
Sedative nonmedical use		
Yes	**1.2 (1.13–1.24)**	**1.2 (1.15–1.30)**
No+	1.0	1.0
Stimulant nonmedical use		
Yes	**1.2 (1.13–1.26)**	**1.2 (1.17–1.33)**
No+	1.0	1.0
Age of first cannabis use		
≤13	–	**1.9 (1.71–2.04)**
14–15		**1.4 (1.32–1.55)**
16–17+		1.0
State legalized commercial sales or personal possession		
Yes	1.0 (0.95–1.15)	0.9 (0.73–1.15)
No+	1.0	1.0
Perceived state legalization of medical cannabis use		
Yes	**1.1 (1.04–1.08)**	**1.1 (1.01–1.13)**
No+	1.0	1.0
Not sure/unknown	**1.0 (0.95–0.99)**	0.9 (0.89–1.01)
Perceived cannabis availability, fairly/very easy		
Yes	**1.6 (1.54–1.63)**	**1.4 (1.29–1.58)**
No+	1.0	1.0

TABLE 2 Other Correlates of Cannabis Use Among Youth as well as Other Correlates of Cannabis Use Disorders Among Youth Users in the US—cont'd

Factors	Cannabis Use among Youth	Cannabis Use Disorders Among Youth Users
	2002–14 NSDUH, $n=288,300^a$	2002–14 NSDUH, $n=41,100^a$
	Adjusted Relative Risk (95% CI)	Adjusted Relative Risk (95% CI)
Perceived risk of smoking cannabis 1–2/week		
No risk+	1.0	1.0
Slight risk	**0.9 (0.82–0.87)**	1.0 (0.95–1.05)
Moderate risk	**0.7 (0.66–0.70)**	**0.9 (0.87–0.98)**
Great risk	**0.5 (0.52–0.56)**	0.9 (0.86–1.02)
Unspecified	**0.7 (0.58–0.76)**	0.9 (0.60–1.46)
Perceived parent disapproval of using cannabis once a month or more		
Strong disapproval+	1.0	1.0
Somewhat disapproval	**1.1 (1.11–1.19)**	1.0 (0.89–1.02)
Neither approval nor disapproval	**1.4 (1.32–1.40)**	1.0 (0.96–1.07)
Perceived peer's disapproval of using cannabis once a month or more		
Strong disapproval+	1.0	1.0
Somewhat disapproval	**1.4 (1.39–1.48)**	1.2 (0.95–1.56)
Neither approval nor disapproval	**1.8 (1.80–1.90)**	**1.2 (1.14–1.31)**
Talked to parents about dangers of tobacco, alcohol, and drugs		
Yes	**1.0 (0.95–0.99)**	**0.9 (0.86–0.94)**
No+	1.0	1.0
Source of cannabis		
Bought it	–	**1.6 (1.50–1.65)**
Traded for it		**1.4 (1.25–1.68)**
Got it for free/shared+		1.0
Grew it yourself		**1.6 (1.34–2.00)**
Method unspecified		**0.4 (0.29–0.47)**
Major depressive episode[b]		
Yes	1.0 (1.00–1.06)	**1.3 (1.21–1.36)**
No+	1.0	1.0

Notes: NH, non-hispanic; CI, confidence interval; +Reference group. Significant relative risks are in bold.
[a] *SAMHSA requires that any description of overall sample sizes based on the restricted-use data files have to be rounded to the nearest 100 to minimize potential disclosure risk.*
[b] *The relative risks of past-year major depressive episode (MDE) were based on separate models using the 2004–14 NSDUH data since MDE among youth was not measured in 2002–03 NSDUH.*

than in 2002 (adjusted relative risks (ARRs)=0.8–0.9). After controlling for other covariates and alcohol use, but not tobacco use, youth were still less likely to use cannabis in 2005–14 than 2002 (ARRs=0.9). Results were similar after controlling for other covariates and tobacco use (with or without alcohol use): youths were less likely to use cannabis only during 2005–2007 and in 2009 compared with 2002 (ARRs=0.9).

Bivariable logistic regression results showed that youth cannabis users were less likely to have cannabis use disorders during 2012–14 than in 2002 (URRs=0.8–0.9) (Table 1). After controlling for other covariates (see Table 1 footnotes and Table 2, entering the risk perceptions of cannabis use and perceived cannabis availability separately did not impact the trend), but not alcohol or tobacco use, youth cannabis users were still less likely to have cannabis use disorders during 2013–14 than in 2002 (ARRs=0.8–0.9). After controlling for other covariates and alcohol use, but not tobacco use, youth cannabis users were less likely to have cannabis use disorders during 2013–14 compared with 2002 (ARRs=0.8–0.9). The results remained similar after controlling for other covariates and tobacco use (with or without alcohol use).

Other Correlates of Cannabis Use and Use Disorders Among Youth

Compared with each corresponding reference group, the adjusted prevalence of cannabis use was higher among youth aged 16–17, non-Hispanic blacks, Hispanics, non-Hispanic youth with more than one race, non-Hispanic Native Americans and Alaska Natives, Medicaid beneficiaries, and youth residing in large metropolitan areas and in the South (Table 2). Higher cannabis use was found among users of tobacco, alcohol, cocaine, hallucinogens, and inhalants, and nonmedical users of prescription pain relievers, sedatives, and stimulants, than among the corresponding nonusers. Compared with each corresponding reference group, cannabis use was higher among youth who perceived no risk of smoking cannabis, youth who perceived cannabis was easy to obtain, and youth who perceived parental or peers' approval of using cannabis as "somewhat disapproval" or "neither approval nor disapproval."

Among cannabis users (Table 2), compared with each corresponding reference group, the adjusted prevalence of cannabis use disorders was higher among those aged 16–17, racial/ethnic minorities, and Medicaid beneficiaries. It was higher among users of tobacco, alcohol, cocaine, hallucinogens, and inhalants, and nonmedical users of prescription pain relievers, sedatives, and stimulants than the corresponding nonusers. Compared with each corresponding reference group, cannabis use disorders were also more prevalent among those who first used cannabis by age 15; users who perceived cannabis was easy to obtain; users who perceived no risk of smoking cannabis; users who perceived peers' approval of using cannabis as "neither approval nor disapproval"; users who bought, traded, or grew cannabis themselves; and users with depression.

Implications of Trends and Correlates of Cannabis Use Among Youth

We found that cannabis use declined among youth in the US during 2005–14 compared with 2002, even after adjusting for sociodemographic and geographic characteristics and substance use factors, except for tobacco use. This decline in cannabis use occurred even in the context of declines in youth risk perceptions of cannabis use, especially during 2007–14. Researchers have suggested that the stable prevalence of parental or peers' disapproval of cannabis use and the decline in perceived cannabis availability may explain the recent stable prevalence of cannabis use among US students despite declining risk perceptions (Johnston, O'Malley, Miech, Bachman, & Schulenberg, 2015). However, we found during 2002–14 that the trends in alcohol use, parental or peers' disapproval of cannabis use, risk perceptions of cannabis use, and perceived cannabis availability were not associated with the decline in cannabis use among youth. However, the decline in tobacco use, with downward trends starting in 2002 and accelerating during 2010–14, was strongly associated with the decline in cannabis use among youth in the US. After adjusting for the prevalence of tobacco use, there were no significant differences in the prevalence of cannabis use in 2010–14 and 2002, suggesting if the prevalence of tobacco use remained unchanged, the prevalence of past-year cannabis use among youth in 2010–14 would have been similar to that in 2002.

These results indicate that the decline in tobacco use among youth may be an important driving factor for the decline in cannabis use. Moreover, the prevalence of cannabis use among tobacco users increased during 2002–14, indicating the importance of tobacco control and prevention among youth (U.S. Department of Health and Human Services, 2014). The gateway theory suggests that tobacco use often precedes the onset of cannabis use (Kandel, 2002; Patton, Coffey, Carlin, Sawyer, & Lynskey, 2005). Furthermore, tobacco use and cannabis use share a common route of administration and genetic liability (Chen et al., 2008; Agrawal, Silberg, Lynskey, Maes, & Eaves, 2010; Peters, Budneyb, & Carrollc, 2012). Future research is needed to monitor trends in tobacco use among youth and to determine whether the prevalence of cannabis use will continue to decline among youth or will begin to parallel the increase among adults.

Cannabis use disorders among youth users were lower in 2013–14 compared with 2002, even after controlling for sociodemographic factors, geographic characteristics, substance-use factors, and risk perceptions of cannabis use. Unlike its association with the decline in cannabis use, tobacco use was not associated with the decline in cannabis use disorders, suggesting that tobacco use may be related to the onset of cannabis use among youth but not its progression to cannabis use disorder.

Our findings that the prevalence rates of cannabis use and use disorders were higher among non-Hispanic blacks than among non-Hispanic whites diverge from previous work (Compton, Grant, Colliver, Glantz, & Stinson, 2004), suggesting a shifting pattern of cannabis use in the US (Hasin, Saha, et al., 2015,

2016). A recent study showed that non-Hispanic black youth tend to view cannabis use favorably (Cavazos-Rehg et al., 2015). Previous studies also found gender differences in pathways to cannabis use among youth (Brook et al., 1998; Schepis et al., 2011; Buu et al., 2014); females tended to be at higher risk for initiating cannabis use at younger ages (Buu et al., 2014), and females had a faster transition from the initiation of cannabis use to regular use (Schepis et al., 2011). However, our study found no gender differences in the prevalence of cannabis use among youth or in the prevalence of cannabis use disorders among youth users, suggesting another shifting pattern of cannabis use. Consistent with previous research (Compton et al., 2004; Peters et al., 2012; Schauer, Berg, Kegler, Donovan, & Windle, 2015), our study identified associations of cannabis use and use disorders with tobacco, alcohol, and other substance use, and an association between cannabis use disorder and depression, suggesting that use of multiple substances and comorbidity with psychiatric illness are common among youth cannabis users. Identification of one psychiatric or behavioral problem should prompt clinicians to carefully probe for other related problems (Center for Substance Abuse Treatment, 1997; Dawson, Compton, & Grant, 2010; Peters et al., 2012; Compton, Blanco, & Wargo, 2015).

Despite some declines, the prevalence rates of cannabis use and cannabis use disorders among youth in the US remain high. Our multivariable results confirmed the existing studies (Johnston et al., 2015; SAMHSA, 2015; Monitoring the Future Study, 2015). Cannabis use has adverse consequences, including deleterious effects on brain development and school performance, mental health problems, and addictions (Silins et al., 2014; Volkow et al., 2014). Our results can help inform ongoing cannabis-related policy discussions and tobacco control efforts, and are important for clinicians providing optimal care for youth using cannabis.

EPIDEMIOLOGY OF CANNABIS USE AND USE DISORDERS AMONG ADULTS

Trends in Cannabis Use, Use Disorders, and Use Frequency Among Adults and Adult Users

Based on 596,500 sampled persons aged 18 or older from the 2002–14 NSDUHs, the 12-month prevalence of cannabis use among adults increased from 10.4% in 2002 to 13.3% in 2014 (absolute difference 2.9%, $p<0.0001$) (Fig. 2). The prevalence of perceiving no risk of smoking cannabis increased from 5.6% in 2002 to 15.1% in 2014 (absolute difference 9.5%, $p<0.0001$). The 12-month prevalence of daily or near daily use increased from 1.9% in 2002 to 3.5% in 2014 (absolute difference 1.6%, $p<0.0001$; data not shown: see Compton, Han, Hughes, Jones, & Blanco, 2016). The 12-month prevalence of perceived state legalization of medical cannabis use increased from 17.9% in 2002 to 32.6% in 2014 (absolute difference 14.7%, $p<0.0001$). However, the prevalence of cannabis use disorders among adults remained stable during 2002–14 at around 1.5%.

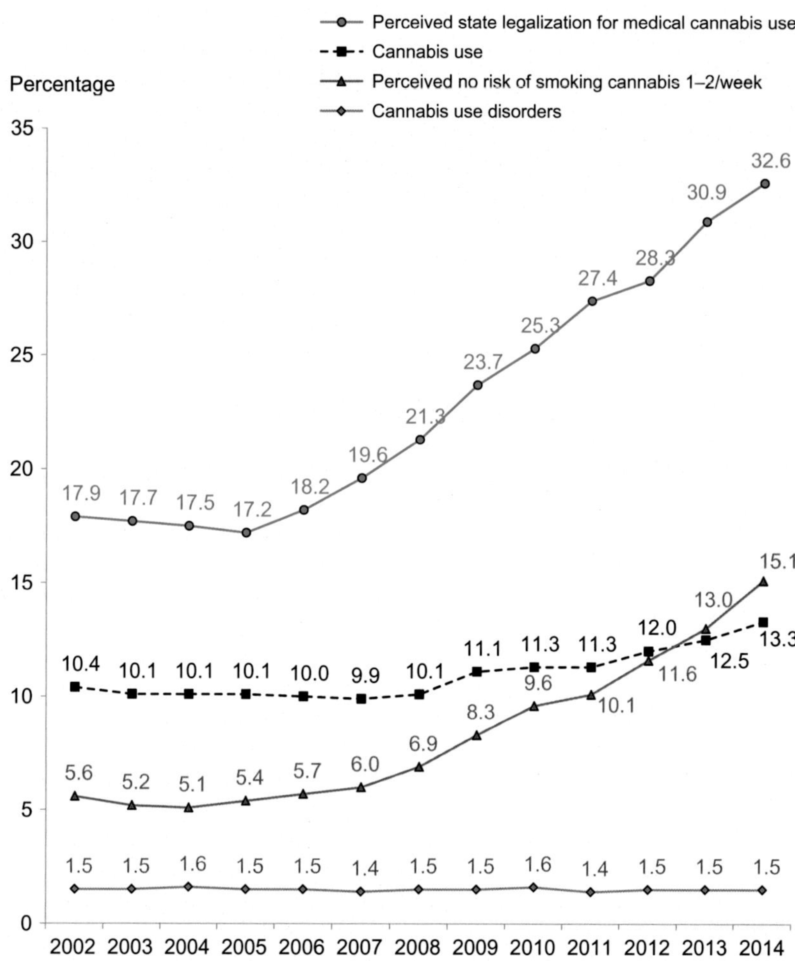

FIG. 2 Twelve-month prevalence of cannabis use, cannabis use disorders, perceived state legalization for medical cannabis use, and perceived no risk of smoking cannabis once or twice a week among adults in the US: 2002–14.

The number of persons aged 18 or older who first used cannabis in the past 12 months increased from 823,000 in 2002 to 1.4 million in 2014 (data not shown). The overall number of cannabis users increased from 21.9 million in 2002 to 31.9 million in 2014. The number of daily or near-daily users increased from 3.9 million in 2002 to 8.4 million in 2014.

Among those who reported using cannabis, the 12-month prevalence of cannabis use disorders decreased from 14.8% in 2002 to 11.0% in 2014 (absolute difference −3.8%, $p<0.0001$) (Fig. 3). The prevalence of perceiving no risk of smoking cannabis once or twice a week increased from 24.8% in 2002 to 49.5% in 2014 (absolute difference 24.7%, $p<0.0001$). The 12-month prevalence of perceived state legalization of medical cannabis use increased from 28.9% in 2002 to 43.8% in 2014 (absolute difference 14.9%, $p<0.0001$).

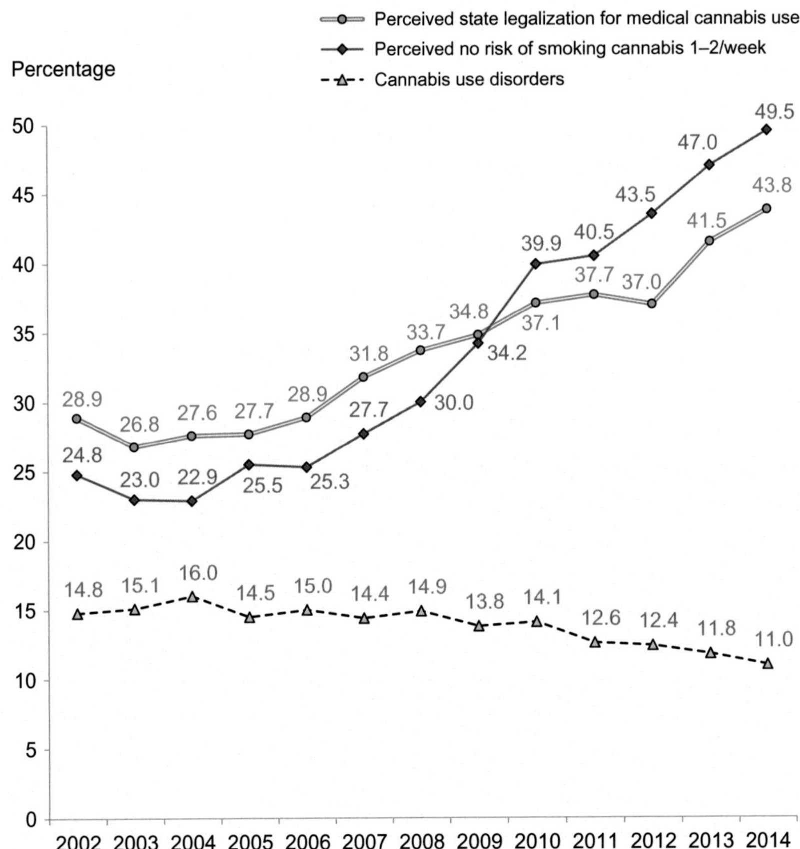

FIG. 3 Twelve-month prevalence of perceived state legalization for medical cannabis use, perceived no risk of smoking cannabis once or twice a week, and cannabis use disorders among adult cannabis users in the US: 2002–14.

Among adults in the US, the mean number of days of cannabis use increased from 10.0 days in 2002 to 16.3 days in 2014 (absolute difference = 6.3 days, $p < 0.0001$; data not shown: see Compton et al., 2016). Among cannabis users, it increased from 97.9 days in 2002 to 124.9 days in 2014 (absolute difference 27.0 days, $p < 0.0001$).

Associations Between Trends in Cannabis Use, Use Disorders, Use Frequency, and Perceived Risk of Smoking Cannabis Among Adults

After adjusting for covariates (except for perceived risk of smoking cannabis), the prevalence of cannabis use in the adult population was lower in 2006–07 than in 2002, but was higher in each year during 2011–14 than in 2002 (Table 3).

TABLE 3 Twelve-Month Unadjusted and Adjusted Relative Risks of Cannabis Use Among Adults in the US

	2002–14 NSDUH, $n=596,500$[a]		
	Bivariable Model Without Controlling for Any Covariates	Multivariable Model Without Perceived Risk of Smoking Cannabis, but With Other Covariates[b]	Multivariable Model With Perceived Risk of Smoking Cannabis and Other Covariates[b]
Factors	Unadjusted Relative Risk (95% CI)	Adjusted Relative Risk (95% CI)	Adjusted Relative Risk (95% CI)
Year			
2002+	1.0	1.0	1.0
2003	1.0 (0.92–1.03)	1.0 (0.93–1.01)	1.0 (0.95–1.02)
2004	1.0 (0.92–1.03)	1.0 (0.94–1.02)	1.0 (0.96–1.04)
2005	1.0 (0.92–1.03)	1.0 (0.94–1.02)	1.0 (0.94–1.02)
2006	1.0 (0.91–1.02)	**0.9 (0.91–0.99)**	**0.9 (0.91–0.99)**
2007	1.0 (0.90–1.01)	**0.9 (0.91–0.99)**	**0.9 (0.89–0.97)**
2008	1.0 (0.92–1.03)	1.0 (0.93–1.02)	**0.9 (0.89–0.97)**
2009	**1.1 (1.01–1.14)**	1.0 (0.99–1.08)	**0.9 (0.91–0.99)**
2010	**1.1 (1.03–1.15)**	1.0 (0.98–1.07)	**0.9 (0.88–0.96)**
2011	**1.1 (1.03–1.15)**	**1.1 (1.02–1.10)**	**0.9 (0.89–0.97)**
2012	**1.2 (1.09–1.22)**	**1.1 (1.03–1.12)**	**0.9 (0.88–0.96)**
2013	**1.2 (1.14–1.27)**	**1.1 (1.08–1.17)**	**0.9 (0.90–0.97)**
2014	**1.3 (1.21–1.35)**	**1.2 (1.14–1.24)**	**0.9 (0.92–0.99)**
Perceived risk of smoking cannabis once/twice weekly			
No risk+			1.0
Slight risk			**0.7 (0.69–0.72)**
Moderate risk			**0.5 (0.44–0.46)**
Great risk			**0.3 (0.26–0.28)**
Unspecified			**0.5 (0.45–0.62)**

Notes: CI, confidence interval. +Reference group. Within each factor, a prevalence that significantly differs from the prevalence of the reference group is in bold.

[a]*SAMHSA requires that any description of overall sample sizes based on the restricted-use data files has to be rounded to the nearest 100 to minimize potential disclosure risk.*

[b]*Each multivariable model adjusted for age, gender, race/ethnicity, education, employment, marital status, health insurance, metropolitan statistical area, region, tobacco use, heavy alcohol use, cocaine use, hallucinogen use, heroin use, inhalant use, nonmedical use of prescription opioids, nonmedical sedative use, nonmedical stimulant use, age at first cannabis use, state legalization of commercial sales or personal possession, perceived state legalization of medical cannabis use, and perceived cannabis availability. Past-year major depressive episode was not associated with cannabis use and was excluded from the final models. (See Table 5 for details about these covariates.)*

However, after additionally controlling for perceived risk, the prevalence of cannabis use was lower in each year during 2006–14 than in 2002. Thus, while the overall trends show an increase in the prevalence of cannabis use in adults from 2011 to 2014, reduced risk perception is an essential correlate of the increases. In contrast, among adult cannabis users (Table 4), the adjusted prevalence of cannabis use disorders was lower in each year during 2011–14 than in 2002 (ARRs = 0.8–0.9). Additional adjustments for perceived risk of smoking cannabis did not change these results. Thus, perceived risk is associated with the use of cannabis itself but not progression to cannabis use disorder.

Other Correlates of Cannabis Use and Use Disorders Among Adults

Among the adult population (Table 5), compared with each corresponding reference group, the adjusted prevalence of cannabis use was higher among younger adults (ages 18–29 and 30–49); men; non-Hispanic blacks; non-Hispanic adults with more than one race; those without a high school diploma; the uninsured; those employed part-time, disabled for work, or unemployed; and those residing in the West. Cannabis use was higher among users of tobacco, cocaine, hallucinogens, and inhalants; nonmedical users of prescription opioids, sedatives, and stimulants; and heavy alcohol users than the corresponding nonusers. The adjusted prevalence of cannabis use was higher among those who perceived legalization of medical cannabis use in their state and those who perceived it to be easy to obtain cannabis compared with each corresponding reference group.

Among those reporting use of cannabis (Table 5), compared with each corresponding reference group, the adjusted prevalence of cannabis use disorders was higher among those aged 18–49; men; non-Hispanic blacks; Hispanics; non-Hispanic Asians; non-Hispanic users with more than one race; adults without a high school diploma; those employed part-time, disabled for work, or unemployed; never-married adults; and those with specific substance use disorders (nicotine, alcohol, prescription opioids, cocaine, hallucinogens, inhalants, sedatives, and stimulants) and MDE. Cannabis use disorders were more prevalent among those who perceived cannabis to be easy to obtain and those who bought/traded cannabis or grew cannabis themselves than each corresponding reference group.

Implications of Trends and Correlates of Cannabis Use and Use Disorders Among Adults

Cannabis use and the frequency of cannabis use among adults increased in the US during 2011–14 compared with a decade earlier, even after adjusting for sociodemographic, geographic, and substance-use factors. Initiation of cannabis use also increased during 2002–14. Furthermore, adults in the US perceived less risk in cannabis use since 2006–07, and these declining risk perceptions were strongly associated with increases in cannabis use and frequency of use among adults.

TABLE 4 Twelve-Month Unadjusted and Adjusted Relative Risks of Cannabis Use Disorders Among Adult Cannabis Users in the US

	2002–14 NSDUH, $n = 114{,}700$[a]		
	Bivariable Model Without Controlling for Any Covariates	Multivariable Model Without Controlling for Perceived Risk of Cannabis Use, but Adjusting for Other Covariates[b]	Multivariable Model Controlling for Perceived Risk of Cannabis Use and Other Covariates[b]
Factors	Unadjusted Relative Risk (95% CI)	Adjusted Relative Risk (95% CI)	Adjusted Relative Risk (95% CI)
Year			
2002+	1.0	1.0	1.0
2003	1.0 (0.91–1.14)	1.0 (0.93–1.15)	1.0 (0.93–1.15)
2004	1.1 (0.97–1.21)	1.1 (0.97–1.19)	1.1 (0.97–1.19)
2005	1.0 (0.87–1.10)	1.0 (0.87–1.08)	1.0 (0.87–1.08)
2006	1.0 (0.90–1.14)	1.0 (0.87–1.09)	1.0 (0.87–1.09)
2007	1.0 (0.86–1.09)	1.0 (0.86–1.07)	1.0 (0.86–1.07)
2008	1.0 (0.90–1.14)	1.0 (0.86–1.08)	1.0 (0.87–1.08)
2009	0.9 (0.83–1.05)	0.9 (0.81–1.01)	0.9 (0.82–1.02)
2010	0.9 (0.84–1.07)	0.9 (0.82–1.02)	0.9 (0.83–1.04)
2011	**0.8 (0.75–0.95)**	**0.8 (0.76–0.94)**	**0.9 (0.77–0.96)**
2012	**0.8 (0.74–0.95)**	**0.8 (0.72–0.90)**	**0.8 (0.73–0.92)**
2013	**0.8 (0.70–0.90)**	**0.8 (0.73–0.93)**	**0.8 (0.74–0.95)**
2014	**0.7 (0.66–0.83)**	**0.8 (0.71–0.78)**	**0.8 (0.73–0.90)**
Perceived risk of smoking cannabis once/twice weekly			
No risk+			1.0
Slight risk			**1.1 (1.07–1.18)**
Moderate risk			**1.1 (1.02–1.16)**
Great risk			1.1 (1.00–1.20)
Unspecified			0.7 (0.36–1.26)

Notes: CI, confidence interval. +Reference group. Within each factor, a prevalence that significantly differs from the prevalence of the reference group is in bold.
[a]*SAMHSA requires that any description of overall sample sizes based on the restricted-use data files have to be rounded to the nearest 100 to minimize potential disclosure risk.*
[b]*Each multivariable model adjusted for age, gender, race/ethnicity, education, employment status, marital status, health insurance, metropolitan statistical area, region, nicotine dependence, alcohol use disorders, cocaine use disorders, hallucinogen use disorders, heroin use disorders, inhalant use disorders, prescription opioid use disorders, sedative use disorders, stimulant use disorders, age at first cannabis use, state legalization of commercial sales or personal possession, perceived state legalization of medical cannabis use, perceived cannabis availability, source of cannabis, and major depressive episode. (See Table 5 for details about these covariates.)*

TABLE 5 Other Correlates of Cannabis Use Among Adults and Other Correlates of Cannabis Use Disorders Among Adult Users in the US

	Cannabis Use Among Adults $n = 596{,}500$[a]	Cannabis Use Disorders Among Adult Cannabis Users $n = 114{,}700$[a]
Factors	**Adjusted Relative Risk**[b] **(95% CI)**	**Adjusted Relative Risk**[c] **(95% CI)**
Age		
18–29	**1.5 (1.41–1.51)**	**1.8 (1.54–2.08)**
30–49	**1.1 (1.08–1.15)**	**1.2 (1.06–1.45)**
≥50+	1.0	1.0
Gender		
Male	**1.1 (1.07–1.10)**	**1.2 (1.14–1.25)**
Female+	1.0	1.0
Race/ethnicity		
NH white+	1.0	1.0
NH black	**1.2 (1.15–1.21)**	**1.4 (1.33–1.49)**
NH Native American/Alaska Native	1.0 (0.93–1.09)	**1.2 (1.01–1.46)**
NH Hawaiian/Other Pacific Islander	1.1 (0.91–1.24)	1.0 (0.76–1.30)
NH Asian	1.0 (0.92–1.02)	**1.3 (1.10–1.49)**
NH more than one race	**1.1 (1.07–1.20)**	**1.2 (1.03–1.36)**
Hispanic	1.0 (0.96–1.02)	**1.3 (1.20–1.37)**
Education		
Less than high school+	1.0	1.0
High school	**0.9 (0.90–0.94)**	**0.8 (0.80–0.90)**
Some college	**0.9 (0.89–0.93)**	**0.8 (0.79–0.89)**
College graduate	**0.9 (0.88–0.93)**	**0.7 (0.68–0.81)**
Employment Status		
Full-time employed+	1.0	1.0
Part-time employed	**1.2 (1.19–1.24)**	**1.2 (1.17–1.31)**
Disabled for work	**1.2 (1.17–1.27)**	**1.2 (1.08–1.40)**
Unemployed	**1.1 (1.10–1.17)**	**1.3 (1.18–1.35)**
Marital Status		
Married+	1.0	1.0
Widowed	**1.1 (1.01–1.23)**	0.7 (0.50–1.11)
Divorced/separated	**1.2 (1.17–1.24)**	1.0 (0.93–1.17)
Never married	**1.4 (1.42–1.48)**	**1.2 (1.13–1.31)**
Health insurance		
Private only+	1.0	1.0
No insurance coverage	**1.1 (1.05–1.09)**	1.0 (0.91–1.01)
Medicaid	1.0 (0.99–1.05)	1.0 (0.96–1.10)
Other	1.0 (0.96–1.02)	1.0 (0.91–1.11)

TABLE 5 Other Correlates of Cannabis Use Among Adults and Other Correlates of Cannabis Use Disorders Among Adult Users in the US—cont'd

Factors	Cannabis Use Among Adults $n = 596{,}500$[a] Adjusted Relative Risk[b] (95% CI)	Cannabis Use Disorders Among Adult Cannabis Users $n = 114{,}700$[a] Adjusted Relative Risk[c] (95% CI)
Metropolitan statistical area		
Large+	1.0	1.0
Small	**1.0 (0.96–0.99)**	1.0 (0.95–1.05)
Nonmetropolitan	**0.9 (0.91–0.94)**	1.0 (0.93–1.05)
Region		
Northeast	**1.0 (1.01–1.06)**	1.0 (0.89–1.02)
Midwest	**0.9 (0.91–0.96)**	1.0 (0.90–1.03)
South	**0.9 (0.90–0.94)**	1.0 (0.92–1.06)
West+	1.0	1.0
Tobacco[d]		
Yes	**1.4 (1.38–1.44)**	**1.1 (1.03–1.14)**
No+	1.0	1.0
Alcohol[d]		
Yes	**1.2 (1.21–1.26)**	**1.8 (1.69–1.85)**
No+	1.0	1.0
Cocaine[d]		
Yes	**1.8 (1.77–1.90)**	**1.5 (1.37–1.64)**
No+	1.0	1.0
Hallucinogen[d]		
Yes	**2.0 (1.91–2.06)**	**2.1 (1.83–2.34)**
No+	1.0	1.0
Heroin[d]		
Yes	0.9 (0.82–1.07)	0.9 (0.71–1.05)
No+	1.0	1.0
Inhalant[d]		
Yes	**1.5 (1.37–1.62)**	**1.7 (1.16–2.45)**
No+	1.0	1.0
Nonmedical prescription opioids[d]		
Yes	**1.3 (1.31–1.38)**	**1.7 (1.53–1.81)**
No+	1.0	1.0
Nonmedical sedative[d]		
Yes	**1.4 (1.31–1.40)**	**1.8 (1.48–2.03)**
No+	1.0	1.0

Continued

TABLE 5 Other Correlates of Cannabis Use Among Adults and Other Correlates of Cannabis Use Disorders Among Adult Users in the US—cont'd

	Cannabis Use Among Adults $n = 596{,}500$[a]	Cannabis Use Disorders Among Adult Cannabis Users $n = 114{,}700$[a]
Factors	**Adjusted Relative Risk**[b] **(95% CI)**	**Adjusted Relative Risk**[c] **(95% CI)**
Nonmedical stimulant[d]		
Yes	**1.4 (1.34–1.46)**	**1.5 (1.31–1.77)**
No+	1.0	1.0
Age at first cannabis use		
<18	**1.1 (1.10–1.15)**	**1.2 (1.18–1.32)**
18–29	1.0	1.0
≥30	**0.1 (0.10–0.12)**	1.2 (0.64–2.20)
State legalized commercial sales or personal possession		
Yes	**1.1 (1.01–1.18)**	1.0 (0.80–1.17)
No+	1.0	1.0
Perceived state legalization of medical cannabis use		
Yes	**1.1 (1.09–1.13)**	1.0 (0.98–1.09)
No+	1.0	1.0
Not sure/unknown	**0.9 (0.86–0.90)**	**0.9 (0.82–0.97)**
Perceived cannabis availability, fairly/very easy		
Yes	**1.4 (1.35–1.42)**	**1.2 (1.13–1.36)**
No+	1.0	1.0
Source of cannabis		
Bought it	–	**2.5 (2.38–2.65)**
Traded for it		**2.4 (2.02–2.93)**
Got it for free/shared+		1.0
Grew it yourself		**2.1 (1.70–2.66)**
Method unspecified		0.8 (0.66–1.12)
Major depressive episode[e]		
Yes	–	**1.5 (1.42–1.65)**
No+		1.0

Notes: NH, non-Hispanic; CI, confidence interval. +Reference group. Within each factor, a model-adjusted prevalence that significantly differs from the model-adjusted prevalence of the reference group is in bold.
[a]*SAMHSA requires that any description of overall sample sizes based on the restricted-use data files have to be rounded to the nearest 100 to minimize potential disclosure risk.*
[b]*Results for each factor are from models that adjust for all the factors listed as well as the survey year and perceived risk of cannabis use (see Table 3). Source of Cannabis and Major Depressive Episode were not associated with cannabis use and so are not included in the model.*
[c]*Results for each factor are from models that adjust for all the factors listed as well as the survey year and perceived risk of cannabis use (see Table 4).*
[d]*For Cannabis Use model, results are for use of each of the corresponding substances, except for heavy alcohol use rather than alcohol use. For cannabis use disorders, results are for substance use disorders (or nicotine dependence) related to each of the corresponding substances.*
[e]*The presented results for Major Depressive Episode are based on a model using just the 2005–14 NSDUH data after controlling for other covariates. Other results are from the 2002–14 NSDUH data.*

While shifts in perceived risk have historically been shown to be important predictors of adolescent cannabis use trends (Bachman, Johnson, & O'Malley, 1998), none of the previous research examined this relationship in adults.

We found neither an increase in cannabis use disorders nor an association between changes in perceived risk and the prevalence of cannabis use disorders. Previous research suggests that the quantity and frequency of cannabis use is highly correlated with criteria of cannabis use disorders (Compton, Saha, Conway, & Grant, 2009). Similarly, previous research has suggested that increases in cannabis potency may have been related to increases in cannabis use disorders (Compton et al., 2004). However, our study did not find increases in the prevalence of cannabis use disorders, despite finding increases in the frequency of use and the known continued increases in potency between 2002 and 2014 (ElSohly, 2015). It is possible that the large number of recent initiators of cannabis use may need time to develop cannabis use disorders (Lopez-Quintero et al., 2011). Future research on trends in cannabis use disorders and their relationships with perceived risk may help elucidate the reasons for the discrepancy between cannabis use patterns and use disorders.

Our results partly confirmed increases in cannabis use suggested in the two-wave National Epidemiologic Survey on Alcohol and Related Conditions (NESARC) study (Hasin, Saha, et al., 2015), but the increases we found are lower in magnitude than the more than doubling reported by Hasin, Saha, and colleagues (2015). Also like NESARC, we found decreases in the proportion of users reporting cannabis use disorders, but the prevalence of cannabis use disorders among users was lower (14.8% in 2002 to 11.0% in 2014) in our study than in NESARC (35.6% in 2001–02 to 30.6% in 2012–13) (Hasin, Saha, et al., 2015). By contrast, our results documented no increase in cannabis use disorders among the adult population, diverging from NESARC, which found a near doubling from 2001–02 to 2012–13 (1.5% to 2.9%) (Hasin, Saha, et al., 2015). Furthermore, we found that increases in the frequency of cannabis use among cannabis users occurred after 2007 and not earlier. Differences in study design and implementation between the NESARC and NSDUH as well as between NESARC waves may explain these differences (Grucza et al., 2007, 2016). First, NSDUHS's Audio Computer-assisted Self-interview process may have resulted in higher prevalence rates than NESARC and NESARC III's Computer-assisted Personal Interview because those using illegal substances may be more likely to self-report accurately when given more privacy (SAMHSA, 2016; Grant et al., 2015). Second, for the 2002–14 NSDUH, each respondent who completed the interview received a $30 cash incentive (SAMHSA, 2016). NESARC did not give any monetary incentive, while NESARC III gave a $90 check incentive for each respondent who completed the interview (Grant et al., 2015). Third, NESARC-III was conducted in six languages, while the NESARC was conducted in English only. Fourth, a federal government agency (US Census Bureau) conducted the NESARC, while a nonprofit agency (Westat) conducted the NESARC-III (SAMHSA, 2016; Grant et al., 2015). This may have affected an individual's willingness to report illegal behaviors (e.g., cannabis use) differently at each NESARC wave and so may be related to the changes seen in NESARC studies that are not seen in the NSDUH data.

The associations of cannabis use and use disorders with younger age, men, low education, and other-than-full-time employment, are consistent with previous research (Chen, Kandel, & Davies, 1997; Compton et al., 2004; Hasin, Wall, et al., 2015, Hasin, Saha, et al., 2015, Hasin et al., 2016). The findings that prevalence is higher among non-Hispanic blacks than among non-Hispanic whites are in contrast to previous work (Compton et al., 2009) and suggest a shifting pattern of cannabis use in the US (Hasin, Saha, et al., 2015). Generally consistent with previous research are the associations of cannabis use disorders with depression and tobacco and other substance use (Merline, O'Malley, Schulenberg, Bachman, & Johnston, 2004; Schauer et al., 2015). Such co-occurrence is a stark reminder that comorbidity between psychiatric illness and use of multiple substances is common. When one psychiatric or behavioral problem is identified, it is essential that clinicians carefully probe for other related problems (Center for Substance Abuse Treatment, 1997; Dawson et al., 2010; Compton et al., 2015). Findings that cannabis use is more prevalent among adults residing in states with legalization of medical cannabis use than adults not residing in those states (Hasin, Wall, et al., 2015), and among adults who believed that the use of medical cannabis was legal in their state, suggest two alternatives: these may be the very locations where residents would favor such laws because of prior experience, or changes in the laws may have been related to increasing use. Evidence supporting both of these pathways has been found, suggesting an overall reciprocal relationship of social attitudes and cannabis use patterns (Cerda, Wall, Keyes, Galea, & Hasin, 2012; Schuermeyer et al., 2014; Pacula, Powell, Heaton, & Sevigny, 2015).

The number of adults in the US reporting past-year cannabis use increased by 10 million and the number with daily/near daily patterns of use increased by 4.4 million between 2002 and 2014. Understanding these trends is relevant for policy makers who continue to consider whether and how to modify laws related to cannabis and for healthcare practitioners who care for patients using cannabis. Lower perceived risk of cannabis use is associated with patterns of high frequency and early onset of use (Volkow et al., 2014; Silins et al., 2014), suggesting an increasing need for clinical intervention. Our results also indicate an increasing need for modifying risk perceptions of cannabis use among adults through effective education and prevention messages.

NEEDED RESEARCH

NSDUH excludes homeless people not living in shelters and those residing in institutions (e.g., those incarcerated), which may lead to underestimates in behaviors such as drug use and drug use disorders. Research should examine the epidemiology of cannabis use and use disorders among homeless people not living in shelters and among those residing in institutions. Moreover, associations between cannabis use and many specific psychiatric disorders could not be considered due to the lack of these measures in NSDUH questionnaires. Studies such as NESARC (e.g., Hasin et al., 2016; Compton et al., 2004) generally show direct associations of cannabis use disorders with mood, anxiety, and personality disorders in both 2001–02 and 2012–13 data. Yet, more studies are needed to assess the reasons for these associations between cannabis use and specific psychiatric disorders, which may be better studied in longitudinal samples (e.g., National Institutes of Health, 2016).

Because of the cross-sectional nature of NSDUH data, this study could not establish either temporal or causal relationships. For example, studying the relationship of tobacco use to cannabis use in longitudinal studies such as the recently launched Adolescent Brain Cognitive Development Study will be useful in determining whether the association represents a causal pathway or is a marker for other factors (National Institutes of Health, 2016). Also, NSDUH does not ascertain the use of electronic cigarettes, which have become common among youth (Kann et al., 2014; Johnston et al., 2015). Although our results based on NSDUH are consistent with trends found in other surveys (Kann et al., 2014; Johnston et al., 2015), longitudinal designs are needed to examine the complex relationships among tobacco use, use of electronic cigarettes, and cannabis use and use disorders among youth.

Key Chapter Points

- Among youth in the US, cannabis use decreased during 2005–14 compared with 2002, and cannabis use disorders declined during 2013–14 compared with 2002.
- Associations between the decline in tobacco use and the decrease in cannabis use among youth suggest the importance of tobacco control and prevention.
- Cannabis use among adults increased in the US during 2011–14 compared with 2002.
- Associations between cannabis use and decreases in perceived risk of smoking cannabis among adults suggest a potential benefit of education and prevention messages.
- For both youth and adults, the co-occurrence of cannabis use and use disorders with other substance use and depression highlights the importance of screening across the full range of behavioral health issues.

REFERENCES

Agrawal, A., Silberg, J. L., Lynskey, M. T., Maes, H. H., & Eaves, L. J. (2010). Mechanisms underlying the lifetime co-occurrence of tobacco and cannabis use in adolescent and young adult twins. *Drug and Alcohol Dependence*, *108*, 49–55.

American Association for Public Opinion Research (AAPOR). (2015). *Standard definitions: Final dispositions of case codes and outcome rates for surveys* (8th ed.). Lenexa, KS: Author. pp. 52–53.

American Psychiatric Association. (1994). *Diagnostic and statistical manual of mental disorders* (4th ed.). Washington, DC: American Psychiatric Association.

Bachman, J. G., Johnson, L. D., & O'Malley, P. M. (1998). Explaining recent increases in students' marijuana use: Impacts of perceived risks and disapproval, 1976 through 1996. *American Journal of Public Health*, *88*, 887–892.

Bieler, G. S., Brown, G. G., Williams, R. L., & Brogan, D. L. (2010). Estimating model-adjusted risks, risk differences, and risk ratio from complex survey data. *American Journal of Epidemiology*, *171*, 618–623.

Brook, J. S., Brook, D. W., De La Rosa, M., Duque, L. F., Rodriguez, E., Montoya, I. D., et al. (1998). Pathways to marijuana use among adolescents: Cultural/ecological, family, peer, and personality influences. *Journal of the American Academy of Child and Adolescent Psychiatry*, *37*, 759–766.

Buu, A., Dabrowska, A., Mygrants, M., Puttler, L. I., Jester, J. M., & Zucker, R. A. (2014). Gender differences in the developmental risk of onset of alcohol, nicotine, and marijuana use and the effects of nicotine and marijuana use on alcohol outcomes. *Journal of Studies on Alcohol and Drugs*, *75*, 850–858.

Cavazos-Rehg, P. A., Krauss, M., Fish, S. L., Salyer, P., Grucza, R. A., & Bierut, L. J. (2015). Twitter chatter about marijuana. *Journal of Adolescent Health*, *56*, 139–145.

Center for Substance Abuse Treatment. (1997). A Guide to Substance Abuse Services for Primary Care Clinicians. In *Treatment Improvement Protocol (TIP) Series, No 24*: Substance Abuse and Mental Health Services Administration. Washington, DC: U.S. Government Printing Office. DHHS Publication No. (SMA) 97–3139.

Cerda, M., Wall, M., Keyes, K. M., Galea, S., & Hasin, D. (2012). Medical marijuana laws in 50 states: investigating the relationship between state legalization of medical marijuana and marijuana use, abuse and dependence. *Drug and Alcohol Dependence*, *120*, 22–27.

Chen, K., Kandel, D. B., & Davies, M. (1997). Relationships between frequency and quantity of marijuana use and last year proxy dependence among adolescents and adults in the United States. *Drug and Alcohol Dependence*, *46*, 53–67.

Chen, X., Williamson, V. S., An, S.-S., Hettema, J. M., Aggen, S. H., Neale, M. C., et al. (2008). Cannabinoid receptor 1 gene association with nicotine dependence. *Archives of General Psychiatry*, *65*, 816–824.

Compton, W. M., Blanco, C., & Wargo, E. M. (2015). Integrating addiction services into general medicine. *JAMA*, *314*, 2401–2402.

Compton, W. M., Grant, B. F., Colliver, J. D., Glantz, M. D., & Stinson, F. S. (2004). Prevalence of marijuana use disorders in the United States: 1991-1992 and 2001-2002. *JAMA*, *291*, 2114–2121.

Compton, W. M., Han, B., Hughes, A., Jones, C. M., & Blanco, B. (2016). Marijuana use and use disorders among adults in the United States, 2002-2014. *Lancet Psychiatry*, *3*, 954–964.

Compton, W. M., Saha, T. D., Conway, K. P., & Grant, B. F. (2009). The role of cannabis use within a dimensional approach to cannabis use disorders. *Drug and Alcohol Dependence*, *100*, 221–227.

Dawson, D. A., Compton, W. M., & Grant, B. F. (2010). Frequency of 5+/4+ Drinks as a screener for drug use and drug use disorders. *Journal of Studies on Alcohol and Drugs*, *71*, 751–760.

ElSohly, M. A. (2015). *Potency Monitoring Program Quarterly Report Number 127: Reporting Period: 9/16/2014-12/15/2014*. Bethesda, MD: National Institute on Drug Abuse.

Grant, B. F., Chu, A., Sigman, R., Amsbary, M., Kali, J., Sugawara, Y., et al. (2015). Source and Accuracy Statement. National Epidemiologic Survey on Alcohol and Related Conditions-III. National Institute on Alcohol Abuse and Alcoholism. https://www.niaaa.nih.gov/sites/default/files/NESARC_Final_Report_FINAL_1_8_15.pdf Accessed 25.04.17.

Grucza, R. A., Abbacchi, A. M., Przybeck, T. R., & Gfroerer, J. C. (2007). Discrepancies in estimates of prevalence and correlates of substance use and disorders between two national surveys. *Addiction*, *102*, 623–629.

Grucza, R. A., Agrawal, A., Krauss, M. J., Cavazos-Rehg, P. A., & Bierut, L. J. (2016). Recent trends in the prevalence of marijuana use and associated disorders in the United States. *JAMA Psychiatry*, *10*. http://dx.doi.org/10.1001/jamapsychiatry.2015.3111.

Han, B., Compton, W. M., Jones, C. M., & Blanco, C. (2017). Cannabis use and cannabis use disorders among youth in the United States: 2002–2014. *Journal of Clinical Psychiatry*, *78*.

Han, B., Compton, W. M., Jones, C. M., & Cai, R. (2015). Nonmedical prescription opioid use and use disorders among adults aged 18 through 64 years in the United States, 2003-2013. *JAMA*, *314*, 1468–1478.

Hasin, D. S., Kerridge, B. T., Saha, T. D., Huang, B., Smith, S. M., Jung, J., et al. (2016). Prevalence and correlates of DSM-5 cannabis use disorder, 2012-2013: Findings from the national epidemiologic survey on alcohol and related conditions-III. *American Journal of Psychiatry*, *173*, 588–599.

Hasin, D. S., Saha, T. D., Kerridge, B. T., Goldstein, R. B., Chou, S. P., Zhang, H., et al. (2015a). Prevalence of marijuana use disorders in the United States between 2001-2002 and 2012-2013. *JAMA Psychiatry*, *72*, 1235–1242.

Hasin, D. S., Wall, M., Keyes, K. M., Cerdá, M., Schulenberg, J., O'Malley, P. M., et al. (2015b). Medical marijuana laws and adolescent marijuana use in the USA from 1991 to 2014: results from annual, repeated cross-sectional surveys. *Lancet Psychiatry*, *2*, 601–608.

Johnston, L. D., O'Malley, P. M., Miech, R. A., Bachman, J. G., & Schulenberg, J. E. (2015). *Monitoring the future national results on adolescent drug use: Overview of key findings, 2014*. Ann Arbor, MI: Institute for Social Research, the University of Michigan.

Jordan, B. K., Karg, R. S., Batts, K. R., Epstein, J. F., & Wiesen, C. (2008). A clinical validation of the National Survey on Drug Use and Health assessment of substance use disorders. *Addictive Behaviors*, *33*, 782–798.

Kandel, D. B. (Ed.), (2002). *Stages and pathways of drug involvement: Examining the gateway hypothesis*. Cambridge: Cambridge University Press.

Kann, L., Kinchen, S., Shanklin, S. L., Flint, K. H., Kawkins, J., Harris, W. A., et al. (2014). Youth risk behavior surveillance—United States, 2013. *MMWR Surveillance Summaries*, *63*(Suppl 4), 1–168.

Lopez, M. F., Compton, W. M., & Volkow, N. D. (2009). Changes in cigarette and illicit drug use among US teenagers. *Archives of Pediatrics and Adolescent Medicine*, *163*, 869–870.

Lopez-Quintero, C., Pérez de los Cobos, J., Hasin, D. S., Okuda, M., Wang, S., Grant, B. F., et al. (2011). Probability and predictors of transition from first use to dependence on nicotine, alcohol, cannabis, and cocaine: Results of the National Epidemiologic Survey on Alcohol and Related Conditions (NESARC). *Drug and Alcohol Dependence*, *115*, 120–130.

Merline, A. C., O'Malley, P. M., Schulenberg, J. E., Bachman, J. G., & Johnston, L. D. (2004). Substance use among adults 35 years of age: Prevalence, adulthood predictors, and impact of adolescent substance use. *American Journal of Public Health*, *94*, 96–102.

Monitoring the Future Study. (2015). Trends in annual prevalence of use of various drugs of for grades 8, 10, and 12 combined. http://www.monitoringthefuture.org/data/15data/15drtbl15.pdf Accessed 01.01.16.

National Institutes of Health (NIH). (2016). Collaborative research on addiction at NIH. Adolescent brain cognitive development study. http://addictionresearch.nih.gov/adolescent-brain-cognitive-development-study Accessed 15.04.16.

Pacula, R. L., Powell, D., Heaton, P., & Sevigny, E. L. (2015). Assessing the effects of medical marijuana laws on marijuana use: the devil is in the details. *Journal of Policy Analysis and Management*, *34*, 7–31.

Pacula, R. L., & Sevigny, E. L. (2014). Marijuana liberalization policies: Why we can't learn much from policy still in motion. *Journal of Policy Analysis and Management*, *33*, 212–221.

Patton, G. C., Coffey, C., Carlin, J. B., Sawyer, S. M., & Lynskey, M. (2005). Reverse gateways? Frequent cannabis use as a predictor of tobacco initiation and nicotine dependence. *Addiction*, *100*, 1518–1525.

Peters, E. N., Budneyb, A. J., & Carrollc, K. M. (2012). Clinical correlates of co-occurring cannabis and tobacco use: A systematic review. *Addiction*, *107*, 1404–1417.

Research Triangle Institute (RTI). (2015). *SUDAAN release 11.0.1*. Research Triangle Park, NC: RTI International.

Roy-Byrne, P., Maynard, C., Bumgardner, K., Krupski, A., Dunn, C., West, I. I., et al. (2015). Are medical marijuana users different from recreational users? The view from primary care. *American Journal on Addictions*, *24*, 599–606.

Schauer, G. L., Berg, C. J., Kegler, M. C., Donovan, D. M., & Windle, M. (2015). Assessing the overlap between tobacco and marijuana: Trends in patterns of co-use of tobacco and marijuana in adults from 2003-2012. *Addictive Behaviors*, *49*, 26–32.

Schepis, T. S., Desai, R. A., Cavallo, D. A., Smith, A. E., McFetridge, A., Liss, T. B., et al. (2011). Gender differences in adolescent marijuana use and associated psychosocial characteristics. *Journal of Addiction Medicine*, *5*, 65–73.

Schuermeyer, J., Salomonsen-Sautel, S., Price, R. K., Balan, S., Thurstone, C., Min, S. J., et al. (2014). Temporal trends in marijuana attitudes, availability and use in Colorado compared to non-medical marijuana states: 2003–11. *Drug and Alcohol Dependence*, *140*, 145–155.

Shiffman, S., Waters, A., & Hickcox, M. (2004). The nicotine dependence syndrome scale: A multidimensional measure of nicotine dependence. *Nicotine & Tobacco Research*, *6*, 327–348.

Silins, E., Horwood, L. J., Patton, G. C., Fergusson, D. M., Olsson, C. A., Hutchinson, D. M., et al. (2014). Young adult sequelae of adolescent cannabis use: An integrative analysis. *Lancet Psychiatry*, *1*, 286–293.

Substance Abuse and Mental Health Services Administration. (2010). In *Methodology series M-8, HHS publication no SMA 09-4425*: Reliability of key measures in the national survey on drug use and health. Rockville, MD: Substance Abuse and Mental Health Services Administration, Office of Applied Studies.

Substance Abuse and Mental Health Services Administration. (2012). In *NSDUH series H-42, HHS publication no (SMA) 11-4667*: Results from the 2010 national survey on drug use and health: Mental health findings. Rockville, MD: Substance Abuse and Mental Health Services Administration.

Substance Abuse and Mental Health Services Administration. (2016). National survey on drug use and health. http://www.samhsa.gov/data/population-data-nsduh/reports Accessed 15.02.16.

Substance Abuse and Mental Health Services Administration, Center for Behavioral Health Statistics and Quality. (2015). Results from the 2014 national survey on drug use and health: Detailed tables. http://www.samhsa.gov/data/sites/default/files/NSDUH-DetTabs2014/NSDUH-DetTabs2014.htm Accessed 15.02.16.

U.S. Department of Health and Human Services. (2014). *The health consequences of smoking—50 Years of progress: A report of the surgeon general*. Atlanta, GA: U.S. Department of Health and Human Services, Centers for Disease Control and Prevention, National Center for Chronic Disease Prevention and Health Promotion, Office on Smoking and Health.

Volkow, N. D., Baler, R. D., Compton, W. M., & Weiss, S. R. B. (2014). Adverse health effects of marijuana use: State of the science. *The New England Journal of Medicine*, *370*, 2219–2227.

Chapter 3

Cannabinoids and the Brain: The Effects of Endogenous and Exogenous Cannabinoids on Brain Systems and Function

David L. Atkinson*, Jeff K. Abbott†
*University of Texas-Southwestern, Dallas, TX, United States
†Northwestern University, Chicago, IL, United States

INTRODUCTION

The endocannabinoid system is important in early development and throughout life. In this chapter, we examine and contrast the natural function of the endocannabinoid system with changes that occur with the exogenous administration of cannabinoids. Moreover, we discuss how these exogenous cannabinoids, many of which are more potent agonists of the cannabinoid system than the endocannabinoids themselves, can accumulate in greater concentrations than naturally occurring endocannabinoids, and disrupt a system that has evolved to fine-tune neurons in early development and throughout adulthood.

The endocannabinoid system mediates multiple processes of neurodevelopment including synaptogenesis, cognition, memory formation, perception, reinforcement, motivation, salience, regulation of fear, and immune processes. Exogenous cannabinoids, whether direct agonists or modulators of the system, potentially disrupt this processes. Aberrant formation of neural connections and the development of psychosis also create dysfunction in these processes. Additionally, endocannabinoids have roles in feeding, movement, and peripheral signaling processes that have no clear link to psychosis, but the full extent of their function and actions on the brain and body are unknown. The key questions to address with regard to the effects of endogenous and exogenous cannabinoids on brain systems include the following:

1. What are the basic mechanisms of endocannabinoid signaling in the brain?
2. What is the role of dopamine and the endocannabinoid system in reward processing and reinforcement learning, and how does dopamine affect psychosis risk?

The Complex Connection between Cannabis and Schizophrenia. http://dx.doi.org/10.1016/B978-0-12-804791-0.00003-3

3. What is the role of the hippocampus in psychosis?
4. What are the effects of cannabinoids on memory?
5. How do endogenous and exogenous cannabinoid activity on memory systems affect the risk of psychosis?
6. What are the developmental effects of early-life cannabis exposure?
7. Is adolescence a critically sensitive developmental period for endocannabinoid signaling?

This chapter explores the natural endocannabinoid system, as well as the effects of exogenous cannabinoids, and how they possibly relate to the development of acute and chronic psychosis. Moreover, we address diverse effects of endocannabinoids and exogenous cannabinoids on perception, memory formation, affective functioning, reinforcement processes, and motivation.

Exogenous cannabinoids have clear effects on the mind, including effects on cognition, euphoric/motivational centers, fear circuits, memory, and motor circuits. The perception of time may be altered, and there can be a tendency for intoxicated individuals to perform actions more slowly. The euphoric effects of intoxication can enhance the subjective experience, color perceptions, as well as salience of other stimuli. Social bonds of friendship may be enhanced, or the individual may remain contented while socially withdrawn. Memory formation in the hippocampus may be reduced, but false associations may be increased. If undetected, these mistaken associations formed in the hippocampus may lead to psychotic thought processes.

Several different and complex hypotheses have attempted to explain psychosis and schizophrenia, ranging from dysregulation of dopamine, γ-aminobutyric acid (GABA), to glutamate (especially *N*-methyl-D-aspartate, NMDA) receptors (Sanchez-Blazquez, Rodriguez-Munoz, & Garzon, 2014). Endocannabinoids are known to interact with neurons that affect each of these neurotransmitters (Busquets Garcia, Soria-Gomez, Bellocchio, & Marsicano, 2016; Gleason, Birnbaum, Shukla, & Ghose, 2012; Volk & Lewis, 2016; Volkow et al., 2010; Zamberletti et al., 2014), as well as other receptors and intracellular pathways. As is described in Chapter 4, Δ^9-tetrahydrocannabinol (THC) and other CB_1 agonists are a factor in the development of schizophrenia and are acute causes of psychosis. The ability of exogenous cannabinoids to cause psychosis does not necessarily mean that pathologies of the endocannabinoid system are a component cause of "endogenous psychosis," but it is worthwhile to describe the normal functioning of the endocannabinoid system to understand how the natural functions may be disturbed to produce psychosis. The endogenous cannabinoids exert a homeostatic role regulating excessive inhibition and excitation by maintaining a constant tone of activity in response to activation of the postsynaptic circuit (Caballero & Tseng, 2012). Rises in activity of some brain areas are met with compensatory increases or decreases in endocannabinoid levels to maintain homeostasis.

To examine the complex question of how cannabinoids might cause psychosis, we begin by looking at the endocannabinoid system, and then examine the actions and disruptions caused by exogenous cannabinoids on these systems in psychosis. We conclude by contrasting the effects of endocannabinoids and exogenous cannabinoids in reference to age-dependent development of the brain.

THE BIOLOGY OF THE ENDOCANNABINOID SYSTEM

The endocannabinoid system is diversely expressed throughout the brain where it regulates many more processes than originally hypothesized. The system involves the regulation and fine-tuning of neurotransmission, memory formation, and response to inflammation, while playing a critical role in neurodevelopment. The endocannabinoid system has wide distribution throughout the cortex, cerebellum, and subcortical areas and is involved in numerous neuropsychological functions across different anatomical locations.

Endogenous Ligands

A ligand is a substance that forms complexes with proteins to effect some biological action. They are activators of various brain receptors. Exogenous cannabinoid ligands, such as THC, were discovered prior to the discovery of endogenous ligands, or endocannabinoids, such as 2-arachidonoylglycerol (2-AG) and anandamide (also known as *N*-arachidonoylethanolamine, or AEA). For this reason, the receptors are called "cannabinoid receptors." The receptors can either be activated at their "regular" binding site, to which endogenous ligands bind, termed *orthosteric modulation*; or, they may bind at a different site, termed *allosteric modulation*. Allosteric modulation can either be positive, enhancing the effects of the natural ligand, or negative, reversing the effects of the natural ligand. These numerous processes can make cannabinoid pharmacology quite complicated.

Ligands can either be full agonists, partial agonists, neutral antagonists, or partial or full inverse agonists. For example, WIN-55 is a full agonist at CB_1 receptors, while THC is only a partial agonist at the receptor. Tetrahydrocannabivarin (THCV) is a neutral antagonist, and may have psychotherapeutic potential, while rimonabant is a partial inverse agonist. Rimonabant is not simply a blocker of endocannabinoid transmission, but produces effects opposite to what endocannabinoid signaling typically does and may cause adverse psychiatric side effects (McPartland, Duncan, Di Marzo, & Pertwee, 2015). Complicating things further is that whether a chemical is a partial agonist or a full agonist at the receptor may depend on which cell type is being innervated (Pertwee, 2008), and the firing rate of the circuit (Roloff & Thayer, 2009). While we will not go into many of these subtleties in this chapter, it is important to understand the complexities of the endocannabinoid system and the regional specificity of the pharmacologic action of any compound (Fig. 1).

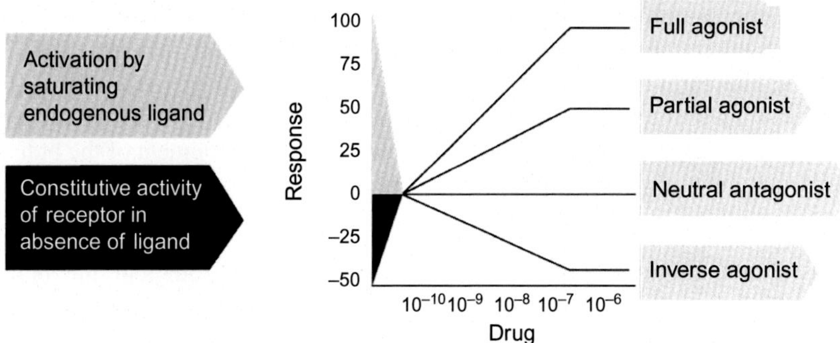

FIG. 1 Spectrum of agonist activity.

CB_1 Receptors

CB_1 receptors are expressed in the hippocampus, mesolimbic dopamine areas (ventral tegmental area and nucleus accumbens), cingulate cortex, prefrontal cortex, and cerebellum. The wide distribution of these receptors may help to explain the diverse effects of manipulating the endocannabinoid signaling system. CB_1 receptors are g-coupled receptors, and in fact are the most prominent $g_{i/0}$ receptors in the human central nervous system (CNS). $G_{i/0}$ proteins downregulate cyclic adenosine monophosphate (cAMP). CB_1 agonism also activates the mitogen-activated protein (MAP) kinase intracellular signaling pathways. The intracellular workings of CB_1 are largely beyond the scope of this chapter, but CB_1 receptors are also present in mitochondria, where they may exert effects on memory formation (Hebert-Chatelain et al., 2016).

CB_1 receptors largely act on presynaptic neurons by decreasing GABA release. This can lead to a net facilitation of dopamine release, as in the ventral tegmental area, when the reduction of inhibitory, GABAergic tone facilitates phasic, burst firing of dopamine (Laricchiuta, Musella, Rossi, & Centonze, 2014). However, in certain brain areas, such as the hippocampus, the CB_1 receptor acts as an excitatory receptor by increasing glutamate. CB_1 receptors are also found in astrocytes and glia, in addition to the signaling neurons, and the receptors may be found inside of the cell (Busquets Garcia et al., 2016). In short, given the widespread distribution in the brain, as well as among different cell types, CB_1 receptors mediate diverse effects. To summarize, CB_1 receptors are responsible for maintaining a delicate balance between neuronal inhibition and excitation, especially in GABAergic, glutamatergic, and dopaminergic transmission (Fig. 2).

CB_2 Receptors

In the brain, CB_2 receptors are mainly localized in the brainstem. CB_2 receptors are also found in abundance, often collocated with CB_1 receptors, on cells involved in the peripheral immune system, where their effects have been known for

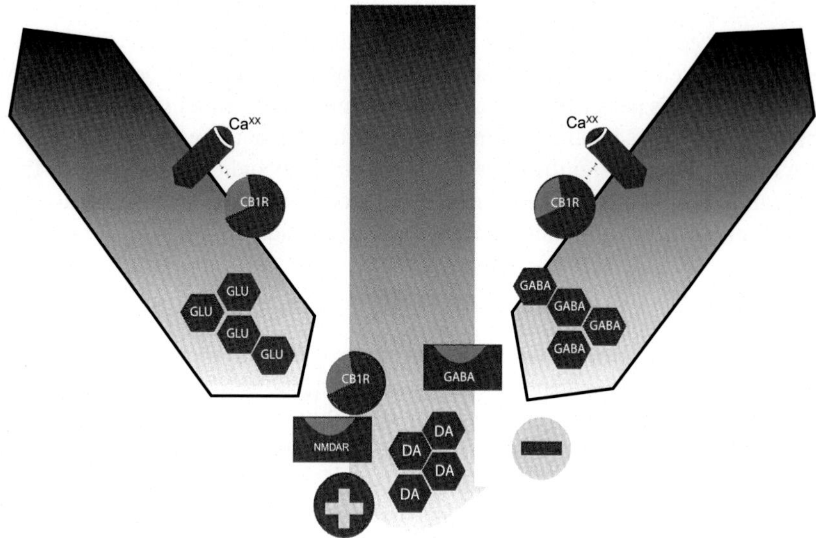

FIG. 2 Schematic of cannabinoid modulation of dopamine release through differential actions of glutamate and GABA.

some time (Karmaus, Chen, Crawford, Kaplan, & Kaminski, 2013). However, CB_2 receptor gene variations have recently been correlated with risk for psychosis (Ishiguro et al., 2010) and the receptor is expressed in the hippocampus (Brusco, Tagliaferro, Saez, & Onaivi, 2008), challenging the assumption that CB_2 receptors are largely unimportant in the brain and for brain processes. CB_2 receptors are also involved in CNS inflammation (Lisboa, Gomes, Guimaraes, & Campos, 2016), which has been an increasingly discussed and researched aspect of the pathophysiology of schizophrenia (Muller, Weidinger, Leitner, & Schwarz, 2015).

TRPV and Other Receptors

A more recently characterized group of receptors are transient receptor potential (TRP) cation channels, or specifically TRPV ("V" for vanilloid) receptors, which are calcium-sensitive channels that are activated by anandamide and other endocannabinoids that may have a significant role in seizures (Devinsky et al., 2014). TRPV receptors may also play a role in hyperdopaminergic states (Tzavara et al., 2006), which could have importance in the pathogenesis of schizophrenia, while other effects are not yet fully characterized. Some researchers have suggested that they be reclassified as CB_3 receptors, but this change to the nomenclature is not currently accepted. It is worth mentioning in order to highlight that the number of "canonical endocannabinoid receptors" has not yet been established.

Related to cannabinoid and endocannabinoid pharmacology, G-protein-coupled receptor 55 (GPR-55) is considered an orphan receptor that may be involved in developmental processes (Cherif et al., 2015) and seizures (Devinsky et al., 2014). It is possible that cannabidiol is an allosteric modulator of GPR-55, but its endogenous ligand is unknown. THC does not act directly on this receptor, but the possible significant interaction with cannabidiol and the theory that cannabidiol may prevent THC-related psychotic processes means that it may ultimately be part of the cannabis and psychosis equation.

MECHANISMS OF ENDOCANNABINOID TRANSMISSION

Endocannabinoids are not as rapid acting as typical neurotransmitters, because rather than being stored in the cell awaiting release, they are synthesized on demand (Sugiura, Kishimoto, Oka, & Gokoh, 2006; van der Stelt et al., 2003). After their activation of presynaptic receptors, they are internalized into the cell and degraded (van der Stelt et al., 2003). The diffusion of endocannabinoid neurotransmitters affects more than one neuron in the presynaptic area, and can spread to the neighboring inhibitory interneurons in the region of the synapse (Kreitzer, Carter, & Regehr, 2002), thus modulating more targets than typical anterograde transmission.

2-Arachidonoylglycerol

2-AG is one of the most important endocannabinoids in the CNS. 2-AG is more tonically active in the adult CNS, whereas anandamide (AEA) is tonically active during development. 2-AG seems to act more as a neuromodulator, whereas AEA acts as a growth factor during development and a stress-responsive modulator during adult life. 2-AG exertsits its effects on CB_1 receptors that can be found in many brain areas, such as the hippocampus, the mesolimbic dopamine area (ventral tegmental area and nucleus accumbens), the cingulate cortex, the prefrontal cortex, and the cerebellum. It has been well established that these receptors lay on the presynaptic regulatory neurons of both GABA and glutamate receptors, and act as a retrograde feedback mechanism via 2-AG, which is synthesized on demand by the postsynaptic neuron (van der Stelt et al., 2003).

2-AG Metabolism

2-AG metabolism is regulated by the enzyme monoacylglycerol lipase (MAGL). MAGL-knockout mice have significantly elevated 2-AG levels, but they have no evidence of behavioral change in many paradigms. However, these mice showed an interesting improvement in a water maze test, indicating an improvement in memory functions (Pan et al., 2011), contrasted with the deficits in memory that exogenous cannabinoids are known to induce (Senn, Keren, Hefetz, & Sarne, 2008). MAGL inhibitors do produce reduction of pain responses in animal models, but do not produce attenuated opioid withdrawal responses. The MAGL knockouts had brain CB_1 receptor adaptations that resulted from altered 2-AG levels, and many peripheral effects were seen, with changes in body temperature and locomotor activity (Ramesh et al., 2011).

Anandamide

Anandamide (AEA) is synthesized in the brain, where levels are controlled by tightly regulated degradation pathways. Its effects in the brain are extremely localized. Like 2-AG, it is a retrograde neurotransmitter whose release can affect many cells in the neighboring synaptic space, but not beyond that space. It does not have the diffuse action of a hormone; its release is locally controlled and does not affect the organism systemically in the way that exogenous cannabinoid administration does. AEA can be characterized as a stress-responsive neurotransmitter, and is largely not as tonically active as 2-AG in the adult brain. However, if AEA is induced to be constantly active, the brain will downregulate its responsiveness to AEA, thus decreasing its neuromodulatory functionality, and the signaling function carried out by stress-responsive variations of anandamide levels will be disrupted. In turn, the endocannabinoid system will not be able to carry out its fine-tuning of neurotransmission (Lu & Mackie, 2016; Lu & Anderson, 2017).

In contrast, AEA is tonically active during development. The function of AEA during this period is not as a retrograde neurotransmitter, but as an important promoter and regulator of neuronal development, where it seems to be critical to processes such as axonal pathfinding and synapse formation. Disruption during neurogenesis could have severe effects (Rivera et al., 2015).

AEA Metabolism

The metabolism of AEA is controlled by fatty acid amide hydrolase (FAAH). After neurogenesis, the inhibition of FAAH seems to lead to effects that trigger rapid adaptation and tolerance. However, FAAH inhibition can lead to sustained and effective pain control when combined with a partial inhibition of MAGL (Ghosh et al., 2015), and a combination targeting these two enzymes would likely yield fewer cannabimimetic side effects. Results such as these demonstrate the potential for modulating the cannabinoid system with greater precision than achieved through administration of an exogenous CB_1 agonist, and highlight the complicated interplay of different endocannabinoids and their metabolisms in the brain.

NEUROPLASTIC CHANGES INDUCED BY CANNABINOIDS

Depolarization-Induced Sustained Inhibition or Excitation

The immediate modulation exhibited by the cannabinoid signaling system can be seen though depolarization-induced sustained inhibition (DSI) or depolarization-induced sustained excitation (DSE). These changes are attributable to the paired calcium channels that open in response to CB_1 activation. In turn, this raises or lowers the threshold for GABAergic or glutamatergic neuron firing, depending on the brain region examined (Zhu & Lovinger, 2010). This underscores some of the more acute changes in neuroplasticity secondary to cannabinoid receptor modulation and helps explain the "fine-tuning" of neurotransmission.

Long-Term Potentiation or Depression

Long-term potentiation (LTP) is a critical neurobiologic underpinning of memory formation. In circuits such as those involving the hippocampus, the effects of the endocannabinoid system on glutamatergic signaling are to suppress excitation (Diana & Marty, 2004). LTP in the hippocampus can be affected by the activation of CB_1 receptors (Abush & Akirav, 2012; Maroso et al., 2016; Misner & Sullivan, 1999). With chronic administration of exogenous cannabinoids, other adaptive changes occur in other areas. Chronic THC exposure induces long-term depression (LTD) of dopamine neurons in the ventral tegmental area (Liu et al., 2010).

LTP and LTD lead to a hypo- or hyper-excitable neuron (Sanchez-Blazquez et al., 2014) and has been shown to be mediated by the NMDA glutamate receptor (Kullmann, Asztely, & Walker, 2000). Activation of NMDA leads to intercellular changes via protein kinase A secondary to adenylyl cyclase/cAMP leading to LTP (MacDonald, Jackson, & Beazely, 2007). In the ventral tegmental area (VTA), chronic activation of exogenous activation of CB1 receptors causes long term depression of the dopamine neurons via activation of NMDA receptors (Liu et al., 2010).

Together, without disruption from exogenous drugs of abuse LTP and LTD achieve a delicate balance of excitation and inhibition that is regulated by activity of dopamine. Endocannbinoid effects on LTD are enhanced by D2 receptor activation, showing the interplay between the two systems (Pan, Hillard, & Liu, 2008). As mentioned before, because 2-AG is synthesized on demand by activation of calcium channels, 2-AG is only utilized while the dopamine neuron is active. However, when the brain is repeatedly flooded with exogenous, untargeted, CB_1 agonists, such as THC, this leads to states that are hypoglutamatergic, with improperly functioning NMDA receptors and dysregulated GABA release, which ultimately leads to both the over- and under-release of dopamine in specific brain regions, similar to what is found in individuals with schizophrenia and other psychotic disorders (Sanchez-Blazquez et al., 2014).

Endocannabinoids and Psychosis

As cannabis use is thought to be a component cause of schizophrenia for some affected individuals, endocannabinoid dysfunction may also be a component "cause" in the development of schizophrenia or psychosis even in the absence of exogenous cannabinoid use. This putative role is possibly more important for some individuals than others. Studies have suggested that individuals with schizophrenia have significant variation in their levels of 2-AG and AEA compared to healthy controls (Muguruza et al., 2013). Elevated 2-AG levels have been shown in individuals with schizophrenia, but it is unknown whether this is part of a causal mechanism or a compensatory reaction to another imbalance in the brain (Fakhoury, 2016; Vigano et al., 2009). AEA levels have also been shown to change when dopamine metabolism is altered. Dopamine transporter knockout mice have shown increased AEA levels in the striatum (Tzavara et al., 2006), and the behavioral alterations of these mice were improved by administering drugs that affect the degradation and reuptake of AEA. The failure of a CB_1 antagonist to affect the phenotype suggests that it is mediated by TRPV receptors.

AEA is degraded to arachidonic acid by an integral phospholipid membrane enzyme, FAAH (Deutsch et al., 2001). Not surprisingly, FAAH-knockout mice have much higher anandamide levels, and these mice have reduced responses to painful stimuli. A similar effect is seen with CB_1 antagonists such as rimonabant and others. FAAH inhibitors have been launched for clinical development as an opportunity for pain signal modulation, and perhaps a relief from chronic pain without the development of tolerance. To determine safety of these compounds, it will be very important to study the effects of changing FAAH metabolism on psychosis, and their safety. Moreover, because inhibition of FAAH would lead to great increases in AEA, this could lead to rapid tolerance, or even possibly major disruptions in neurogenesis if administered during key developmental stages (Basavarajappa, 2015).

Enzymatic production of endocannabinoids is mediated through a phospholipase C enzyme, diacylglycerol lipase (DAGL). There are two forms of enzymes, alpha and beta, which are both expressed in the brain, alpha being expressed in neurons and beta expressed in microglia. The disruption of DAGL in the neuron disrupts multiple synaptic plasticity pathways. Excess 2-AG could possibly be linked to psychosis through its involvement in inflammation (Szafran, Borazjani, Lee, Ross, & Kaplan, 2015), which has been increasingly hypothesized to play a role in schizophrenia (Muller et al., 2015). It is possible that modulation of DAGL-beta might target the inflammatory response in the CNS, which could be part of the pathophysiologic process in schizophrenia; but this is still untested. Peripherally, DAGL has been shown to be dysregulated in first-episode psychosis patients (Bioque et al., 2013), and DAGL has been shown to be upregulated in the dopaminergic circuits following social isolation, which produces numerous alterations in the brain similar to those found in schizophrenia (Robinson, Loiacono, Christopoulos, Sexton, & Malone, 2010).

Individuals' differences in endocannabinoid genetics, both in receptors and in metabolic pathways, may be clinically significant in determining an individual's risk of developing psychosis with or without the addition of exogenous cannabinoids. Differences in endocannabinoid signaling are associated with elevated rates of alcohol and marijuana abuse, such as variations in CNR1 (the gene encoding the CB_1 receptor) and FAAH polymorphisms. Specifically, FAAH genotype is associated with increased craving and negative affect in withdrawal, and certain CNR1 genotypes are more related to the development of psychosis than others (Fernandez-Espejo, Viveros, Nunez, Ellenbroek, & Rodriguez de Fonseca, 2009), as well as to craving and structural brain differences in marijuana-using individuals (Filbey, Schacht, Myers, Chavez, & Hutchison, 2010; Haughey, Marshall, Schacht, Louis, & Hutchison, 2008; Schacht, Hutchison, & Filbey, 2012).

MECHANISMS OF PSYCHOSIS AND THE EFFECTS OF CANNABINOIDS

Perception and Psychosis

Building upon the scientific foundation discussed earlier, it is possible to examine the biological plausibility of a connection between cannabis and psychosis by further evaluating the roles of various biological systems under the effects of exogenous cannabinoids. Here we examine alterations of the dopamine system and mesolimbic circuits, the NMDA receptor system, the hippocampal formation, as well as prefrontal and other cognitive control circuits.

Dopamine, Cannabinoids, and Schizophrenia

Owing to the historical prominence and wide familiarity of the dopamine hypothesis of schizophrenia, a natural question to ask is whether the psychotomimetic effects of cannabis are mediated by dopamine (Kuepper et al., 2010). Dopamine response is clearly blunted among chronic cannabis users, showing a dose-dependent effect (Albrecht et al., 2013; DiNieri et al., 2011; Kowal, Colzato, & Hommel, 2011; Volkow et al., 2014). However, the degree of this effect is not correlated with the level of psychotic symptoms in these users (Bloomfield et al., 2014), and the magnitude of the dopamine release also does not correlate with the degree of psychotic symptoms that are acutely produced by cannabinoids (Sherif, Radhakrishnan, D'Souza, & Ranganathan, 2016). These findings suggest that the effects of cannabis on psychosis are not explained by these alterations in the dopaminergic system. The modern dopaminergic hypothesis of schizophrenia is much more complex than hypothesizing cortical hypofunctioning and mesolimbic hyperfunctioning, and readers wanting a more in-depth analysis of the dopamine hypothesis of schizophrenia are directed to recent reviews on the subject (Brisch et al., 2014; Eyles, Feldon, & Meyer, 2012; Perez & Lodge, 2014).

Endocannabinoids, Psychosis, and Perception and Understanding of Self

The Research Domain Criteria (RDoC) framework has defined a subconstruct of the Perception and Understanding of Self construct, “Agency,” as the ability to recognize one's self as the agent of one's actions and thoughts, including the recognition of one's own body and body parts. Agency can be impaired by disrupting some of the frontotemporal connections in the brain, which seems to occur in both schizophrenia (Wible, 2012) and cannabis use (Stone et al., 2012). When there are breakdowns in the "Agency" system of the brain, actions planned and carried out in the brain often feel foreign to the individual, who loses the necessary brain synchrony to be aware of one's own agency in generating these actions. This can lead to individuals believing that their actions, words, and thoughts arose from an external source. Hallucinations are generated from Broca's area (McGuire, Shah, & Murray, 1993), but are thought of as emanating from a foreign source to the person with schizophrenia. The subconstruct of "Agency" is likely to be important in the development of cannabis-induced psychosis as well as all other hallucinatory psychoses.

Shared Biomarkers of Schizophrenia and Cannabinoid Activity

There have been numerous research paradigms utilizing biomarkers of schizophrenia, and there are several biomarkers in animal models of schizophrenia that are thought to be correlates of human biomarkers of the disorder. Cannabis administration shows effects similar to those found in schizophrenia, such as P50 suppression, P300 potential alterations, and mismatch negativity (MMN) (Steeds, Carhart-Harris, & Stone, 2015). P50 suppression is the term for a process by which a stimulus is presented multiple times, leading to a suppression of amplitude in the P50 auditory-evoked potential after repeated exposure of the stimulus. An analogous finding in rodents may be the N40 auditory-evoked potential (Swerdlow et al., 2012), which is reduced in amplitude after neonatal lesioning of the ventral hippocampus (Chen et al., 2012). The P300 potential is a positive wave on EEG (electroencephalogram) that is usually elicited after the presentation of an "oddball" stimulus, which reaches its peak about 300 ms after the presentation of non-matching "oddball," or novel, stimulus. A reduction in amplitude is not only demonstrated in individuals with schizophrenia, but also in their unaffected relatives (Earls, Curran, & Mittal, 2016) and high-risk groups (del Re et al., 2015). The P300 wave amplitude is affected by the CNR1 genotype in individuals who have been administered THC (Stadelmann et al., 2011).

MMN is a laboratory investigational paradigm where a series of repeated tones are given, and then the tones change in temporal frequency, volume, duration, or pitch. MMN is one of the best-established electrophysiologic correlates of chronic schizophrenia (Salisbury, Shenton, Griggs, Bonner-Jackson, & McCarley, 2002), but with little difference shown in first-episode patients (Haigh, Coffman, & Salisbury, 2016) or in unaffected relatives when using most EEG channels (Earls et al., 2016). This suggests that it could be a marker of a later-stage disruption. Disturbances in these paradigms are thought to involve the lack of persistence of memory for the previous stimulus, and a lack of implicit, associative learning. The findings may involve one of the working memory subconstructs of the RDoC framework. All of these disruptions are seen in individuals who have been using cannabis chronically (Gallinat, Rentzsch, & Roser, 2012). However, it remains uncertain whether these findings are due to the types of individuals who use cannabis chronically, because these disruptions have also been seen in other individuals; e.g., those with attention-deficit/hyperactivity disorder (ADHD) (Cheng, Chan, Hsieh, & Chen, 2016; Micoulaud-Franchi et al., 2016; Micoulaud-Franchi et al., 2015), which has been associated prospectively with greater risk of cannabis use even after controlling for conduct disorder (Elkins, McGue, & Iacono, 2007; Ottosen, Petersen, Larsen, & Dalsgaard, 2016). Controlling for ADHD symptoms and status in the users does not give one a clear picture of the effects of cannabis, because the cannabis use is also associated with greater persistence of ADHD symptoms (Fergusson & Boden, 2008), and could lead to an under-identification of negative effects of drug use. Complications such as these underscore the necessity of longitudinal studies, and phenotyping of childhood presentation of ADHD and preexisting temperamental factors.

The neonatal ventral hippocampus lesion is a model of schizophrenia in animals (Lipska, 2004). Sensory gating is impaired by CB_1 receptor activation (Hajos, Hoffmann, & Kocsis, 2008), which is related to deficits in prepulse inhibition both in individuals with schizophrenia (Swerdlow et al., 2006) and in mice that are given neonatal hippocampal lesions (Swerdlow et al., 2012). There is also a progressive decrease in prepulse inhibition and fear conditioning seen in animals exposed to cannabis (Gleason et al., 2012), as well as in individuals with psychosis (Schneider & Koch, 2003).

Prefrontal Cortex Anomalies

Given the importance of appropriate prefrontal cortex functioning in regulating multiple aspects of human behavior, it would be surprising if the prefrontal cortex was not involved in any given psychopathology. Prefrontal cortex dysfunction is, in fact, seen in schizophrenia (Minzenberg, Laird, Thelen, Carter, & Glahn, 2009; Tan, Callicott, & Weinberger, 2007), and is also a hallmark of drug use (Koob & Volkow, 2010). The prefrontal cortex has multiple important roles in attentional processing, goal selection, and salience processing, and is involved in the previously mentioned domain of working memory and paradigms such as prepulse inhibition.

Adolescence is thought to be a critical time of prefrontal cortical development, with developmentally sensitive periods of these higher-order neural circuits occurring later than in other cortical areas (Tau & Peterson, 2010). Optimal prefrontal functioning depends on proper functioning and balance of dopaminergic and GABAergic systems, which reorganize during adolescence (Bossong & Niesink, 2010). Endogenous cannabinoids and activation of NMDA receptors determine whether synaptic connections will be strengthened or pruned in the cortical pyramidal neurons (Bossong & Niesink, 2010). Exogenous administration of either an NMDA antagonist (phencyclidine/ketamine) or a CB_1 agonist might greatly disrupt the intricate and delicate process, and repeated administration could lead to abnormal structural changes. As brain circuits and systems are pushed to their limits either in excess or deficiency, they lose the ability to momentarily modulate and fine-tune responses that have evolved over eons.

There are similar differences in cortical oscillations in the theta and gamma range among cannabis users and those with schizophrenia (Caballero & Tseng, 2012), and there are characteristic deficits found in event-related potentials (see above). The remodeling of the prefrontal cortex is actually dependent upon hippocampal glutamatergic efferent neurons synapsing on the prefrontal cortex (Tseng, Chambers, & Lipska, 2009), so parsing psychosis into component brain areas proves a difficult exercise as circuitry becomes more important than neuroanatomic location. Prefrontal CB_1 activity may be important because CB_1 receptors inhibit excitatory synaptic transmission from these glutamatergic inputs from the hippocampus. Ultimately, this would lead to decreased functional maturation of GABAergic interneurons, resulting in decreased synchrony of the prefrontal network (Caballero & Tseng, 2012).

Interindividual Vulnerability to Cannabis-Induced Psychosis

The liability to cannabis-related psychosis is a dimension that can be measured through sensitivity to THC in the laboratory (D'Souza, Sewell, & Ranganathan, 2009), which has shown great interindividual variability. Moreover, individuals who are screened and determined to be healthy controls are still found to be vulnerable to the psychotomimetic effects of THC (D'Souza et al., 2004). General psychosis liability impacts this risk of developing THC-related psychotic disorders (Verdoux, Gindre, Sorbara, Tournier, & Swendsen, 2003), but the extent to which vulnerability to THC-related psychotic disorders overlaps with general psychosis risk is unknown. It is uncertain whether the sensitivity to THC-induced psychosis is independent of general psychosis risk, and it is also uncertain as to how many individuals who developed psychosis related to THC use would not have had any problems without exposure to the drug.

Lack of a family history of schizophrenia may reduce the risk for THC-induced psychosis, but not eliminate it (Arendt, Mortensen, Rosenberg, Pedersen, & Waltoft, 2008). However, a very interesting possibility has been raised by the Bipolar-Schizophrenia Network on Intermediate Phenotypes (B-SNIP) group, which collected data on subtypes of psychosis, searching for an endophenotypic correlation across individuals with bipolar disorder, schizoaffective disorder, and schizophrenia. One phenotype did emerge with less baseline cognitive impairment and greater lifetime marijuana use, and less family history of psychosis (Carol Tamminga, personal communication). This data has not been published, as it is awaiting replication from other sites. The presence of a cannabis-related phenotype would be extremely interesting, and would help elucidate potential causal mechanisms.

Catechol-*O*-methyltransferase (COMT) is an important brain enzyme in the degradation of monoamine neurotransmitters. A functional polymorphism at position 158 in the gene encoding COMT substitutes a Val-encoding codon for the "wild-type" Met-encoding allele. This creates a more active version of the enzyme, degrading dopamine more rapidly. The Met-encoding allele produces an enzyme that only has 25% of the metabolic activity of the enzyme deriving from the Val allele. The enzymatic phenotype is present whether the individual is homozygous or heterozygous, and the genotype has been associated with increased risk for cannabis-induced psychosis in some (Caspi et al., 2005; Ermis et al., 2015; Estrada et al., 2011), but not all studies (Costas et al., 2011; Morgan, Freeman, Powell, & Curran, 2016; Nawaz & Siddiqui, 2015; Zammit et al., 2007). Other studies have partially replicated the finding, as in the study where individuals with the Val allele had greater hallucinations in the flow of day-to-day life with cannabis exposure. However, this was only seen in individuals who evidenced some psychosis vulnerability on psychometric tests (Henquet et al., 2009). There are also alterations shown in the caudate nucleus in Val allele carriers that have been moderated, at least statistically, by cannabis use status (Batalla et al., 2014). It is important to question whether this is an additive effect or a true gene-by-environment interaction, as the COMT Val allele has been associated with elevated psychosis risk independent of cannabis use (Chen, Wang, O'Neill, Walsh, & Kendler, 2004; Molero, Ortuno, Zalacain, & Patino-Garcia, 2007; Nawaz & Siddiqui, 2015). Another study failed to demonstrate that COMT polymorphisms were a distinct risk (Morgan et al., 2016).

PSYCHOSIS AND CANNABIS: INTERSECTION IN THE HIPPOCAMPUS

A Working Model of Cannabis-Induced Psychosis Based on Declarative Memory Dysfunction

False memory formation can be explained by the fact that multiple possible associations can be made from a given set of stimuli. If associations are made more easily, then the individual could have insight or creativity; on the other hand, associations could be false and improper associations might be made which would need to be weeded out. The establishment of an experience as real does not follow the rules of strict logical deduction, but it seems to work fairly well, and fairly consistently in most of us when we are healthy. The individual must integrate numerous inputs simultaneously, and it may be that for every individual a different level of coherence is required before the association is accepted as valid and encoded into memory. Some may make associations very easily, and some not easily enough, such that causally related events escape detection. Coordination of multiple inputs, with the formation and checking of proper associations, may be the function of the endocannabinoid system that is disrupted in psychosis. It may be that psychotic delusions are the result of inappropriate associations being laid down as memories, which would explain the notoriously recalcitrant nature of delusions, readily apparent to mental health professionals in the clinical setting.

The hippocampus has been identified as a critical brain area in schizophrenia, and has numerous highly important functions. Proper hippocampal functioning is critical in generating normal memories. Memory formation occurs through the connection of certain pieces of information in the environment being associated, and then those associations being checked before being laid down as a memory (Tamminga, Stan, & Wagner, 2010).

This requires the proper connectivity of the dentate gyrus and the CA1 and CA3 hippocampal regions (Liu et al., 2012). The dentate gyrus sends mossy fiber pathways to the CA3 area. The level of stimulation from the dentate gyrus affects categorical changes in the CA3 area (Pelkey & McBain, 2008). In the CA3 area, a delicate balance of inputs is required to produce accurate "pattern completion" of associations (Kremin & Hasselmo, 2007). It is thought that schizophrenia could be a disorder of aberrant pattern completion without proper error checking, correlated with inadequate formation of accurate memory tracings, causing inaccurate declarative memories to be encoded, leading to the clinical manifestation of delusions.

Outside the hippocampus, the majority of CB_1 receptors are associated with GABAergic inhibitory interneurons, whereas the synaptic association areas of the hippocampus are densely populated with CB_1 receptors and are mostly glutamatergic neurons (Fig. 3).

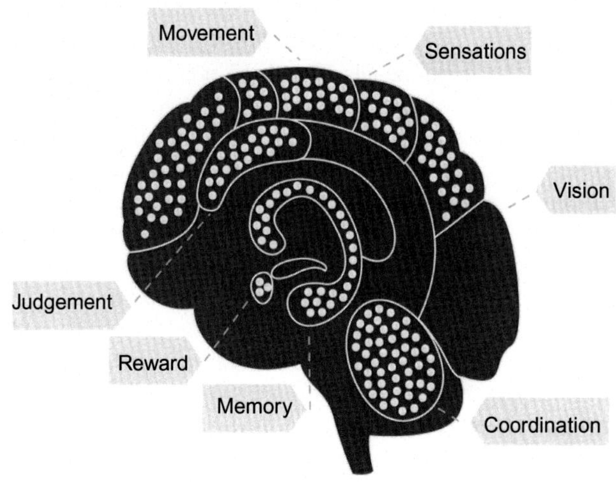

FIG. 3 CB_1 receptor distribution in the brain.

Effects of Cannabis on the Hippocampus and the Relationship to Schizophrenia

The effects of cannabis on these areas of the hippocampus are well documented, giving plausibility to the hypothesis of a central role for endocannabinoid transmission in the hippocampus in the pathogenesis of cannabis-related psychosis or schizophrenia. Murine research has shown that cannabis exposure led to significant reductions in dendritic length in hippocampal pyramidal neurons (Scallet et al., 1987) and reduced neuronal density (Landfield, Cadwallader, & Vinsant, 1988). Critical to the question of cannabis-induced psychosis through long-term hippocampal dysfunction, animal studies have suggested that the effects of cannabis on LTP persist after discontinuation of the drug (Hill, Froc, Fox, Gorzalka, & Christie, 2004). The effects of cannabis exposure on the hippocampus are robustly demonstrated in human volumetric studies (Bolla, Eldreth, Matochik, & Cadet, 2005; Yucel et al., 2008), and recreational users have been found to have lower total hippocampal gray matter (Demirakca et al., 2011). Even individuals studied after being abstinent for months showed reduced volume (Yucel et al., 2012; Ashtari et al., 2011), indicating that this area might be more vulnerable to at least "medium-term" changes from exposure to the drug. A positron emission tomography study (Hirvonen et al., 2012) showed that reductions in CB_1 receptor density returned to normal in all brain areas *except the hippocampus* after 28 days of abstinence, meaning that this is an area particularly vulnerable to disruption that continues after drug cessation (Hirvonen et al., 2012). Lorenzetti, Solowij, and Yucel (2016) summarized the deficits documented with neuroanatomic imaging, and they reviewed the difficulties in the field at its current state. In general, there was an association with dose and gray matter volumes, as well as a dose-dependent effect on the hippocampus. The article pointed out some methodological differences leading to limitations in meta-analytic conclusions, such as the clear need for better quantification of drug use, and consistent reporting of time abstinent (Fig. 4). The same findings are also reported in nonpsychotic individuals (Rocchetti et al., 2013).

An increased forming of associations in the hippocampus with a decreased error-checking of these associations may lead to the free-flowing creativity which some users report with cannabinoids, but the lack of an intact "editing mechanism" may mean that invalid associations are laid down as false memories: the delusional thought content of psychosis. As Tamminga wrote in *Neuropsychopharmacology Reviews*, "It is plausible that psychosis is dependent on a pathologically increased level of neuronal function in CA3, which exceeds the associational capacity of this subfield and results in mistaken and false associations, some with psychotic content, which then get consolidated, as normal memory, albeit with psychotic content. These memories utilize normal declarative memory neural pathways, including limbic and prefrontal cortical regions, even though they have psychotic content" (Tamminga, 2013). Here, again, we see a potential role for cannabis in the pathophysiology of schizophrenia: chronic cannabinoid exposure causes alterations of

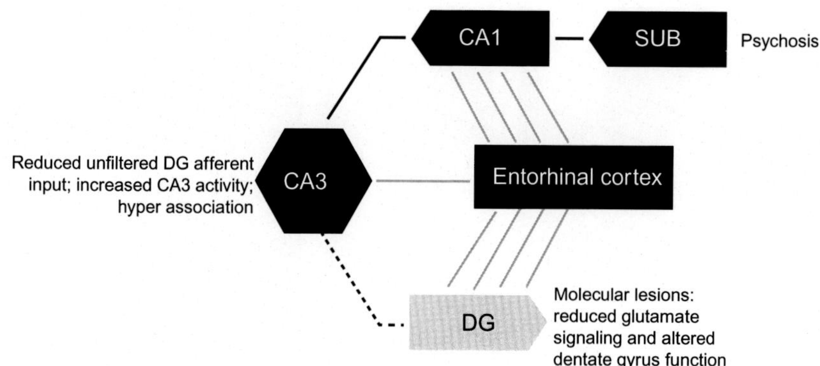

FIG. 4 This figure shows theoretical points of breakdown in the hippocampus that could lead to psychosis. Actions of THC and other exogenous cannabinoids would be likely to disrupt the balance in the dentate gyrus and CA3 areas. *(Adapted from Tamminga, C. A. (2013). Psychosis is emerging as a learning and memory disorder.* Neuropsychopharmacology, 38*(1), 247. http://dx.doi.org/10.1038/npp.2012.187).*

the hippocampal dentate gyrus of rats (Candelaria-Cook & Hamilton, 2014; Rubino et al., 2009), which could lead to decreased inhibition of CA3 (Alle & Geiger, 2007; Kobayashi, 2010). The centrality of the hippocampus to psychosis, and the specific damaging effects on cannabis on this brain region, might explain the psychotogenic potential of exogenous cannabinoids, despite the relatively low release of dopamine from cannabis smoking (Barkus et al., 2011). This is in contrast to the robust release of dopamine from amphetamine use, which is not as widely associated with psychosis as is cannabis or phencyclidine (Sherif et al., 2016).

These effects may explain some of the differences that have been reported in the literature between the persistence of cannabis-related psychosis compared to amphetamine-induced psychosis. Specifically, individuals initially diagnosed with cannabis-induced psychotic illnesses are more likely to convert to schizophrenia (46% of cases), as compared to those with amphetamine-induced psychosis (30%) (Niemi-Pynttari et al., 2013). The greater propensity for cannabis to induce permanent psychosis would not be expected if the sole mechanism of generating psychosis was the release of dopamine.

Cannabis is not alone among illicit drugs in its ability to produce psychotic symptoms: NMDA antagonist drugs (e.g., PCP and ketamine) have shown similar effects on psychotic symptoms and processes (Javitt, Zukin, Heresco-Levy, & Umbricht, 2012; Steeds et al., 2015). The overlap between NDMA receptor antagonists and cannabinoids is not a mystery, and the two can cause similar deficiencies in cognitive processes and positive symptoms of psychosis. Exogenous cannabinoids reduce the release of glutamate in the hippocampus, and this may be below the level needed to activate NMDA receptors (Haj-Dahmane & Shen, 2010; Iversen, 2012; Misner & Sullivan, 1999) leading to a convergence of the two mechanisms of exogenous cannabinoid agonism and

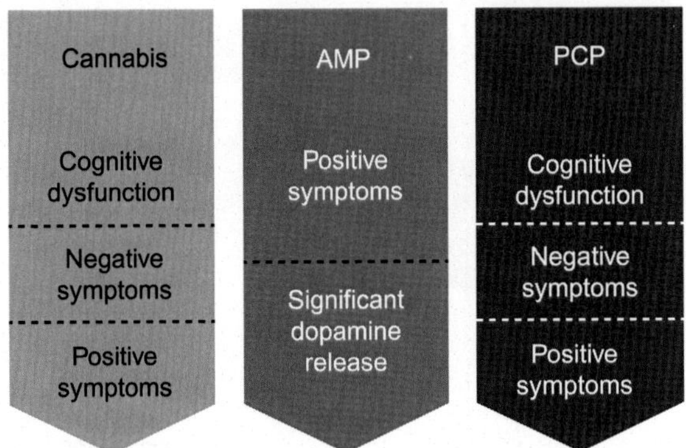

FIG. 5 Different psychotic dimensions with different drug models of psychosis (AMP = amphetamine, PCP = phencyclidine).

NMDA antagonism. NMDA antagonism will likely have effects in many brain regions that are less heavily modulated by cannabinoids, and vice versa, so the two drugs are not identical and overlapping in their other areas of activity. The hippocampal effects of the two drugs are very similar, raising the possibility of overlapping risks, as well as potential therapeutic benefits, from modulating these two different systems (Fig. 5).

Summarizing a Working Model

Cannabis is specifically toxic to the hippocampus. Hippocampal lesions in early life are known to produce changes that biologically mimic schizophrenia. Cannabis use disrupts hippocampal circuit formation, especially with adolescent exposure, and leads to deficits in the checking of associations, and the formation of accurate memories. These alterations in memory formation and the disrupted hippocampal involvement in cognition lead to the clinical phenotype of schizophrenia. The exogenous cannabinoids affect some people more than others, but it is currently unknown which individuals will be affected more (Fig. 6).

CANNABINOIDS AND PSYCHOSIS: NEUROPSYCHOLOGICAL CONSTRUCTS

Cannabis, Mood and Affect, and Psychosis

As affective and mood disturbance are known to cause psychosis in depressive and bipolar disorders, it is important to consider a psychosis mediated by cannabis-induced alterations in mood. Validity of this concept is suggested by a study of acute intravenous THC administration, showing that aberrant experiences and negative affect statistically mediated the increase in psychosis from THC (Freeman et al., 2014). But, when wondering whether mania and depression

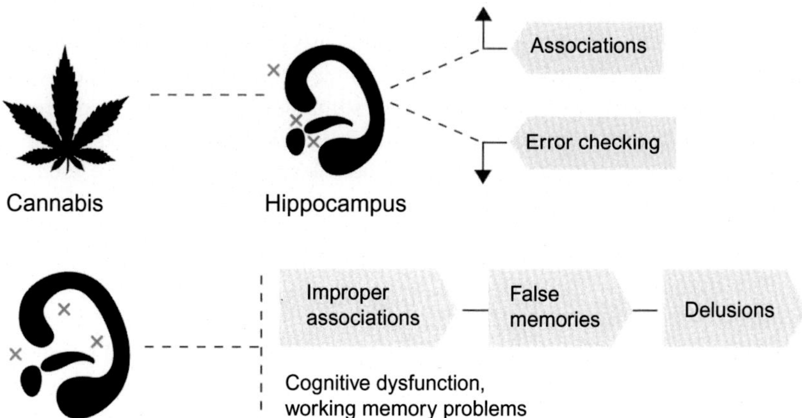

FIG. 6 A working model of cannabis-induced hippocampal damage leading to psychosis.

are significant drivers of the cannabis link to psychosis, it is important to reference the epidemiologic literature (Moore et al., 2007; McGrath et al., 2010). The increased risk of psychosis among those who have excessively used cannabis during adolescence does not seem to stem solely from an affective (bipolar or depressive) psychosis. Blunted affect is a symptom of schizophrenia, and hedonic deficits are also seen in schizophrenia (Dowd & Barch, 2010). There also may be blunted hedonic responses in cannabis users with first-episode psychosis relative to nonusers and community controls who have not used cannabis (Cassidy, Lepage, Harvey, & Malla, 2012), but this affective blunting is better characterized as a part of the schizophrenia spectrum of disorder and is not an independent mood disorder.

Endocannabinoids, Cannabis, Reward Processing, and Psychosis

A deficit in reward processing is seen in psychosis and schizophrenia, and alterations of reward processing are seen in both psychotic and nonpsychotic cannabis users as well (Cassidy et al., 2012). Whether this is a causal effect of psychosis or a more downstream effect is an important question. An interesting question might be whether long-term drug use would induce reward dysfunction and thereby create psychosis. The relatively early onset of cannabis-induced psychosis tends to argue against such a process.

Cannabis, Psychosis, and Salience

Psychosis also involves aberrations in how the brain determines the salience of stimuli (Kapur, 2003; Kapur, Mizrahi, & Li, 2005). It has been proposed that psychosis is a state of aberrant salience (Kapur et al., 2005); however, not all studies support this association (Abboud et al., 2016). Cannabis use causes aberrations in attentional salience processing (Bhattacharyya et al., 2012; Bhattacharyya et al., 2015). During attentional salience processing, cannabis alters prefrontal and caudate activation, and the activity in the caudate (which can suppress the approach motivation of a stimulus) is related to the degree of psychotic symptoms (Bhattacharyya et al., 2012). The inability of the brain to properly filter out and select attentional focus leads to an overactive focus on stimuli that is difficult to redirect, explaining the extreme focus on delusional thought content in psychosis.

Cannabis, Psychosis, Attention, and Cognition

Attentional processes are difficult to disentangle from other mental functions, but young adults who are recently abstinent from heavy cannabis use show poor executive function and working memory (Fontes et al., 2011), with attention regulation problems being manifest in these individuals. The fact that cannabis use is correlated with decreased sustained attention does not mean that it is an effect of the drug itself, as differences between users and nonusers may be based on individual premorbid characteristics, such as childhood ADHD (Scholes-Balog & Martin-Iverson, 2011; Tamm et al., 2013).

The neuroimaging findings around cannabis use also suggest differences in activation of attentional networks, but functional magnetic resonance imaging (fMRI) findings are often seemingly contradictory between studies, and the reader is directed to reviews on this particular subject (Lisdahl, Wright, Kirchner-Medina, Maple, & Shollenbarger, 2014). The findings could be summarized by saying that there is often increased activation during memory tasks, and decreased deactivation during memory tasks, and for those unfamiliar with fMRI tasks, it is worth mentioning that "more activation" is not "better." Greater activation of a specific brain region often represents inefficient processing and/or alternate brain strategies being employed to complete the task, which causes the increase in blood flow measured by the scanner. The literature is complicated by the fact that there are inconsistent definitions of levels of cannabis use between studies, as well as different age groups, ages at initiation of use, and amounts of time abstinent prior to image acquisition (Bossong, Jager, Bhattacharyya, & Allen, 2014). Construction of longitudinal studies of youth before and after cannabis initiation would be very helpful to disentangling questions of causality of cannabis use on these findings.

Effects of Cannabis Use on Motivation—A Possible Dimension of Psychosis

Occasionally termed "avolition," motivational deficits are one of the core features of schizophrenia. The RDoC construct of Approach Motivation offers two subconstructs affecting the commonly conceived concept of motivation, called "Reward Valuation" and "Effort Valuation" (NIH, 2013). In the short term, decreased locomotor activity is evident in both humans and animals with CB_1 agonism (Milstein, MacCannell, Karr, & Clark, 1975; Wiley, 2003). Both dopaminergic and extra-dopaminergic circuits are responsible for the effects of cannabis on motivation, and CB_1 activation can uniquely affect motivation without the involvement of dopamine in the anterior cingulate (Khani et al., 2015).

Cannabis has long been thought to have effects on motivation in chronic users. The term *amotivational syndrome* has been used colloquially and in media, but the term has not been extensively used in research. Motivation may clinically manifest itself through decreased academic, financial, and social achievement. Difficulty arises in differentiating psychotic-like effects of cannabis that might produce negative symptoms of schizophrenia, versus a motivational deficit unrelated to psychotic processes. Addiction also affect motivation, not through a global decrease in motivation, but through a decreased range of activities that become motivating; i.e., those pertaining to use, acquisition, and protection of the preferred drug.

The interaction between cannabis use and motivation has been directly studied in animal models and in humans (Lane & Cherek, 2002; Lane, Cherek, Pietras, & Steinberg, 2005). The effects of cannabis on reactivity to dopamine have been studied in a recent positron emission tomography study (Volkow et al., 2014), and previous studies have established dopamine's important role in motivation (Volkow, Fowler, Wang, & Swanson, 2004). As mentioned above, the effects of CB_1 agonism selective to the anterior cingulate suggest that the drug can affect motivation. Animals given CB_1 agonists have changes in the amount of effort they are willing to exert to obtain a reward, and develop a greater preference for larger, immediate rewards (Khani et al., 2015). Adolescents, like such animal models, show a tendency to be selectively motivated for large rewards in their decision-making under the influence of cannabis (Mokrysz, Freeman, Korkki, Griffiths, & Curran, 2016; Solowij et al., 2012). It is uncertain whether any of these deficits in motivation with cannabis use are related to the pathophysiology of cannabis-induced psychosis, or whether schizophrenia and cannabis-induced deficits in motivation are distinct, with deficits in motivation being a "downstream effect" of schizophrenia.

THE ENDOCANNABINOID SYSTEM'S ROLE IN DEVELOPMENT AND THE DISRUPTIVE EFFECTS OF EXOGENOUS CANNABINOIDS ON NEUROGENESIS

Cannabinoids, Psychosis, and the Genome

In human studies, it is very difficult to differentiate the effects of parental behavioral characteristics from the effects of the drug itself. This makes it difficult to understand the exact effects of disrupting endocannabinoid transmission during early development. Although the field is young, it is interesting that there have been significant effects found in animal models of maternal-use-induced epigenetic changes affecting the risk of substance use (DiNieri et al., 2011). More observational data over time are needed, particularly to answer questions regarding potential intergenerational effects of cannabis use on DNA raised by a recent narrative review (Albert Stuart Reece & Hulse, 2016). It is also worth investigating whether endocannabinoid signaling has any role in the activation of DNA repair and mutagenesis, possibly through inflammatory cascades (Bileck et al., 2016), as THC alone does not seem to be a mutagen. The effects of this DNA damage are also currently unknown, but DNA damage is a well-established finding in schizophrenia, and research has indicated that DNA damage is possibly part of the pathogenesis of schizophrenia and psychosis (Markkanen, Meyer, & Dianov, 2016; Muraleedharan, Menon, Rajkumar, & Chand, 2015).

Effects of Cannabinoids on Embryonic Development Related to Psychosis

The process of organizing a human brain is the most complicated in all of nature, and it would be unsurprising if perturbations in the complex chemical machinery would have consequences during development. The endocannabinoid anandamide is critically important during embryonic brain development and for sustaining the pregnancy (Karasu, Marczylo, Maccarrone, & Konje, 2011). The administration of a THC analogue, a CB_1 agonist, can result in disrupted neural development for some animals (Psychoyos, Hungund, Cooper, & Finnell, 2008). The colocalization of CB_1 receptors and GABAergic neurons suggests the possibility of the endocannabinoid system playing a significant role in GABAergic development and GABA regulation of glutamatergic neurons (Van Waes, Beverley, Siman, Tseng, & Steiner, 2012), which are important in both the hippocampal and striatal pathways.

THC has been shown to have enduring effects on the fetus in a variety of outcomes through animal studies (Dinieri & Hurd, 2012; Spano, Ellgren, Wang, & Hurd, 2007; Spano, Fadda, Fratta, & Fattore, 2010) and there are impairments in cognitive functioning and certain neuropsychological tasks in human studies (Fried, 2002a, 2002b; Fried & Smith, 2001; Fried, Watkinson, & Gray, 2003; Porath & Fried, 2005; Smith, Fried, Hogan, & Cameron, 2004, 2006; Spano et al., 2007). Specifically, executive dysfunction is one of the common effects thought to occur from exposure during development. Few studies have looked at the effect of in utero cannabis use on the subsequent development of psychosis, but Zammit (Zammit et al., 2009) failed to find an association for cannabis use, while finding an association with in utero nicotine exposure upon subsequent development of psychosis. The potential for significant effects on developmental systems without existing research establishing clear safety of the drug is a cause for great concern, particularly when pregnant women are using marijuana to control nausea (Alpar, Di Marzo, & Harkany, 2016; Reece, 2009). It is unknown what the effect of prenatal cannabis exposure is on schizophrenia risk, though this should be a critical area of interest (Alpar et al., 2016).

Effects of Cannabinoids on Childhood Development

There is a lack of research on regular exposure of children to cannabis. Investigation is needed into the developmentally neurotoxic effects of cannabis on processes such as synaptic pruning (Bossong & Niesink, 2010; Malone, Hill, & Rubino, 2010) and axonal pathfinding (Berghuis et al., 2007; Zalesky et al., 2012), before we can elucidate how endocannabinoid disruption during this stage might affect the adult brain in regards to psychosis.

Effects of Cannabinoids on Adolescent Development and Disruptions Related to Psychosis

Adolescence is a critical time of brain development during which the individual learns new social roles, gathers useful experiences, and acquires knowledge for future productivity. Physically, the adolescent brain is still undergoing development, with ongoing myelination, synaptic pruning, and shifts in functional connectivity. As will be discussed in further chapters, longitudinal epidemiologic studies have identified adolescent cannabis use as being linked to the development of psychosis. Are adolescent cannabis users manifesting an increased risk due to a correlated susceptibility to both psychosis and early cannabis use? Or does cannabis cause true age-dependent effects on the brain?

The mesolimbic system is involved in schizophrenia, as previously discussed, and the mesolimbic system of adolescents is highly responsive to reward. Heightened reward response in the context of an incompletely developed prefrontal cortex places the adolescent at higher risk of impulsive decision-making and choosing highly rewarding stimuli over safer alternatives (Casey & Jones, 2010). Prospective studies have shown that adolescent onset of substance use is associated with a greater risk of developing dependence (Butterworth, Slade, & Degenhardt, 2014). For cannabis use and the subsequent risks of other substance use disorders, certain genetically informed twin studies have questioned whether the age at initiation is critical in twins (Grant et al., 2010), but other studies have suggested there is a link with age at initiation (Lynskey et al., 2003; Lynskey, Vink, & Boomsma, 2006). Furthermore, preclinical data demonstrates that adolescent animals are at greater risk of acquisition of self-administration (Ellgren, Spano, & Hurd, 2007; Spano et al., 2010), and stress-induced relapse (Stopponi et al., 2014). As there is an indication that the reward system may be highly vulnerable to long-term plastic changes from exposure during adolescence, the question of whether this increased vulnerability of mesolimbic reward-related pathology extends beyond addiction and pertains to the adolescent vulnerability to develop cannabis-related psychosis remains to be answered.

The prefrontal cortex and limbic areas more widely express CB_1 receptors during adolescence (Heng, Beverley, Steiner, & Tseng, 2011). Adolescent THC exposure in animal models causes permanent affective changes in the adult animal (Realini, Rubino, & Parolaro, 2009). It should be noted that an adolescent's increased vulnerability to the development of schizophrenia stemming from cannabis use is not reflected by an increased vulnerability to acute psychotic effects from cannabis consumption: in a placebo-controlled clinical laboratory experiment, adolescents were *less* vulnerable to acute cannabis-induced psychosis (Mokrysz et al., 2016). This is a potentially important lesson, that the vulnerability to develop schizophrenia from cannabis use and the liability to acute psychosis with THC administration may not entirely overlap.

The adolescent period is also critical to synaptic pruning, which is important for the correct regulation of brain circuits involving cortical areas. CB_1 agonism induces premature cell death through the induction of an aberrant cell signaling cascade (Tomiyama & Funada, 2011), and during adolescence there is a high rate of change in synaptic refinement and CB_1 receptor density. Synaptic refinement seems to be disrupted by CB_1 agonism, and it is possible that normal brain development involves a shift in the regulatory inputs of hippocampal activity at this age, which are disrupted by endocannabinoid activation (Gleason et al., 2012). Cognitive effects in adolescents have been noted, and have consistently been found in multiple studies, with differences in list learning, story memory, recall, number-letter sequencing, digit-symbol encoding, sequencing errors and intrusion errors, and continued deficits in attention, memory, and spatial functioning (Jacobus & Tapert, 2014; Randolph, Brock, Margolis, & Tau, 2013). THC administration to primates demonstrates an age-related vulnerability in spatial perception with rhesus monkeys (Verrico et al., 2012), a finding also discovered in rodents (Rubino et al., 2009). The age dependence of the effects of ventral hippocampal lesioning in rodents (Chen et al., 2012) also suggests that there is a relative stability to the system in adulthood, but an enhanced vulnerability when the organism is young and undergoing development.

Neural Progenitor Cells

Neural progenitor cells have been shown to be important to brain development, and are still active in adult neurogenesis, particularly in the hippocampus. The endocannabinoid system modifies the function of adult progenitor cells in a manner dependent upon CB_1 expression. Absence of CB_1 receptors in mice leads to abnormal radial migration, progenitor cell proliferation, and deficits in axonal pathfinding (Mulder et al., 2008). The CB_1 receptor is also important to cell signaling in development which helps the formation of synapses and the selection of proper targets for the formation of synapses (Berghuis et al., 2007). CB_1-receptor-deficient mice have lower levels of brain-derived neurotrophic factor (Aso et al., 2008).

The effect of CB_1 receptor activation is not always the facilitation of neurogenesis, and late in gestation, CB_1 activation leads to the collapse of growth cones (Berghuis et al., 2007). This collapse of the axonal growth cone during development is a normal developmental process that keeps the neurons from excessively branching, but the question of *to grow or not to grow* is regulated by local signaling, and systemic application of a substance that either globally promotes or inhibits growth cones would be detrimental to the natural local signaling and regulation of axonal growth (Kaplan, Kent, Charron, & Fournier, 2014).

Key Chapter Points

- The endocannabinoid system interacts with brain systems implicated in psychosis and schizophrenia in a number of ways. Perhaps most convincing are the mesolimbic system and the hippocampal system, both showing strong evidence of vulnerability to cannabis-induced changes, as well as pathophysiologic plausibility in the generation of psychotic disorders.
- Endocannabinoids are important regulators of development, memory, reward system functioning, and homeostatic brain processes.
- Exogenous cannabinoids may unbalance the endocannabinoid system.
- Clinical laboratory studies show that nonpsychotic individuals can develop positive and negative symptoms with controlled exposure to THC. The preclinical evidence suggests mechanisms that would explain a more detrimental effect of adolescent cannabis initiation.
- The brains of individuals exposed to cannabis and those with schizophrenia share biomarkers, which may also be partially shared by other syndromes.
- The hippocampal formation is disturbed in schizophrenia, and cannabis use has profound effects on the hippocampus.
- The effects of cannabis use on cognition may contribute to psychosis and schizophrenia. The effects of adolescent use are clinically significant and are suggested to persist following cessation.
- Endocannabinoids have critical roles in embryonic and later development, and exogenous cannabinoids could have profound effects on these processes.
- Adolescence may be a critical period of vulnerability for the development of psychosis with cannabis exposure.

ACKNOWLEDGMENTS

The authors acknowledge Maria Haag, Maria Haag Art, for the graphics provided in this chapter.

REFERENCES

Abboud, R., Roiser, J. P., Khalifeh, H., Ali, S., Harrison, I., Killaspy, H. T., et al. (2016). Are persistent delusions in schizophrenia associated with aberrant salience? *Schizophrenia Research Cognition*, *4*, 32–38. http://dx.doi.org/10.1016/j.scog.2016.04.002.

Abush, H., & Akirav, I. (2012). Short- and long-term cognitive effects of chronic cannabinoids administration in late-adolescence rats. *PloS ONE*, *7*(2), e31731. http://dx.doi.org/10.1371/journal.pone.0031731.

Albrecht, D. S., Skosnik, P. D., Vollmer, J. M., Brumbaugh, M. S., Perry, K. M., Mock, B. H., et al. (2013). Striatal D(2)/D(3) receptor availability is inversely correlated with cannabis consumption in chronic marijuana users. *Drug and Alcohol Dependence*, *128*(1-2), 52–57. http://dx.doi.org/10.1016/j.drugalcdep.2012.07.016.

Alle, H., & Geiger, J. R. (2007). GABAergic spill-over transmission onto hippocampal mossy fiber boutons. *Journal of Neuroscience*, *27*(4), 942–950. http://dx.doi.org/10.1523/jneurosci.4996-06.2007.

Alpar, A., Di Marzo, V., & Harkany, T. (2016). At the tip of an iceberg: Prenatal marijuana and its possible relation to neuropsychiatric outcome in the offspring. *Biological Psychiatry*, *79*(7), e33–e45. http://dx.doi.org/10.1016/j.biopsych.2015.09.009.

Arendt, M., Mortensen, P. B., Rosenberg, R., Pedersen, C. B., & Waltoft, B. L. (2008). Familial predisposition for psychiatric disorder: Comparison of subjects treated for cannabis-induced psychosis and schizophrenia. *Archives of General Psychiatry*, *65*(11), 1269–1274. http://dx.doi.org/10.1001/archpsyc.65.11.1269.

Ashtari, M., Avants, B., Cyckowski, L., Cervellione, K. L., Roofeh, D., Cook, P., et al. (2011). Medial temporal structures and memory functions in adolescents with heavy cannabis use. *Journal of Psychiatric Research*, *45*(8), 1055–1066. http://dx.doi.org/10.1016/j.jpsychires.2011.01.004.

Aso, E., Ozaita, A., Valdizan, E. M., Ledent, C., Pazos, A., Maldonado, R., et al. (2008). BDNF impairment in the hippocampus is related to enhanced despair behavior in CB1 knockout mice. *Journal of Neurochemistry*, *105*(2), 565–572. http://dx.doi.org/10.1111/j.1471-4159.2007.05149.x.

Barkus, E., Morrison, P. D., Vuletic, D., Dickson, J. C., Ell, P. J., Pilowsky, L. S., et al. (2011). Does intravenous Delta9-tetrahydrocannabinol increase dopamine release? A SPET study. *Journal of Psychopharmacology*, *25*(11), 1462–1468. http://dx.doi.org/10.1177/0269881110382465.

Basavarajappa, B. S. (2015). Fetal alcohol spectrum disorder: Potential role of endocannabinoids signaling. *Brain Sciences*, *5*(4), 456–493. http://dx.doi.org/10.3390/brainsci5040456.

Batalla, A., Soriano-Mas, C., Lopez-Sola, M., Torrens, M., Crippa, J. A., Bhattacharyya, S., et al. (2014). Modulation of brain structure by catechol-O-methyltransferase Val(158) Met polymorphism in chronic cannabis users. *Addiction Biology*, *19*(4), 722–732. http://dx.doi.org/10.1111/adb.12027.

Berghuis, P., Rajnicek, A. M., Morozov, Y. M., Ross, R. A., Mulder, J., Urban, G. M., et al. (2007). Hardwiring the brain: Endocannabinoids shape neuronal connectivity. *Science*, *316*(5828), 1212–1216. http://dx.doi.org/10.1126/science.1137406.

Bhattacharyya, S., Crippa, J. A., Allen, P., Martin-Santos, R., Borgwardt, S., Fusar-Poli, P., et al. (2012). Induction of psychosis by Delta9-tetrahydrocannabinol reflects modulation of prefrontal and striatal function during attentional salience processing. *Archives of General Psychiatry*, *69*(1), 27–36. http://dx.doi.org/10.1001/archgenpsychiatry.2011.161.

Bhattacharyya, S., Falkenberg, I., Martin-Santos, R., Atakan, Z., Crippa, J. A., Giampietro, V., et al. (2015). Cannabinoid modulation of functional connectivity within regions processing attentional salience. *Neuropsychopharmacology*, *40*(6), 1343–1352. http://dx.doi.org/10.1038/npp.2014.258.

Bileck, A., Ferk, F., Al-Serori, H., Koller, V. J., Muqaku, B., Haslberger, A., et al. (2016). Impact of a synthetic cannabinoid (CP-47,497-C8) on protein expression in human cells: Evidence for induction of inflammation and DNA damage. *Archives of Toxicology*, *90*(6), 1369–1382. http://dx.doi.org/10.1007/s00204-015-1569-7.

Bioque, M., Garcia-Bueno, B., Macdowell, K. S., Meseguer, A., Saiz, P. A., Parellada, M., et al. (2013). Peripheral endocannabinoid system dysregulation in first-episode psychosis. *Neuropsychopharmacology*, *38*(13), 2568–2577. http://dx.doi.org/10.1038/npp.2013.165.

Bloomfield, M. A., Morgan, C. J., Egerton, A., Kapur, S., Curran, H. V., & Howes, O. D. (2014). Dopaminergic function in cannabis users and its relationship to cannabis-induced psychotic symptoms. *Biological Psychiatry*, *75*(6), 470–478. http://dx.doi.org/10.1016/j.biopsych.2013.05.027.

Bolla, K. I., Eldreth, D. A., Matochik, J. A., & Cadet, J. L. (2005). Neural substrates of faulty decision-making in abstinent marijuana users. *NeuroImage*, *26*(2), 480–492. http://dx.doi.org/10.1016/j.neuroimage.2005.02.012.

Bossong, M. G., Jager, G., Bhattacharyya, S., & Allen, P. (2014). Acute and non-acute effects of cannabis on human memory function: A critical review of neuroimaging studies. *Current Pharmaceutical Design*, *20*(13), 2114–2125.

Bossong, M. G., & Niesink, R. J. (2010). Adolescent brain maturation, the endogenous cannabinoid system and the neurobiology of cannabis-induced schizophrenia. *Progress in Neurobiology*, *92*(3), 370–385. http://dx.doi.org/10.1016/j.pneurobio.2010.06.010.

Brisch, R., Saniotis, A., Wolf, R., Bielau, H., Bernstein, H. G., Steiner, J., et al. (2014). The role of dopamine in schizophrenia from a neurobiological and evolutionary perspective: Old fashioned, but still in vogue. *Frontiers in Psychiatry*, *5*, 47. http://dx.doi.org/10.3389/fpsyt.2014.00047.

Brusco, A., Tagliaferro, P., Saez, T., & Onaivi, E. S. (2008). Postsynaptic localization of CB2 cannabinoid receptors in the rat hippocampus. *Synapse*, *62*(12), 944–949. http://dx.doi.org/10.1002/syn.20569.

Busquets Garcia, A., Soria-Gomez, E., Bellocchio, L., & Marsicano, G. (2016). Cannabinoid receptor type-1: Breaking the dogmas. *F1000Res*, *5*, F1000. http://dx.doi.org/10.12688/f1000research.8245.1.

Butterworth, P., Slade, T., & Degenhardt, L. (2014). Factors associated with the timing and onset of cannabis use and cannabis use disorder: Results from the 2007 Australian National Survey of Mental Health and Well-Being. *Drug and Alcohol Review*, *33*(5), 555–564. http://dx.doi.org/10.1111/dar.12183.

Caballero, A., & Tseng, K. Y. (2012). Association of cannabis use during adolescence, prefrontal CB1 receptor signaling, and schizophrenia. *Frontiers in Pharmacology*, *3*, 101. http://dx.doi.org/10.3389/fphar.2012.00101.

Candelaria-Cook, F. T., & Hamilton, D. A. (2014). Chronic cannabinoid agonist (WIN 55,212-2) exposure alters hippocampal dentate gyrus spine density in adult rats. *Brain Research*, *1542*, 104–110. http://dx.doi.org/10.1016/j.brainres.2013.10.039.

Casey, B. J., & Jones, R. M. (2010). Neurobiology of the adolescent brain and behavior: Implications for substance use disorders. *Journal of the American Academy of Child and Adolescent Psychiatry*, *49*(12), 1189–1201. http://dx.doi.org/10.1016/j.jaac.2010.08.017. quiz 1285.

Caspi, A., Moffitt, T. E., Cannon, M., McClay, J., Murray, R., Harrington, H., et al. (2005). Moderation of the effect of adolescent-onset cannabis use on adult psychosis by a functional polymorphism in the catechol-O-methyltransferase gene: Longitudinal evidence of a gene X environment interaction. *Biological Psychiatry*, *57*(10), 1117–1127. http://dx.doi.org/10.1016/j.biopsych.2005.01.026.

Cassidy, C. M., Lepage, M., Harvey, P. O., & Malla, A. (2012). Cannabis use and anticipatory pleasure as reported by subjects with early psychosis and community controls. *Schizophrenia Research*, *137*(1-3), 39–44. http://dx.doi.org/10.1016/j.schres.2012.02.028.

Chen, X., Wang, X., O'Neill, A. F., Walsh, D., & Kendler, K. S. (2004). Variants in the catechol-o-methyltransferase (COMT) gene are associated with schizophrenia in Irish high-density families. *Molecular Psychiatry*, *9*(10), 962–967. http://dx.doi.org/10.1038/sj.mp.4001519.

Chen, X. S., Zhang, C., Xu, Y. F., Zhang, M. D., Lou, F. Y., Chen, C., et al. (2012). Neonatal ventral hippocampal lesion as a valid model of schizophrenia: Evidence from sensory gating study. *Chinese Medical Journal*, *125*(15), 2752–2754.

Cheng, C. H., Chan, P. Y., Hsieh, Y. W., & Chen, K. F. (2016). A meta-analysis of mismatch negativity in children with attention deficit-hyperactivity disorders. *Neuroscience Letters*, *612*, 132–137. http://dx.doi.org/10.1016/j.neulet.2015.11.033.

Cherif, H., Argaw, A., Cecyre, B., Bouchard, A., Gagnon, J., Javadi, P., et al. (2015). Role of GPR55 during axon growth and target innervation. *eNeuro*, *2*(5). http://dx.doi.org/10.1523/eneuro.0011-15.2015.

Costas, J., Sanjuan, J., Ramos-Rios, R., Paz, E., Agra, S., Tolosa, A., et al. (2011). Interaction between COMT haplotypes and cannabis in schizophrenia: A case-only study in two samples from Spain. *Schizophrenia Research*, *127*(1-3), 22–27. http://dx.doi.org/10.1016/j.schres.2011.01.014.

del Re, E. C., Spencer, K. M., Oribe, N., Mesholam-Gately, R. I., Goldstein, J., Shenton, M. E., et al. (2015). Clinical high risk and first episode schizophrenia: Auditory event-related potentials. *Psychiatry Research*, *231*(2), 126–133. http://dx.doi.org/10.1016/j.pscychresns.2014.11.012.

Demirakca, T., Sartorius, A., Ende, G., Meyer, N., Welzel, H., Skopp, G., et al. (2011). Diminished gray matter in the hippocampus of cannabis users: Possible protective effects of cannabidiol. *Drug and Alcohol Dependence*, *114*(2-3), 242–245. http://dx.doi.org/10.1016/j.drugalcdep.2010.09.020.

Deutsch, D. G., Glaser, S. T., Howell, J. M., Kunz, J. S., Puffenbarger, R. A., Hillard, C. J., et al. (2001). The cellular uptake of anandamide is coupled to its breakdown by fatty-acid amide hydrolase. *The Journal of Biological Chemistry*, *276*(10), 6967–6973. http://dx.doi.org/10.1074/jbc.M003161200.

Devinsky, O., Cilio, M. R., Cross, H., Fernandez-Ruiz, J., French, J., Hill, C., et al. (2014). Cannabidiol: Pharmacology and potential therapeutic role in epilepsy and other neuropsychiatric disorders. *Epilepsia*, *55*(6), 791–802. http://dx.doi.org/10.1111/epi.12631.

Diana, M. A., & Marty, A. (2004). Endocannabinoid-mediated short-term synaptic plasticity: Depolarization-induced suppression of inhibition (DSI) and depolarization-induced suppression of excitation (DSE). *British Journal of Pharmacology*, *142*(1), 9–19. http://dx.doi.org/10.1038/sj.bjp.0705726.

Dinieri, J. A., & Hurd, Y. L. (2012). Rat models of prenatal and adolescent cannabis exposure. *Methods in Molecular Biology*, *829*, 231–242. http://dx.doi.org/10.1007/978-1-61779-458-2_14.

DiNieri, J. A., Wang, X., Szutorisz, H., Spano, S. M., Kaur, J., Casaccia, P., et al. (2011). Maternal cannabis use alters ventral striatal dopamine D2 gene regulation in the offspring. *Biological Psychiatry*, *70*(8), 763–769. http://dx.doi.org/10.1016/j.biopsych.2011.06.027.

Dowd, E. C., & Barch, D. M. (2010). Anhedonia and emotional experience in schizophrenia: Neural and behavioral indicators. *Biological Psychiatry*, *67*(10), 902–911. http://dx.doi.org/10.1016/j.biopsych.2009.10.020.

D'Souza, D. C., Perry, E., MacDougall, L., Ammerman, Y., Cooper, T., Wu, Y. T., et al. (2004). The psychotomimetic effects of intravenous delta-9-tetrahydrocannabinol in healthy individuals: Implications for psychosis. *Neuropsychopharmacology*, *29*(8), 1558–1572. http://dx.doi.org/10.1038/sj.npp.1300496.

D'Souza, D. C., Sewell, R. A., & Ranganathan, M. (2009). Cannabis and psychosis/schizophrenia: Human studies. *European Archives of Psychiatry and Clinical Neuroscience*, *259*(7), 413–431. http://dx.doi.org/10.1007/s00406-009-0024-2.

Earls, H. A., Curran, T., & Mittal, V. (2016). A meta-analytic review of auditory event-related potential components as endophenotypes for perspectives from first-degree relatives. *Schizophrenia Bulletin*, *42*(6), 1504–1516. http://dx.doi.org/10.1093/schbul/sbw047.

Elkins, I. J., McGue, M., & Iacono, W. G. (2007). Prospective effects of attention-deficit/hyperactivity disorder, conduct disorder, and sex on adolescent substance use and abuse. *Archives of General Psychiatry*, *64*(10), 1145–1152. http://dx.doi.org/10.1001/archpsyc.64.10.1145.

Ellgren, M., Spano, S. M., & Hurd, Y. L. (2007). Adolescent cannabis exposure alters opiate intake and opioid limbic neuronal populations in adult rats. *Neuropsychopharmacology*, *32*(3), 607–615. http://dx.doi.org/10.1038/sj.npp.1301127.

Ermis, A., Erkiran, M., Dasdemir, S., Turkcan, A. S., Ceylan, M. E., Bireller, E. S., et al. (2015). The relationship between catechol-O-methyltransferase gene Val158Met (COMT) polymorphism and premorbid cannabis use in Turkish male patients with schizophrenia. *In Vivo*, *29*(1), 129–132.

Estrada, G., Fatjo-Vilas, M., Munoz, M. J., Pulido, G., Minano, M. J., Toledo, E., et al. (2011). Cannabis use and age at onset of psychosis: Further evidence of interaction with COMT Val158Met polymorphism. *Acta Psychiatrica Scandinavica*, *123*(6), 485–492. http://dx.doi.org/10.1111/j.1600-0447.2010.01665.x.

Eyles, D., Feldon, J., & Meyer, U. (2012). Schizophrenia: Do all roads lead to dopamine or is this where they start? Evidence from two epidemiologically informed developmental rodent models. *Translational Psychiatry*, *2*, e81. http://dx.doi.org/10.1038/tp.2012.6.

Fakhoury, M. (2016). Role of the endocannabinoid system in the pathophysiology of schizophrenia. *Molecular Neurobiology*, *54*(1), 768–778. http://dx.doi.org/10.1007/s12035-016-9697-5.

Fergusson, D. M., & Boden, J. M. (2008). Cannabis use and adult ADHD symptoms. *Drug and Alcohol Dependence*, *95*(1-2), 90–96. http://dx.doi.org/10.1016/j.drugalcdep.2007.12.012.

Fernandez-Espejo, E., Viveros, M. P., Nunez, L., Ellenbroek, B. A., & Rodriguez de Fonseca, F. (2009). Role of cannabis and endocannabinoids in the genesis of schizophrenia. *Psychopharmacology*, *206*(4), 531–549. http://dx.doi.org/10.1007/s00213-009-1612-6.

Filbey, F. M., Schacht, J. P., Myers, U. S., Chavez, R. S., & Hutchison, K. E. (2010). Individual and additive effects of the CNR1 and FAAH genes on brain response to marijuana cues. *Neuropsychopharmacology*, *35*(4), 967–975. http://dx.doi.org/10.1038/npp.2009.200.

Fontes, M. A., Bolla, K. I., Cunha, P. J., Almeida, P. P., Jungerman, F., Laranjeira, R. R., et al. (2011). Frontal Assessment Battery (FAB) is a simple tool for detecting executive deficits in chronic cannabis users. *Journal of Clinical and Experimental Neuropsychology*, *33*(5), 523–531. http://dx.doi.org/10.1080/13803395.2010.535505.

Freeman, D., Dunn, G., Murray, R. M., Evans, N., Lister, R., Antley, A., et al. (2014). How cannabis causes paranoia: Using the intravenous administration of 9-tetrahydrocannabinol (THC) to identify key cognitive mechanisms leading to paranoia. *Schizophrenia Bulletin*, http://dx.doi.org/10.1093/schbul/sbu098.

Fried, P. A. (2002a). Adolescents prenatally exposed to marijuana: Examination of facets of complex behaviors and comparisons with the influence of in utero cigarettes. *The Journal of Clinical Pharmacology*, *42*(11), 97s–102s.

Fried, P. A. (2002b). Conceptual issues in behavioral teratology and their application in determining long-term sequelae of prenatal marihuana exposure. *The Journal of Child Psychology and Psychiatry*, *43*(1), 81–102.

Fried, P. A., & Smith, A. M. (2001). A literature review of the consequences of prenatal marihuana exposure. An emerging theme of a deficiency in aspects of executive function. *Neurotoxicology and Teratology*, *23*(1), 1–11.

Fried, P. A., Watkinson, B., & Gray, R. (2003). Differential effects on cognitive functioning in 13- to 16-year-olds prenatally exposed to cigarettes and marihuana. *Neurotoxicology and Teratology*, *25*(4), 427–436.

Gallinat, J., Rentzsch, J., & Roser, P. (2012). Neurophysiological effects of cannabinoids: Implications for psychosis research. *Current Pharmaceutical Design*, *18*(32), 4938–4949.

Ghosh, S., Kinsey, S. G., Liu, Q. S., Hruba, L., McMahon, L. R., Grim, T. W., et al. (2015). Full fatty acid amide hydrolase inhibition combined with partial monoacylglycerol lipase inhibition: Augmented and sustained antinociceptive effects with reduced cannabimimetic side effects in mice. *The Journal of Pharmacology and Experimental Therapeutics*, *354*(2), 111–120. http://dx.doi.org/10.1124/jpet.115.222851.

Gleason, K. A., Birnbaum, S. G., Shukla, A., & Ghose, S. (2012). Susceptibility of the adolescent brain to cannabinoids: Long-term hippocampal effects and relevance to schizophrenia. *Translational Psychiatry*, *2*, e199. http://dx.doi.org/10.1038/tp.2012.122.

Grant, J. D., Lynskey, M. T., Scherrer, J. F., Agrawal, A., Heath, A. C., & Bucholz, K. K. (2010). A cotwin-control analysis of drug use and abuse/dependence risk associated with early-onset cannabis use. *Addictive Behaviors*, *35*(1), 35–41. http://dx.doi.org/10.1016/j.addbeh.2009.08.006.

Haigh, S. M., Coffman, B. A., & Salisbury, D. F. (2016). Mismatch negativity in first-episode schizophrenia: A meta-analysis. *Clinical EEG and Neuroscience*. http://dx.doi.org/10.1177/1550059416645980.

Haj-Dahmane, S., & Shen, R. Y. (2010). Regulation of plasticity of glutamate synapses by endocannabinoids and the cyclic-AMP/protein kinase a pathway in midbrain dopamine neurons. *The Journal of Physiology*, *588*(Pt 14), 2589–2604. http://dx.doi.org/10.1113/jphysiol.2010.190066.

Hajos, M., Hoffmann, W. E., & Kocsis, B. (2008). Activation of cannabinoid-1 receptors disrupts sensory gating and neuronal oscillation: Relevance to schizophrenia. *Biological Psychiatry*, *63*(11), 1075–1083. http://dx.doi.org/10.1016/j.biopsych.2007.12.005.

Haughey, H. M., Marshall, E., Schacht, J. P., Louis, A., & Hutchison, K. E. (2008). Marijuana withdrawal and craving: Influence of the cannabinoid receptor 1 (CNR1) and fatty acid amide hydrolase (FAAH) genes. *Addiction*, *103*(10), 1678–1686. http://dx.doi.org/10.1111/j.1360-0443.2008.02292.x.

Hebert-Chatelain, E., Desprez, T., Serrat, R., Bellocchio, L., Soria-Gomez, E., Busquets-Garcia, A., et al. (2016). A cannabinoid link between mitochondria and memory. *Nature*, *539*(7630), 555–559. http://dx.doi.org/10.1038/nature20127.

Heng, L., Beverley, J. A., Steiner, H., & Tseng, K. Y. (2011). Differential developmental trajectories for CB1 cannabinoid receptor expression in limbic/associative and sensorimotor cortical areas. *Synapse*, *65*(4), 278–286. http://dx.doi.org/10.1002/syn.20844.

Henquet, C., Rosa, A., Delespaul, P., Papiol, S., Fananas, L., van Os, J., et al. (2009). COMT ValMet moderation of cannabis-induced psychosis: A momentary assessment study of 'switching on' hallucinations in the flow of daily life. *Acta Psychiatrica Scandinavica*, *119*(2), 156–160. http://dx.doi.org/10.1111/j.1600-0447.2008.01265.x.

Hill, M. N., Froc, D. J., Fox, C. J., Gorzalka, B. B., & Christie, B. R. (2004). Prolonged cannabinoid treatment results in spatial working memory deficits and impaired long-term potentiation in the CA1 region of the hippocampus in vivo. *The European Journal of Neuroscience*, *20*(3), 859–863. http://dx.doi.org/10.1111/j.1460-9568.2004.03522.x.

Hirvonen, J., Goodwin, R. S., Li, C. T., Terry, G. E., Zoghbi, S. S., Morse, C., et al. (2012). Reversible and regionally selective downregulation of brain cannabinoid CB1 receptors in chronic daily cannabis smokers. *Molecular Psychiatry*, *17*(6), 642–649. http://dx.doi.org/10.1038/mp.2011.82.

Ishiguro, H., Horiuchi, Y., Ishikawa, M., Koga, M., Imai, K., Suzuki, Y., et al. (2010). Brain cannabinoid CB2 receptor in schizophrenia. *Biological Psychiatry*, *67*(10), 974–982. http://dx.doi.org/10.1016/j.biopsych.2009.09.024.

Iversen, L. (2012). How cannabis works in the brain. In D. Castle, R. Murray, & D. C. D'Souza (Eds.), *Marijuana and madness*. (2nd ed.). Cambridge: Cambridge University Press.

Jacobus, J., & Tapert, S. F. (2014). Effects of cannabis on the adolescent brain. *Current Pharmaceutical Design*, *20*(13), 2186–2193.

Javitt, D. C., Zukin, S. R., Heresco-Levy, U., & Umbricht, D. (2012). Has an angel shown the way? Etiological and therapeutic implications of the PCP/NMDA model of schizophrenia. *Schizophrenia Bulletin*, *38*(5), 958–966. http://dx.doi.org/10.1093/schbul/sbs069.

Kaplan, A., Kent, C. B., Charron, F., & Fournier, A. E. (2014). Switching responses: Spatial and temporal regulators of axon guidance. *Molecular Neurobiology*, *49*(2), 1077–1086. http://dx.doi.org/10.1007/s12035-013-8582-8.

Kapur, S. (2003). Psychosis as a state of aberrant salience: A framework linking biology, phenomenology, and pharmacology in schizophrenia. *The American Journal of Psychiatry*, *160*(1), 13–23. http://dx.doi.org/10.1176/appi.ajp.160.1.13.

Kapur, S., Mizrahi, R., & Li, M. (2005). From dopamine to salience to psychosis—Linking biology, pharmacology and phenomenology of psychosis. *Schizophrenia Research*, *79*(1), 59–68. http://dx.doi.org/10.1016/j.schres.2005.01.003.

Karasu, T., Marczylo, T. H., Maccarrone, M., & Konje, J. C. (2011). The role of sex steroid hormones, cytokines and the endocannabinoid system in female fertility. *Human Reproduction Update*, *17*(3), 347–361. http://dx.doi.org/10.1093/humupd/dmq058.

Karmaus, P. W., Chen, W., Crawford, R., Kaplan, B. L., & Kaminski, N. E. (2013). Delta9-tetrahydrocannabinol impairs the inflammatory response to influenza infection: Role of antigen-presenting cells and the cannabinoid receptors 1 and 2. *Toxicological Sciences*, *131*(2), 419–433. http://dx.doi.org/10.1093/toxsci/kfs315.

Khani, A., Kermani, M., Hesam, S., Haghparast, A., Argandona, E. G., & Rainer, G. (2015). Activation of cannabinoid system in anterior cingulate cortex and orbitofrontal cortex modulates cost-benefit decision making. *Psychopharmacology*, *232*(12), 2097–2112. http://dx.doi.org/10.1007/s00213-014-3841-6.

Kobayashi, K. (2010). Hippocampal mossy fiber synaptic transmission and its modulation. *Vitamins and Hormones*, *82*, 65–85. http://dx.doi.org/10.1016/s0083-6729(10)82004-7.

Koob, G. F., & Volkow, N. D. (2010). Neurocircuitry of addiction. *Neuropsychopharmacology*, *35*(1), 217–238. http://dx.doi.org/10.1038/npp.2009.110.

Kowal, M. A., Colzato, L. S., & Hommel, B. (2011). Decreased spontaneous eye blink rates in chronic cannabis users: Evidence for striatal cannabinoid-dopamine interactions. *PloS One*, *6*(11), e26662. http://dx.doi.org/10.1371/journal.pone.0026662.

Kreitzer, A. C., Carter, A. G., & Regehr, W. G. (2002). Inhibition of interneuron firing extends the spread of endocannabinoid signaling in the cerebellum. *Neuron*, *34*(5), 787–796.

Kremin, T., & Hasselmo, M. E. (2007). Cholinergic suppression of glutamatergic synaptic transmission in hippocampal region CA3 exhibits laminar selectivity: Implication for hippocampal network dynamics. *Neuroscience*, *149*(4), 760–767. http://dx.doi.org/10.1016/j.neuroscience.2007.07.007.

Kuepper, R., Morrison, P. D., van Os, J., Murray, R. M., Kenis, G., & Henquet, C. (2010). Does dopamine mediate the psychosis-inducing effects of cannabis? A review and integration of findings across disciplines. *Schizophrenia Research*, *121*(1-3), 107–117. http://dx.doi.org/10.1016/j.schres.2010.05.031.

Kullmann, D. M., Asztely, F., & Walker, M. C. (2000). The role of mammalian ionotropic receptors in synaptic plasticity: LTP, LTD and epilepsy. *Cellular and Molecular Life Sciences*, *57*(11), 1551–1561.

Landfield, P. W., Cadwallader, L. B., & Vinsant, S. (1988). Quantitative changes in hippocampal structure following long-term exposure to delta 9-tetrahydrocannabinol: Possible mediation by glucocorticoid systems. *Brain Research*, *443*(1-2), 47–62.

Lane, S. D., & Cherek, D. R. (2002). Marijuana effects on sensitivity to reinforcement in humans. *Neuropsychopharmacology*, *26*(4), 520–529. http://dx.doi.org/10.1016/s0893-133x(01)00375-x.

Lane, S. D., Cherek, D. R., Pietras, C. J., & Steinberg, J. L. (2005). Performance of heavy marijuana-smoking adolescents on a laboratory measure of motivation. *Addictive Behaviors*, *30*(4), 815–828. http://dx.doi.org/10.1016/j.addbeh.2004.08.026.

Laricchiuta, D., Musella, A., Rossi, S., & Centonze, D. (2014). Behavioral and electrophysiological effects of endocannabinoid and dopaminergic systems on salient stimuli. *Frontiers in Behavioral Neuroscience*, *8*, 183. http://dx.doi.org/10.3389/fnbeh.2014.00183.

Lipska, B. K. (2004). Using animal models to test a neurodevelopmental hypothesis of schizophrenia. *Journal of Psychiatry and Neuroscience*, *29*(4), 282–286.

Lisboa, S. F., Gomes, F. V., Guimaraes, F. S., & Campos, A. C. (2016). Microglial cells as a link between cannabinoids and the immune hypothesis of psychiatric disorders. *Frontiers in Neurology*, *7*, 5. http://dx.doi.org/10.3389/fneur.2016.00005.

Lisdahl, K. M., Wright, N. E., Kirchner-Medina, C., Maple, K. E., & Shollenbarger, S. (2014). The effects of regular cannabis use on neurocognition in adolescents and young adults. *Current Addiction Reports*, *1*(2), 144–156. http://dx.doi.org/10.1007/s40429-014-0019-6.

Liu, Z., Han, J., Jia, L., Maillet, J. C., Bai, G., Xu, L., et al. (2010). Synaptic neurotransmission depression in ventral tegmental dopamine neurons and cannabinoid-associated addictive learning. *PloS One*, *5*(12), e15634. http://dx.doi.org/10.1371/journal.pone.0015634.

Liu, X., Ramirez, S., Pang, P. T., Puryear, C. B., Govindarajan, A., Deisseroth, K., et al. (2012). Optogenetic stimulation of a hippocampal engram activates fear memory recall. *Nature*, *484*(7394), 381–385. http://dx.doi.org/10.1038/nature11028.

Lorenzetti, V., Solowij, N., & Yucel, M. (2016). The role of cannabinoids in neuroanatomic alterations in cannabis users. *Biological Psychiatry*, *79*(7), e17–e31. http://dx.doi.org/10.1016/j.biopsych.2015.11.013.

Lu, Y., & Anderson, H. D. (2017). Cannabinoid signaling in health and disease. *Canadian Journal of Physiology and Pharmacology*, *95*(4), 311–327. http://dx.doi.org/10.1139/cjpp-2016-0346.

Lu, H. C., & Mackie, K. (2016). An introduction to the endogenous cannabinoid system. *Biological Psychiatry*, *79*(7), 516–525. http://dx.doi.org/10.1016/j.biopsych.2015.07.028.

Lynskey, M. T., Heath, A. C., Bucholz, K. K., Slutske, W. S., Madden, P. A., Nelson, E. C., et al. (2003). Escalation of drug use in early-onset cannabis users vs co-twin controls. *JAMA*, *289*(4), 427–433.

Lynskey, M. T., Vink, J. M., & Boomsma, D. I. (2006). Early onset cannabis use and progression to other drug use in a sample of Dutch twins. *Behavior Genetics*, *36*(2), 195–200. http://dx.doi.org/10.1007/s10519-005-9023-x.

MacDonald, J. F., Jackson, M. F., & Beazely, M. A. (2007). G protein-coupled receptors control NMDARs and metaplasticity in the hippocampus. *Biochimica et Biophysica Acta*, *1768*(4), 941–951. http://dx.doi.org/10.1016/j.bbamem.2006.12.006.

Malone, D. T., Hill, M. N., & Rubino, T. (2010). Adolescent cannabis use and psychosis: Epidemiology and neurodevelopmental models. *British Journal of Pharmacology*, *160*(3), 511–522. http://dx.doi.org/10.1111/j.1476-5381.2010.00721.x.

Markkanen, E., Meyer, U., & Dianov, G. L. (2016). DNA damage and repair in schizophrenia and autism: Implications for cancer comorbidity and beyond. *International Journal of Molecular Sciences*, *17*(6), 856. http://dx.doi.org/10.3390/ijms17060856.

Maroso, M., Szabo, G. G., Kim, H. K., Alexander, A., Bui, A. D., Lee, S. H., et al. (2016). Cannabinoid control of learning and memory through HCN channels. *Neuron*, *89*(5), 1059–1073. http://dx.doi.org/10.1016/j.neuron.2016.01.023.

McGrath, J., Welham, J., Scott, J., Varghese, D., Degenhardt, L., Hayatbakhsh, M. R., et al. (2010). Association between cannabis use and psychosis-related outcomes using sibling pair analysis in a cohort of young adults. *Archives of General Psychiatry*, *67*(5), 440–447. http://dx.doi.org/10.1001/archgenpsychiatry.2010.6.

McGuire, P. K., Shah, G. M., & Murray, R. M. (1993). Increased blood flow in Broca's area during auditory hallucinations in schizophrenia. *Lancet*, *342*(8873), 703–706.

McPartland, J. M., Duncan, M., Di Marzo, V., & Pertwee, R. G. (2015). Are cannabidiol and delta(9) -tetrahydrocannabivarin negative modulators of the endocannabinoid system? a systematic review. *British Journal of Pharmacology*, *172*(3), 737–753. http://dx.doi.org/10.1111/bph.12944.

Micoulaud-Franchi, J. A., Lopez, R., Cermolacce, M., Vaillant, F., Peri, P., Boyer, L., et al. (2016). Sensory gating capacity and attentional function in adults with ADHD: A preliminary neurophysiological and neuropsychological study. *Journal of Attention Disorders*. http://dx.doi.org/10.1177/1087054716629716.

Micoulaud-Franchi, J. A., Vaillant, F., Lopez, R., Peri, P., Baillif, A., Brandejsky, L., et al. (2015). Sensory gating in adult with attention-deficit/hyperactivity disorder: Event-evoked potential and perceptual experience reports comparisons with schizophrenia. *Biological Psychology*, *107*, 16–23. http://dx.doi.org/10.1016/j.biopsycho.2015.03.002.

Milstein, S. L., MacCannell, K., Karr, G., & Clark, S. (1975). Marijuana-produced impairments in coordination. Experienced and nonexperienced subjects. *The Journal of Nervous and Mental Disease*, *161*(1), 26–31.

Minzenberg, M. J., Laird, A. R., Thelen, S., Carter, C. S., & Glahn, D. C. (2009). Meta-analysis of 41 functional neuroimaging studies of executive function in schizophrenia. *Archives of General Psychiatry*, *66*(8), 811–822. http://dx.doi.org/10.1001/archgenpsychiatry.2009.91.

Misner, D. L., & Sullivan, J. M. (1999). Mechanism of cannabinoid effects on long-term potentiation and depression in hippocampal CA1 neurons. *Journal of Neuroscience*, *19*(16), 6795–6805.

Mokrysz, C., Freeman, T. P., Korkki, S., Griffiths, K., & Curran, H. V. (2016). Are adolescents more vulnerable to the harmful effects of cannabis than adults? A placebo-controlled study in human males. *Translational Psychiatry*, *6*(11), e961. http://dx.doi.org/10.1038/tp.2016.225.

Molero, P., Ortuno, F., Zalacain, M., & Patino-Garcia, A. (2007). Clinical involvement of catechol-O-methyltransferase polymorphisms in schizophrenia spectrum disorders: Influence on the severity of psychotic symptoms and on the response to neuroleptic treatment. *The Pharmacogenomics Journal*, *7*(6), 418–426. http://dx.doi.org/10.1038/sj.tpj.6500441.

Moore, T. H., Zammit, S., Lingford-Hughes, A., Barnes, T. R., Jones, P. B., Burke, M., et al. (2007). Cannabis use and risk of psychotic or affective mental health outcomes: a systematic review. *Lancet*, *370*(9584), 319–328.

Morgan, C. J., Freeman, T. P., Powell, J., & Curran, H. V. (2016). AKT1 genotype moderates the acute psychotomimetic effects of naturalistically smoked cannabis in young cannabis smokers. *Translational Psychiatry*, *6*, e738. http://dx.doi.org/10.1038/tp.2015.219.

Muguruza, C., Lehtonen, M., Aaltonen, N., Morentin, B., Meana, J. J., & Callado, L. F. (2013). Quantification of endocannabinoids in postmortem brain of schizophrenic subjects. *Schizophrenia Research*, *148*(1-3), 145–150. http://dx.doi.org/10.1016/j.schres.2013.06.013.

Mulder, J., Aguado, T., Keimpema, E., Barabas, K., Ballester Rosado, C. J., Nguyen, L., et al. (2008). Endocannabinoid signaling controls pyramidal cell specification and long-range axon patterning. *Proceedings of the National Academy of Sciences of the United States of America*, *105*(25), 8760–8765. http://dx.doi.org/10.1073/pnas.0803545105.

Muller, N., Weidinger, E., Leitner, B., & Schwarz, M. J. (2015). The role of inflammation in schizophrenia. *Frontiers in Neuroscience*, *9*, 372. http://dx.doi.org/10.3389/fnins.2015.00372.

Muraleedharan, A., Menon, V., Rajkumar, R. P., & Chand, P. (2015). Assessment of DNA damage and repair efficiency in drug naive schizophrenia using comet assay. *Journal of Psychiatric Research*, *68*, 47–53. http://dx.doi.org/10.1016/j.jpsychires.2015.05.018.

Nawaz, R., & Siddiqui, S. (2015). Association of single nucleotide polymorphisms in catechol-O-methyltransferase and serine-threonine protein kinase genes in the Pakistani schizophrenic population: A study with special emphasis on cannabis and smokeless tobacco. *CNS & Neurological Disorders Drug Targets*, *14*(8), 1086–1095.

Niemi-Pynttari, J. A., Sund, R., Putkonen, H., Vorma, H., Wahlbeck, K., & Pirkola, S. P. (2013). Substance-induced psychoses converting into schizophrenia: A register-based study of 18,478 Finnish inpatient cases. *The Journal of Clinical Psychiatry*, *74*(1), e94–e99. http://dx.doi.org/10.4088/JCP.12m07822.

NIH, N.I.o.H. (2013). *The research domain criteria*. Retrieved from http://www.nimh.nih.gov/research-priorities/rdoc/rdoc-constructs.shtml#reward_valuation.

Ottosen, C., Petersen, L., Larsen, J. T., & Dalsgaard, S. (2016). Gender differences in associations between attention-deficit/hyperactivity disorder and substance use disorder. *Journal of the American Academy of Child and Adolescent Psychiatry*, *55*(3), 227–234. http://dx.doi.org/10.1016/j.jaac.2015.12.010.

Pan, B., Hillard, C. J., & Liu, Q. S. (2008). D2 dopamine receptor activation facilitates endocannabinoid-mediated long-term synaptic depression of GABAergic synaptic transmission in midbrain dopamine neurons via cAMP-protein kinase A signaling. *J Neurosci*, *28*(52), 14018–14030. http://dx.doi.org/10.1523/JNEUROSCI.4035-08.2008.

Pan, B., Wang, W., Zhong, P., Blankman, J. L., Cravatt, B. F., & Liu, Q. S. (2011). Alterations of endocannabinoid signaling, synaptic plasticity, learning, and memory in monoacylglycerol lipase knock-out mice. *Journal of Neuroscience*, *31*(38), 13420–13430. http://dx.doi.org/10.1523/jneurosci.2075-11.2011.

Pelkey, K. A., & McBain, C. J. (2008). Target-cell-dependent plasticity within the mossy fibre-CA3 circuit reveals compartmentalized regulation of presynaptic function at divergent release sites. *The Journal of Physiology*, *586*(6), 1495–1502. http://dx.doi.org/10.1113/jphysiol.2007.148635.

Perez, S. M., & Lodge, D. J. (2014). New approaches to the management of schizophrenia: Focus on aberrant hippocampal drive of dopamine pathways. *Drug Design, Development and Therapy*, *8*, 887–896. http://dx.doi.org/10.2147/dddt.s42708.

Pertwee, R. G. (2008). The diverse CB1 and CB2 receptor pharmacology of three plant cannabinoids: Delta9-tetrahydrocannabinol, cannabidiol and delta9-tetrahydrocannabivarin. *British Journal of Pharmacology*, *153*(2), 199–215. http://dx.doi.org/10.1038/sj.bjp.0707442.

Porath, A. J., & Fried, P. A. (2005). Effects of prenatal cigarette and marijuana exposure on drug use among offspring. *Neurotoxicology and Teratology*, *27*(2), 267–277. http://dx.doi.org/10.1016/j.ntt.2004.12.003.

Psychoyos, D., Hungund, B., Cooper, T., & Finnell, R. H. (2008). A cannabinoid analogue of Delta9-tetrahydrocannabinol disrupts neural development in chick. *Birth Defects Research Part B, Developmental and Reproductive Toxicology*, *83*(5), 477–488. http://dx.doi.org/10.1002/bdrb.20166.

Ramesh, D., Ross, G. R., Schlosburg, J. E., Owens, R. A., Abdullah, R. A., Kinsey, S. G., et al. (2011). Blockade of endocannabinoid hydrolytic enzymes attenuates precipitated opioid withdrawal symptoms in mice. *The Journal of Pharmacology and Experimental Therapeutics*, *339*(1), 173–185. http://dx.doi.org/10.1124/jpet.111.181370.

Randolph, K., Brock, P., Margolis, A., & Tau, G. (2013). Cannabis and cognitive systems in adolescents. *Adolescent Psychiatry*, *2013*(3), 13.

Realini, N., Rubino, T., & Parolaro, D. (2009). Neurobiological alterations at adult age triggered by adolescent exposure to cannabinoids. *Pharmacological Research*, *60*(2), 132–138. http://dx.doi.org/10.1016/j.phrs.2009.03.006.

Reece, A. S. (2009). Chronic toxicology of cannabis. *Clinical Toxicology (Philadelphia, Pa)*, *47*(6), 517–524. http://dx.doi.org/10.1080/15563650903074507.

Reece, A. S., & Hulse, G. K. (2016). Chromothripsis and epigenomics complete causality criteria for cannabis- and addiction-connected carcinogenicity, congenital toxicity and heritable genotoxicity. *Mutation Research, Fundamental and Molecular Mechanisms of Mutagenesis*, *789*, 15–25. http://dx.doi.org/10.1016/j.mrfmmm.2016.05.002.

Rivera, P., Bindila, L., Pastor, A., Perez-Martin, M., Pavon, F. J., Serrano, A., et al. (2015). Pharmacological blockade of the fatty acid amide hydrolase (FAAH) alters neural proliferation, apoptosis and gliosis in the rat hippocampus, hypothalamus and striatum in a negative energy context. *Frontiers in Cellular Neuroscience*, *9*, 98. http://dx.doi.org/10.3389/fncel.2015.00098.

Robinson, S. A., Loiacono, R. E., Christopoulos, A., Sexton, P. M., & Malone, D. T. (2010). The effect of social isolation on rat brain expression of genes associated with endocannabinoid signaling. *Brain Research*, *1343*, 153–167. http://dx.doi.org/10.1016/j.brainres.2010.04.031.

Rocchetti, M., Crescini, A., Borgwardt, S., Caverzasi, E., Politi, P., Atakan, Z., et al. (2013). Is cannabis neurotoxic for the healthy brain? A meta-analytical review of structural brain alterations in non-psychotic users. *Psychiatry and Clinical Neurosciences*, *67*(7), 483–492. http://dx.doi.org/10.1111/pcn.12085.

Roloff, A. M., & Thayer, S. A. (2009). Modulation of excitatory synaptic transmission by delta 9-tetrahydrocannabinol switches from agonist to antagonist depending on firing rate. *Molecular Pharmacology*, *75*(4), 892–900. http://dx.doi.org/10.1124/mol.108.051482.

Rubino, T., Realini, N., Braida, D., Guidi, S., Capurro, V., Vigano, D., et al. (2009). Changes in hippocampal morphology and neuroplasticity induced by adolescent THC treatment are associated with cognitive impairment in adulthood. *Hippocampus*, *19*(8), 763–772. http://dx.doi.org/10.1002/hipo.20554.

Salisbury, D. F., Shenton, M. E., Griggs, C. B., Bonner-Jackson, A., & McCarley, R. W. (2002). Mismatch negativity in chronic schizophrenia and first-episode schizophrenia. *Archives of General Psychiatry*, *59*(8), 686–694.

Sanchez-Blazquez, P., Rodriguez-Munoz, M., & Garzon, J. (2014). The cannabinoid receptor 1 associates with NMDA receptors to produce glutamatergic hypofunction: Implications in psychosis and schizophrenia. *Frontiers in Pharmacology*, *4*, 169. http://dx.doi.org/10.3389/fphar.2013.00169.

Scallet, A. C., Uemura, E., Andrews, A., Ali, S. F., McMillan, D. E., Paule, M. G., et al. (1987). Morphometric studies of the rat hippocampus following chronic delta-9-tetrahydrocannabinol (THC). *Brain Research*, *436*(1), 193–198.

Schacht, J. P., Hutchison, K. E., & Filbey, F. M. (2012). Associations between cannabinoid receptor-1 (CNR1) variation and hippocampus and amygdala volumes in heavy cannabis users. *Neuropsychopharmacology*, *37*(11), 2368–2376. http://dx.doi.org/10.1038/npp.2012.92.

Schneider, M., & Koch, M. (2003). Chronic pubertal, but not adult chronic cannabinoid treatment impairs sensorimotor gating, recognition memory, and the performance in a progressive ratio task in adult rats. *Neuropsychopharmacology*, *28*(10), 1760–1769. http://dx.doi.org/10.1038/sj.npp.1300225.

Scholes-Balog, K. E., & Martin-Iverson, M. T. (2011). Cannabis use and sensorimotor gating in patients with schizophrenia and healthy controls. *Human Psychopharmacology*, *26*(6), 373–385. http://dx.doi.org/10.1002/hup.1217.

Senn, R., Keren, O., Hefetz, A., & Sarne, Y. (2008). Long-term cognitive deficits induced by a single, extremely low dose of tetrahydrocannabinol (THC): Behavioral, pharmacological and biochemical studies in mice. *Pharmacology, Biochemistry, and Behavior*, *88*(3), 230–237. http://dx.doi.org/10.1016/j.pbb.2007.08.005.

Sherif, M., Radhakrishnan, R., D'Souza, D. C., & Ranganathan, M. (2016). Human laboratory studies on cannabinoids and psychosis. *Biological Psychiatry*, *79*(7), 526–538. http://dx.doi.org/10.1016/j.biopsych.2016.01.011.

Smith, A. M., Fried, P. A., Hogan, M. J., & Cameron, I. (2004). Effects of prenatal marijuana on response inhibition: An fMRI study of young adults. *Neurotoxicology and Teratology*, *26*(4), 533–542. http://dx.doi.org/10.1016/j.ntt.2004.04.004.

Smith, A. M., Fried, P. A., Hogan, M. J., & Cameron, I. (2006). Effects of prenatal marijuana on visuospatial working memory: An fMRI study in young adults. *Neurotoxicology and Teratology*, *28*(2), 286–295. http://dx.doi.org/10.1016/j.ntt.2005.12.008.

Solowij, N., Jones, K. A., Rozman, M. E., Davis, S. M., Ciarrochi, J., Heaven, P. C., et al. (2012). Reflection impulsivity in adolescent cannabis users: A comparison with alcohol-using and non-substance-using adolescents. *Psychopharmacology*, *219*(2), 575–586. http://dx.doi.org/10.1007/s00213-011-2486-y.

Spano, M. S., Ellgren, M., Wang, X., & Hurd, Y. L. (2007). Prenatal cannabis exposure increases heroin seeking with allostatic changes in limbic enkephalin systems in adulthood. *Biological Psychiatry*, *61*(4), 554–563. http://dx.doi.org/10.1016/j.biopsych.2006.03.073.

Spano, M. S., Fadda, P., Fratta, W., & Fattore, L. (2010). Cannabinoid-opioid interactions in drug discrimination and self-administration: Effect of maternal, postnatal, adolescent and adult exposure to the drugs. *Current Drug Targets*, *11*(4), 450–461.

Stadelmann, A. M., Juckel, G., Arning, L., Gallinat, J., Epplen, J. T., & Roser, P. (2011). Association between a cannabinoid receptor gene (CNR1) polymorphism and cannabinoid-induced alterations of the auditory event-related P300 potential. *Neuroscience Letters*, *496*(1), 60–64. http://dx.doi.org/10.1016/j.neulet.2011.04.003.

Steeds, H., Carhart-Harris, R. L., & Stone, J. M. (2015). Drug models of schizophrenia. *Therapeutic Advances in Psychopharmacology*, *5*(1), 43–58. http://dx.doi.org/10.1177/2045125314557797.

Stone, J. M., Morrison, P. D., Brugger, S., Nottage, J., Bhattacharyya, S., Sumich, A., et al. (2012). Communication breakdown: Delta-9 tetrahydrocannabinol effects on pre-speech neural coherence. *Molecular Psychiatry*, *17*(6), 568–569. http://dx.doi.org/10.1038/mp.2011.141.

Stopponi, S., Soverchia, L., Ubaldi, M., Cippitelli, A., Serpelloni, G., & Ciccocioppo, R. (2014). Chronic THC during adolescence increases the vulnerability to stress-induced relapse to heroin seeking in adult rats. *European Neuropsychopharmacology*, *24*(7), 1037–1045. http://dx.doi.org/10.1016/j.euroneuro.2013.12.012.

Sugiura, T., Kishimoto, S., Oka, S., & Gokoh, M. (2006). Biochemistry, pharmacology and physiology of 2-arachidonoylglycerol, an endogenous cannabinoid receptor ligand. *Progress in Lipid Research*, *45*(5), 405–446. http://dx.doi.org/10.1016/j.plipres.2006.03.003.

Swerdlow, N. R., Light, G. A., Breier, M. R., Shoemaker, J. M., Saint Marie, R. L., Neary, A. C., et al. (2012). Sensory and sensorimotor gating deficits after neonatal ventral hippocampal lesions in rats. *Developmental Neuroscience*, *34*(2-3), 240–249. http://dx.doi.org/10.1159/000336841.

Swerdlow, N. R., Light, G. A., Cadenhead, K. S., Sprock, J., Hsieh, M. H., & Braff, D. L. (2006). Startle gating deficits in a large cohort of patients with schizophrenia: Relationship to medications, symptoms, neurocognition, and level of function. *Archives of General Psychiatry*, *63*(12), 1325–1335. http://dx.doi.org/10.1001/archpsyc.63.12.1325.

Szafran, B., Borazjani, A., Lee, J. H., Ross, M. K., & Kaplan, B. L. (2015). Lipopolysaccharide suppresses carboxylesterase 2 g activity and 2-arachidonoylglycerol hydrolysis: A possible mechanism to regulate inflammation. *Prostaglandins & Other Lipid Mediators*, *121*, 199–206. http://dx.doi.org/10.1016/j.prostaglandins.2015.09.005.

Tamm, L., Epstein, J. N., Lisdahl, K. M., Molina, B., Tapert, S., Hinshaw, S. P., et al. (2013). Impact of ADHD and cannabis use on executive functioning in young adults. *Drug and Alcohol Dependence*, *133*(2), 607–614. http://dx.doi.org/10.1016/j.drugalcdep.2013.08.001.

Tamminga, C. A. (2013). Psychosis is emerging as a learning and memory disorder. *Neuropsychopharmacology*, *38*(1), 247. http://dx.doi.org/10.1038/npp.2012.187.

Tamminga, C. A., Stan, A. D., & Wagner, A. D. (2010). The hippocampal formation in schizophrenia. *The American Journal of Psychiatry*, *167*(10), 1178–1193. http://dx.doi.org/10.1176/appi.ajp.2010.09081187.

Tan, H. Y., Callicott, J. H., & Weinberger, D. R. (2007). Dysfunctional and compensatory prefrontal cortical systems, genes and the pathogenesis of schizophrenia. *Cerebral Cortex*, *17*(Suppl 1), i171–i181. http://dx.doi.org/10.1093/cercor/bhm069.

Tau, G. Z., & Peterson, B. S. (2010). Normal development of brain circuits. *Neuropsychopharmacology*, *35*(1), 147–168. http://dx.doi.org/10.1038/npp.2009.115.

Tomiyama, K., & Funada, M. (2011). Cytotoxicity of synthetic cannabinoids found in "spice" products: The role of cannabinoid receptors and the caspase cascade in the NG 108-15 cell line. *Toxicology Letters*, *207*(1), 12–17. http://dx.doi.org/10.1016/j.toxlet.2011.08.021.

Tseng, K. Y., Chambers, R. A., & Lipska, B. K. (2009). The neonatal ventral hippocampal lesion as a heuristic neurodevelopmental model of schizophrenia. *Behavioural Brain Research*, *204*(2), 295–305. http://dx.doi.org/10.1016/j.bbr.2008.11.039.

Tzavara, E. T., Li, D. L., Moutsimilli, L., Bisogno, T., Di Marzo, V., Phebus, L. A., et al. (2006). Endocannabinoids activate transient receptor potential vanilloid 1 receptors to reduce hyperdopaminergia-related hyperactivity: Therapeutic implications. *Biological Psychiatry*, *59*(6), 508–515. http://dx.doi.org/10.1016/j.biopsych.2005.08.019.

van der Stelt, M., Hansen, H. H., Veldhuis, W. B., Bar, P. R., Nicolay, K., Veldink, G. A., et al. (2003). Biosynthesis of endocannabinoids and their modes of action in neurodegenerative diseases. *Neurotoxicity Research*, *5*(3), 183–200.

Van Waes, V., Beverley, J. A., Siman, H., Tseng, K. Y., & Steiner, H. (2012). CB1 cannabinoid receptor expression in the striatum: Association with corticostriatal circuits and developmental regulation. *Frontiers in Pharmacology*, *3*, 21. http://dx.doi.org/10.3389/fphar.2012.00021.

Verdoux, H., Gindre, C., Sorbara, F., Tournier, M., & Swendsen, J. D. (2003). Effects of cannabis and psychosis vulnerability in daily life: An experience sampling test study. *Psychological Medicine*, *33*(1), 23–32.

Verrico, C. D., Liu, S., Bitler, E. J., Gu, H., Sampson, A. R., Bradberry, C. W., et al. (2012). Delay- and dose-dependent effects of Delta(9)-tetrahydrocannabinol administration on spatial and object working memory tasks in adolescent rhesus monkeys. *Neuropsychopharmacology*, *37*(6), 1357–1366. http://dx.doi.org/10.1038/npp.2011.321.

Vigano, D., Guidali, C., Petrosino, S., Realini, N., Rubino, T., Di Marzo, V., et al. (2009). Involvement of the endocannabinoid system in phencyclidine-induced cognitive deficits modelling schizophrenia. *The International Journal of Neuropsychopharmacology*, *12*(5), 599–614. http://dx.doi.org/10.1017/s1461145708009371.

Volk, D. W., & Lewis, D. A. (2016). The role of endocannabinoid signaling in cortical inhibitory neuron dysfunction in schizophrenia. *Biological Psychiatry*, *79*(7), 595–603. http://dx.doi.org/10.1016/j.biopsych.2015.06.015.

Volkow, N. D., Fowler, J. S., Wang, G. J., & Swanson, J. M. (2004). Dopamine in drug abuse and addiction: Results from imaging studies and treatment implications. *Molecular Psychiatry*, *9*(6), 557–569. http://dx.doi.org/10.1038/sj.mp.4001507.

Volkow, N. D., Wang, G. J., Fowler, J. S., Tomasi, D., Telang, F., & Baler, R. (2010). Addiction: Decreased reward sensitivity and increased expectation sensitivity conspire to overwhelm the brain's control circuit. *Bioessays*, *32*(9), 748–755. http://dx.doi.org/10.1002/bies.201000042.

Volkow, N. D., Wang, G. J., Telang, F., Fowler, J. S., Alexoff, D., Logan, J., et al. (2014). Decreased dopamine brain reactivity in marijuana abusers is associated with negative emotionality and addiction severity. *Proceedings of the National Academy of Sciences of the United States of America*, *111*(30), E3149–E3156. http://dx.doi.org/10.1073/pnas.1411228111.

Wible, C. G. (2012). Hippocampal temporal-parietal junction interaction in the production of psychotic symptoms: A framework for understanding the schizophrenic syndrome. *Frontiers in Human Neuroscience*, *6*, 180. http://dx.doi.org/10.3389/fnhum.2012.00180.

Wiley, J. L. (2003). Sex-dependent effects of delta 9-tetrahydrocannabinol on locomotor activity in mice. *Neuroscience Letters*, *352*(2), 77–80.

Yucel, M., Bora, E., Lubman, D. I., Solowij, N., Brewer, W. J., Cotton, S. M., et al. (2012). The impact of cannabis use on cognitive functioning in patients with schizophrenia: A meta-analysis of existing findings and new data in a first-episode sample. *Schizophrenia Bulletin*, *38*(2), 316–330. http://dx.doi.org/10.1093/schbul/sbq079.

Yucel, M., Solowij, N., Respondek, C., Whittle, S., Fornito, A., Pantelis, C., et al. (2008). Regional brain abnormalities associated with long-term heavy cannabis use. *Archives of General Psychiatry*, *65*(6), 694–701. http://dx.doi.org/10.1001/archpsyc.65.6.694.

Zalesky, A., Solowij, N., Yucel, M., Lubman, D. I., Takagi, M., Harding, I. H., et al. (2012). Effect of long-term cannabis use on axonal fibre connectivity. *Brain*, *135*(Pt 7), 2245–2255. http://dx.doi.org/10.1093/brain/aws136.

Zamberletti, E., Beggiato, S., Steardo, L., Jr., Prini, P., Antonelli, T., Ferraro, L., et al. (2014). Alterations of prefrontal cortex GABAergic transmission in the complex psychotic-like phenotype induced by adolescent delta-9-tetrahydrocannabinol exposure in rats. *Neurobiology of Disease*, *63*, 35–47. http://dx.doi.org/10.1016/j.nbd.2013.10.028.

Zammit, S., Spurlock, G., Williams, H., Norton, N., Williams, N., O'Donovan, M. C., et al. (2007). Genotype effects of CHRNA7, CNR1 and COMT in schizophrenia: Interactions with tobacco and cannabis use. *The British Journal of Psychiatry*, *191*, 402–407. http://dx.doi.org/10.1192/bjp.bp.107.036129.

Zammit, S., Thomas, K., Thompson, A., Horwood, J., Menezes, P., Gunnell, D., et al. (2009). Maternal tobacco, cannabis and alcohol use during pregnancy and risk of adolescent psychotic symptoms in offspring. *The British Journal of Psychiatry*, *195*(4), 294–300. http://dx.doi.org/10.1192/bjp.bp.108.062471.

Zhu, P. J., & Lovinger, D. M. (2010). Developmental alteration of endocannabinoid retrograde signaling in the hippocampus. *Journal of Neurophysiology*, *103*(2), 1123–1129. http://dx.doi.org/10.1152/jn.00327.2009.

Chapter 4

Psychotomimetic and Cognitive Effects of Δ^9-Tetrahydrocannabinol in Laboratory Settings

John D. Cahill*^,†^, Swapnil Gupta*^,†^, Jose Cortes-Briones*^,‡^, Rajiv Radhakrishnan*^,†,‡^, Mohamed Sherif*^,‡^, Deepak C. D'Souza*^,†,‡^

**Yale School of Medicine, New Haven, CT, United States †Connecticut Mental Health Center, New Haven, CT, United States ‡VA Connecticut, West Haven, CT, United States*

INTRODUCTION

A drug, in the context of human laboratory studies (HLS), is said to be psychotomimetic when its actions mimic signs and symptoms characteristic of psychosis, namely perceptual abnormalities (illusions and hallucinations), delusional beliefs and feelings, disorganization of thoughts and speech, and altered perceptions of self (such as dissociation). In addition, drugs may produce effects that resemble negative psychotic symptoms (apathy, anhedonia, alogia, asociality, avolition). Finally, drugs may also impair several aspects of cognition (memory, attention, executive function). Psychosis may be viewed as a core cluster of deficits in a number of functional domains, notably perception, cognition, and reward, and is observable in a range of acute (e.g., substance induced) and chronic (e.g., schizophrenia) conditions. Each condition may possess a unique admixture of additional deficits, for example in schizophrenia, where further, specific cognitive, affective, and functional impairments may also manifest. In this way, psychotomimetic drugs simulate core elements of psychosis in the absence of some of these collateral deficits, rather than a complete clinical picture.

This chapter will first discuss the properties of Δ^9-tetrahydrocannabinol (THC) and related cannabinoids used in HLS alongside key considerations when designing and interpreting these studies. Second, it will outline the psychotomimetic and cognitive effects in healthy humans and clinical populations that are relevant to schizophrenia, followed by factors that modulate those effects. Lastly, it will summarize biomarkers associated with these clinical effects with views toward future work (Fig. 1).

The Complex Connection between Cannabis and Schizophrenia. http://dx.doi.org/10.1016/B978-0-12-804791-0.00004-5

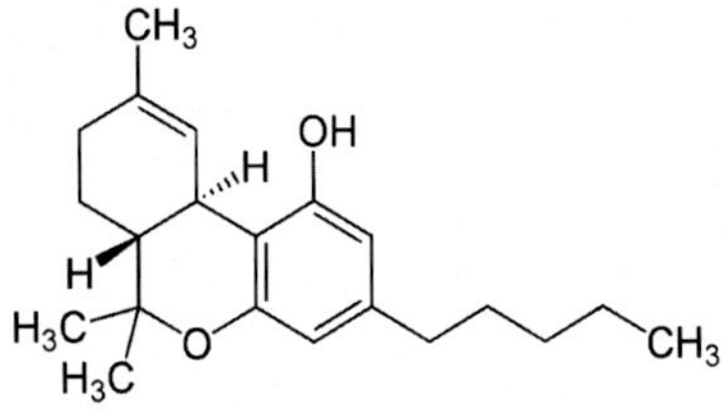

FIG. 1 Chemical structure of THC.

DESIGNING AND INTERPRETING HLS OF CANNABINOIDS

Properties of Cannabinoid Formulations Used in Laboratory Studies

THC, the principal psychoactive constituent of cannabis, is a partial agonist of type 1 (CB_1) and type 2 (CB_2) cannabinoid receptors. Various formulations of THC and related cannabinoids have been used in HLS in the United States (US): oral Dronabinol (THC), oral Nabilone (Cesamet), smoked National Institute of Drug Abuse (NIDA) standard cigarettes, and intravenous (IV) THC. Oral Dronabinol (Marinol) (10mg capsule) is a US Food and Drug Administration (FDA)-approved synthetic form of THC that is typically used to treat anorexia in AIDS (acquired immunodeficiency syndrome) (and other wasting diseases), emesis in cancer patients undergoing chemotherapy, and chronic pain. Similarly, the 9-trans-ketocannabinoid Nabilone (Cesamet), a synthetic analog of THC, is indicated in the US for the treatment of chemotherapy-induced nausea. Of significance for investigators, these two oral agents are classed as schedule 3 (opposed to 1) substances due to their FDA-approved clinical indication, and thus have lesser regulatory requirements around use. NIDA manufactures and supplies standard cannabis cigarettes for medical research purposes in a range of THC potencies and more recently, cannabidiol (CBD) content. IV THC can be formulated for research use under the purview of the FDA as an investigational new drug, and stored under a schedule 1 license with the Drug Enforcement Administration (DEA). However, with the legalization of cannabis for medical and recreational purposes in the US, some states are in the process of making it possible to conduct research with cannabinoids, circumventing the schedule 1 license.

A key consideration in designing HLS is whether the pharmacokinetic and pharmacodynamic properties of the cannabinoid provide an adequate time window of neuropharmacological activity during which measurements can be made to answer the research question of interest. As noted by numerous studies, herbal marijuana contains over 400 chemicals belonging to 18 different classes (ElSohly et al., 2016; Mehmedic et al., 2010; Turner, Elsohly, & Boeren, 1980). The other lesser constituents that vary widely in concentration may have individual, interactive, or possibly entourage effects, which are not well understood and may confound the effects of the principle psychoactive constituent, THC. The use of THC alone hence provides a more precise and controlled probe of the principal psychoactive effects of cannabis. However, it is important to recognize that THC may not accurately capture the full range of effects of herbal cannabis.

Another factor that determines the ecological validity of HLS is the degree to which the pharmacokinetic and pharmacodynamic properties of the administered THC parallel that of socially normative use. While smoking (and vaporizing) cannabis remains the most common method of consumption, the use of "edibles" or oral cannabinoid preparations has been steadily increasing in recent years. Homemade cannabis products (e.g., brownies and "cannabutter") are popular, and an array of commercial "edibles" including cannabis-infused drinks, candies, baked goods, and "dissolvables" (e.g., Cannastrips, Cannalixir) are widely available. Oral, IV, and vaporized cannabinoids have different pharmacodynamic and pharmacokinetic properties that provide unique opportunities for HLS to capture more ecologically valid scenarios of cannabis use.

THC is metabolized primarily in the liver by the cytochrome P450 (CYP) enzymes CYP2C9, CYP2C19, and CYP3A (Watanabe, Matsunaga, Yamamoto, Funae, & Yoshimura, 1995; Watanabe, Yamaori, Funahashi, Kimura, & Yamamoto, 2007). The primary metabolite, 11-hydroxy-delta-9-tetrahydrocannobinol (11-OH-THC), is at least as potent as THC, has a similar pharmacokinetic profile, and is thought to contribute significantly to the effects of THC (Perez-Reyes, Timmons, Lipton, Davis, & Wall, 1972). 11-OH-THC is further metabolized into the inactive metabolite 11-nor-9-carboxy-tetrahydrocannobinol (THC-COOH). It is THC-COOH that is detected in urine drug tests. Tissue distribution of THC and its major metabolites is dependent on its high lipophilicity (Garrett & Hunt, 1974; Hunt & Jones, 1980), causing it to distribute readily into highly vascularized tissue (Ho et al., 1970) and lipophilic tissue such as body fat (Garrett & Hunt, 1974; Johansson, Noren, Sjovall, & Halldin, 1989). THC also demonstrates significant plasma-protein binding (95%–99%) (Garrett & Hunt, 1974; Hunt & Jones, 1980). In plasma, low THC concentrations can be measured for extended periods after administration due to initial rapid redistribution and subsequent slow release ("reabsorption") from fatty tissue into the bloodstream (Hunt & Jones, 1980; Leuschner, Harvey, Bullingham, & Paton, 1986; Ohlsson et al., 1982); however, whether this release is of pharmacological relevance is not clear. Several models for the pharmacokinetics of THC have been published in studies with humans after IV (Lemberger, Tamarkin, Axelrod, & Kopin, 1971), oral (Wall, Sadler, Brine, Taylor, & Perez-Reyes, 1983), and intrapulmonary administration via smoking (Chiang & Barnett, 1984; Cocchetto, Owens, Perez-Reyes, DiGuiseppi, & Miller, 1981; Harder & Rietbrock, 1997) and using a vaporizer (Strougo et al., 2008).

Oral Administration

Orally consumed cannabinoids have a pharmacokinetic profile that is very different from inhaled or IV cannabinoids. The effects have a slower onset of action but last much longer. Absorption is slower when cannabinoids are ingested, with lower, more-delayed peak THC concentrations (Agurell et al., 1986; Hampson, Grimaldi, Axelrod, & Wink, 1998; Harder & Rietbrock, 1997). Dose, route of administration, vehicle, and physiological factors such as absorption and rates of metabolism and excretion can influence drug concentrations in circulation. A high octanol/water partition coefficient (P) of THC (estimated to be between 6000 and $>9\times10^6$) is thought to be responsible for its high rate of absorption. Perez-Reyes et al. examined the efficacy of five different vehicles (ethanol, sesame, glycocholate, emulsion of Tween-80, combination of ethanol and glycocholate) for oral administration of THC in gelatin capsules (Perez-Reyes et al., 1973). They found that glycocholate and sesame oil improved bioavailability although there was considerable variability in absorption and peak serum levels even within vehicles. Wall et al. have reported the bioavailability of oral THC to be 10%–20% (Wall et al., 1983). This complements a more accurate assessment of the oral bioavailability of THC by Ohlsson et al. based on gas chromatography/mass spectroscopy (GC/MS) experiments (Ohlsson et al., 1980). The peak THC concentration following ingestion of 20 mg of THC in a chocolate cookie occurred 1–5 h later and ranged from 4.4 to 11 ng/mL; the estimated oral bioavailability was 6%. Factors that may be responsible for the observed pharmacokinetics include variability in absorption, degradation of THC in the stomach, and significant first-pass metabolism to active 11-OH-THC as well as inactive metabolites in the liver.

In a study of 17 volunteers, following oral administration of a single capsule of Dronabinol (containing 10 mg of THC), mean peak plasma concentrations obtained 1–2 h later were as follows: THC = 3.8 ng/mL (range 1.1–12.7 ng/mL), 11-OH-THC = 3.4 ng/mL (range 1.2–5.6 ng/mL), and THC-COOH = 26 ng/mL (range 14–46 ng/mL) (Hampson et al., 1998). THC and 11-OH-THC concentrations were comparable, while consistently higher THC-COOH concentrations were observed. Also, interestingly, two THC peaks were frequently observed, possibly due to enterohepatic circulation. Compared to the smoked route, following ingestion of Dronabinol the onset of effects is delayed, peak THC concentrations are lower, and duration of pharmacodynamic effects is generally prolonged (Hampson et al., 1998; Ohlsson et al., 1980; Perez-Reyes et al., 1973; Plasse et al., 1991; Wall et al., 1983). Oral Nabilone has also been specifically shown to have a delayed onset of action (1–2 h) and long duration of effects (8–12 h) (Lemberger & Rowe, 1975). There are significant implications for recreational use as well as study dosing protocols and procedures. With the inhaled route, the dose can be titrated in real time. In contrast, given the delayed onset and longer duration of effects when administered orally, little can be done to alleviate negative effects (e.g., panic or psychosis), once they emerge. Furthermore, since the duration of effects is longer with oral consumption, individuals who are not familiar with the effects of cannabis may feel overwhelmed. Moreover, the inability to titrate effects with oral consumption may contribute to a greater sense of loss of control.

Intravenous Administration

The use of IV THC to probe the endocannabinoid system in the laboratory offers several advantages: it standardizes drug delivery and minimizes inter- and intraindividual variability in plasma THC levels, has rapid onset of action (10–15 min), and has a predictable peak (30–60 min), which enables researchers to capture time-locked psychophysiological measures relevant to the study. The pharmacokinetics of oral, inhaled (smoked), and IV cannabinoids were reviewed by Agurell et al. (1986) (Fig. 2). They reported that the plasma profiles of IV THC were comparable to those of smoked THC, although bioavailability was higher following smoking among heavy cannabis users versus light cannabis users. Ohlsson et al. compared the effects of oral, inhaled (smoked), and IV (5 mg over 2 min) THC in 11 healthy male volunteers. Compared to THC plasma levels after smoking and IV injection, THC plasma levels after oral doses were lower and irregular, indicating slow and erratic absorption (Ohlsson et al., 1980). As discussed, THC has low oral bioavailability due to degradation in the gut and extensive first-pass metabolism in the liver (Wall et al., 1983), which may account for these observations. Plasma levels of THC after smoking were similar to plasma levels after IV administration but were about 50% lower. Based on $AUC_{0\text{–}360\,min}$, systemic availability of THC after smoking and after oral administration were 18% ± 6% and 6% ± 3%, respectively. Wall et al. compared the pharmacokinetics of IV and oral THC in male and female volunteers (Wall et al., 1983). They found no significant differences in dynamic activity,

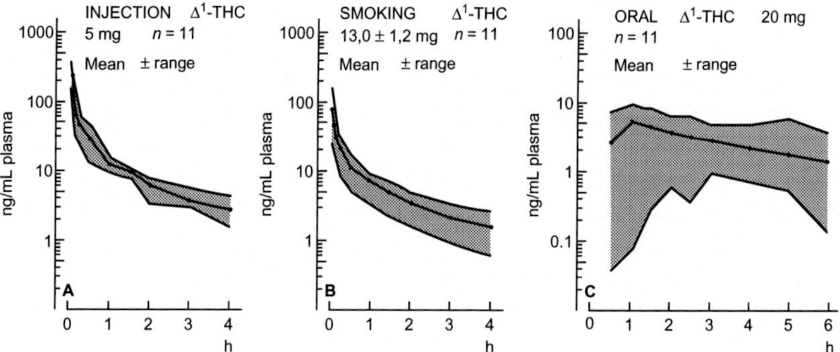

FIG. 2 Comparison of plasma Δ^9-tetrahydrocannabinol concentrations over time (in hours) following intravenous injection, smoking, and oral administration. *Reprinted with permission from Agurell, S., Halldin, M., Lindgren, J. E., Ohlsson, A., Widman, M., Gillespie, H., et al. (1986). Pharmacokinetics and metabolism of delta 1-tetrahydrocannabinol and other cannabinoids with emphasis on man.* Pharmacological Reviews, 38, *21–43 and originally adapted from Ohlsson, A., Lindgren, J. E., Wahlen, A., Agurell, S., Hollister, L. E., & Gillespie, H. K. (1980). Plasma delta-9 tetrahydrocannabinol concentrations and clinical effects after oral and intravenous administration and smoking.* Clinical Pharmacology and Therapeutics, 28, *409–416.*

metabolism, excretion, and kinetics between the two groups; however, the ratio of the concentrations of 11-OH-Δ-9-THC to Δ-9-THC in plasma after IV dosing was 1:10–20 versus 0.5–1:1 after oral dosing (Wall et al., 1983).

Smoking and Vaporization

Although smoking is a well-established route of administration, vaporization is a relatively new technology that allows for cannabinoids to be heated to a temperature below the point of combustion (which occurs at 230°C and above), where active cannabinoid vapors are rapidly formed while suppressing the production of pyrolytic toxic compounds and eliminating the harmful effects of secondhand smoke. In a number of studies, the vaporization of cannabis samples was systematically tested to show its advantages over smoking. A variety of vaporization and smoking devices (including water pipes) used for cannabis consumption have been systematically studied. When compared with direct smoking of cannabis, only vaporizers were found to be capable of achieving reductions in tar intake (McPartland & Pruitt, 1997). Furthermore, prior studies have shown that cannabis vaporization is more efficient than burning the plant material, due to reduced THC degradation (Abrams et al., 2007), and when compared to standard smoking, participants have shown preference for vaporization (Fischedick, Van Der Kooy, & Verpoorte, 2010). One study evaluated the Volcano vaporizer to test the variability of vaporizing parameters with respect to using pure cannabinoid preparations (THC and tetrahydrocannabinolic acid (THCA) diluted as ethanolic solutions) and cannabis plant material (200 mg at 12% THCA) (Hazekamp, Ruhaak, Zuurman, van Gerven, & Verpoorte, 2006). Their findings indicated that the vaporization of the cannabis plant material resulted in a maximum THC delivery of 29%, while the vaporization of pure THC resulted in a maximum THC delivery of about 53.9%, at a temperature of 226 ± 5°C. In addition, their results indicated that 30%–40% of inhaled THC was not absorbed by the lungs. In light of these findings, the difference in THC delivery yield between pure THC vaporization and cannabis plant material vaporization must be taken into account while deciding on an inhaled THC dose. The vaporization of 0.45 g of NIDA cannabis plant material (at a THC concentration of 1.7%) has been reported to produce a peak mean "high" effect of ~60, measured on a 0–100 visual analog scale (VAS) (Abrams et al., 2007). This effect is comparable to the effects of 2 mg of IV THC, which was reported to produce a peak "high" effect of ~50, measured on a 0–100 VAS (Borg, Gershon, & Alpert, 1975).

The rate at which THC enters the brain largely determines its effects. Thus, for smoking and vaporizer studies, the administration of each puff cycle, including the durations of inhalation, holding the smoke/vapor in lungs, and exhalation, has to be standardized by using a cued puff procedure such as the "Foltin Puff Procedure" (Fischman, Foltin, & Brady, 1988; Foltin, Fischman, & Byrne, 1988). In this procedure, participants are prompted to inhale for approximately

5 s, hold vapor in their lungs for approximately 10 s, and exhale and wait for approximately 45 s, after which the entire process is repeated.

A previous study comparing the administration of cannabis by vaporization to standard cannabis smoking found that subjective effects and plasma THC concentrations were generally similar in both administration techniques (Abrams et al., 2007). Their results indicated that the self-reported "high" did not differ during vaporization compared to smoking overall (6-h AUC), or at any observation point after consumption of cannabis. Furthermore, their findings indicated that the two modalities were not significantly different from one another at any of the three THC strengths (1.7%, 3.4%, and 6.8%) in the 6-h area under the plasma THC concentration-time curve (AUC), or for the peak THC plasma concentration measured at 2 min (Abrams et al., 2007).

Safety and Tolerability

The unpleasant psychoactive effects of THC include anxiety, panic, paranoia, and rarely, psychosis; all of these may be extremely distressing. However, the extent to which THC effects are perceived as unpleasant is context-dependent and can be reduced by preparing individuals in advance for the possible effects of THC. The acute physical effects of THC include motor incoordination, tremulousness, muscle weakness, hypo or hypertension, tachycardia, conjunctival injection, dry mouth, and increased appetite, which generally resolve spontaneously (Carbuto et al., 2012b). The estimated lethal dose of THC is 30 mg/kg (2100 mg in a 70-kg individual), which is several thousand times the amount of THC in a standard cannabis joint. It should be noted that no fatalities have been reported with cannabis overdose.

THC has been administered in very large oral doses (50–712 mg) and modest IV doses (1–10 mg) to cannabis users without any serious adverse events (Hollister, 1986). D'Souza et al. have conducted a series of studies with varying doses of THC (1.75–5 mg) administered over varying times (2–20 min) to several groups of subjects ($n > 400$) including cannabis users, nonusers, and patients with schizophrenia (D'Souza et al., 2005), and have shown that it was safe and well tolerated (Ranganathan & D'Souza, 2006b).

There are no major differences between the type of unpleasant behavioral and acute physical effects induced by IV THC, smoked cannabis, and oral Dronabinol. Hypotension may be more likely to be observed with IV THC (D'Souza: unpublished observations). However, as discussed earlier, the slower onset and longer duration of effects of oral THC may have implications if adverse effects are experienced during a study.

The majority of the studies that make up the existing literature on the effects of cannabis consumption use standard smoking as their drug administration method, which exposes subjects to a variety of toxins produced by the combustion of the cannabis plant material. Previous studies that investigated the smoke harm reduction associated with vaporization of cannabis showed that the vaporizers produced qualitative reduction in carbon monoxide and particulates, and complete elimination of benzene, toluene, and naphthalene, which are three toxic hydrocarbons produced by standard cannabis smoking. Another study that investigated vaporization as a smokeless means of delivery of inhaled *Cannabis sativa* reported no adverse events and indicated that vaporization of cannabis is a safe and effective mode of delivery of cannabinoids (Abrams et al., 2007).

Many recent studies that have investigated the administration of vaporized THC point to the conclusion that vaporized THC administration may be a safer method, producing cognitive and psychotropic effects similar to oral and IV THC, with no adverse events reported. In terms of somatic effects produced by vaporization, previous studies using THC dissolved in an ethanolic solution reported side effects such as slight irritation of the throat and upper respiratory tract as well as mild coughing (Hazekamp et al., 2006; Naef, Russmann, Petersen-Felix, & Brenneisen, 2004). These side effects were reversible within 30 min of finishing the inhalation. One study found that these complaints were also observed during inhalation of placebo and were attributed to effects of residual ethanol. No serious adverse events were reported.

HLS in clinical populations have been controversial due to theoretical safety concerns and ethical considerations. However, it has been demonstrated that the administration of THC in laboratory contexts induces symptoms only transiently and has no effect on the participants' desire to use cannabis for up to a year after study participation (Carbuto et al., 2012a). One study reported an event of hospitalization for exacerbation of hypertension, which the subject failed to disclose at the time of screening. As with healthy controls, nausea and dizziness were the common minor side effects at the higher end of IV dosing. Study participation did not impact the course of illness or cannabis use in patients with schizophrenia in a 6-month follow-up study (Carbuto et al., 2012a).

Some Considerations on Study Design

Multiple lines of evidence have indicated a role for the endocannabinoid system in the pathophysiology of schizophrenia (Bossong, Jansma, Bhattacharyya, & Ramsey, 2014; Gupta, Cahill, Ranganathan, & Correll, 2014). Therefore, in addition to inducing nonphysiological states in healthy individuals, laboratory-based THC challenge studies can be used to probe abnormalities in the endocannabinoid systems of individuals with schizophrenia in order to identify markers of disease. Further, conducting similar studies in individuals at genetic risk for disease but without psychosis may help elucidate pathophysiology without confounds such as collateral effects of the illness and the presence of antipsychotic medications. THC challenge studies in individuals with different degrees of genetic and clinical risk also form a paradigm for probing gene–gene interaction related to the role of cannabis in the pathophysiology of schizophrenia. THC challenges in individuals with a genetic diathesis for psychosis may also serve as a test for the extent of their vulnerability for psychosis and may assist in identification of at-risk individuals who might benefit from early interventions to prevent the development of schizophrenia (Gupta, Ranganathan, & D'Souza, 2016b).

It is recognized that individuals with psychotic disorders and comorbid cannabis use disorder have a significantly worse prognosis than those with psychotic disorders alone (Swendsen, Ben-Zeev, & Granholm, 2011). Their cannabis use leads to symptom exacerbation (Buckley, Miller, Lehrer, & Castle, 2009), an increased likelihood of psychotic relapse, and an overall worse global functioning (Zammit et al., 2008). However, subjective reports from some psychotic individuals who use cannabis indicate that they experience a relief of symptoms when they use cannabis (Schofield et al., 2006). Further, a small case series of the use of dronabinol in the treatment of schizophrenia has indicated some efficacy (Schwarcz, Karajgi, & McCarthy, 2009). In the situation where subjective reports contradict large-scale empirical data, a controlled laboratory-based measurement of the effects of THC on individuals with psychosis is of immense value. Studies of the effects of THC in schizophrenia have helped clarify these contradictory findings. Further, these challenge studies can serve as a paradigm to test specific treatments for comorbid cannabis use and schizophrenia, targeting either the cognitive or the subjective euphoric effects of cannabis (similar to the use of naltrexone in alcohol or opioid dependence).

The acute administration of cannabinoids in tightly controlled laboratory settings allows for both a clearer inference of causality of effects and a more detailed characterization of those effects. In the laboratory, the high variability in cannabinoid content (specifically THC and CBD) of street cannabis products available and the manner and route of consumption (greatly affecting bioavailability) can be controlled, facilitating more precise delineation of dose-response relationships. Similarly, the biological, psychological, and the social context of cannabinoid administration can be better described and controlled through a careful selection and characterization of subjects, the enforcement of study restrictions, and the rigorous standardization of the study environment and procedures. Furthermore, given recent evidence questioning the validity of self-reported cannabinoid effects, HLS allow for the ongoing investigation of a potential dissociation between subjective and objective effects. Lastly, although HLS have direct inferential value in studying the nature of acute, intoxication-related cannabinoid-induced psychosis, the degree to which psychotomimesis reflects any other type of clinical psychosis remains subject to debate. Nevertheless, the induction of these transient psychosis-like (and cognitive) effects in healthy controls allows for the isolation of "state" (opposed to "trait") features and biomarkers that are, at the very least, phenomenologically relevant to psychosis. This differentiation of *state* and *trait* may further facilitate the discovery of potential overlaps in the seemingly heterogeneous pathoetiologies of psychotic states.

HLS with cannabinoids are not without limitations. Tight inclusion and exclusion criteria, operationalized cannabinoid administration, and the inevitable influence of the laboratory environment on effects experienced constrain the generalizability and/or limit the interpretation of this type of study. Subjects may self-select due to past positive or negative associations with cannabis and/ or laboratory-based research in general. Most notably, those who have experienced past adverse effects from cannabinoids may be unlikely to volunteer for such studies. Despite providing a useful laboratory model of psychosis in humans, HLS using cannabinoids are unlikely to ever adequately replicate the full range of deficits observed in psychotic disorders and therefore can provide only a restricted, though penetrating, view. They remain a vital complement to epidemiological and other more ecologically valid study designs.

EFFECTS OF Δ^9-TETRAHYDROCANNABINOL IN HUMAN LABORATORY STUDIES

Effects in Healthy Individuals

Experimental studies have shown that the acute effects of cannabinoids resemble, in an attenuated way, some core phenomenology of psychosis and schizophrenia; these effects have been captured using standardized scales such as the Positive and Negative Symptom Scale for Schizophrenia (PANSS) (Fig. 3), Clinician Administered Dissociative Symptoms Scale (CADSS), Psychotomimetic States Inventory (PSI), and the Brief Psychiatric Rating Scale (BPRS). These psychotomimetic effects include positive- and disorganization-like symptoms such as suspiciousness, paranoid and grandiose delusions, perceptual alterations, conceptual disorganization, and fragmented thinking. Furthermore, they also include negative-like symptoms such as blunted affect, reduced rapport, emotional withdrawal, psychomotor retardation, and lack of spontaneity (D'Souza et al., 2004; Kleinloog et al., 2012b; Liem-Moolenaar et al., 2010; Morrison & Stone, 2011; Morrison et al., 2009), which are not an effect of the sedating and cataleptic effects of THC (Morrison & Stone, 2011). In addition, cannabinoids have been shown to induce dissociative symptoms

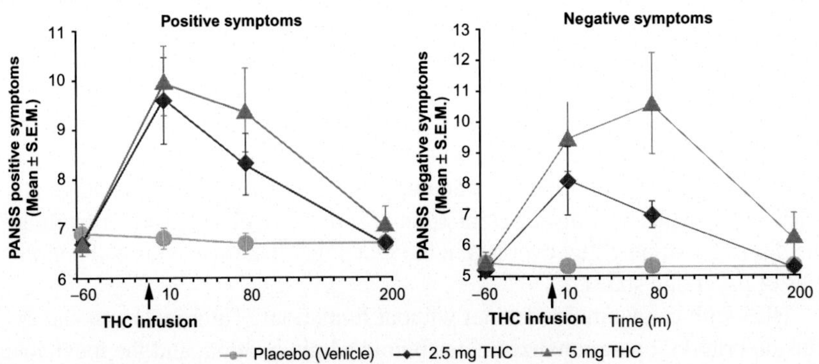

FIG. 3 Comparison of pharmacokinetics of IV smoked and oral THC.

TABLE 1 The Quality of Psychotomimetic Effects Reported by Healthy Individuals Following Intravenous Infusion of 5 or 2.5 mg of THC as It Relates to Typical Signs and Symptoms of Schizophrenia

Subject Quote	Symptom
I thought you could read my mind, that's why I didn't answer	Suspiciousness/paranoia
I thought you all were trying to trick me by changing the rules of the tests to make me fail	
I could hear someone on typing on the computer... and I thoughts you all were trying to program me	
I thought you all were giving me THC through the BP machines and the sheets	
I couldn't keep track of my thoughts... they'd suddenly disappear	Formal thought disorder
It seemed as if all the questions were coming to me at once... everything was happening in staccato	
My thoughts were fragmented... the past, present and future all seemed to be happening at once	
I felt I could see into the future... I thoughts I was God	Grandiosity
The AC that I couldn't hear before suddenly became deafening	Perceptual abnormalities
I thought I could hear the dripping of the iv and it was louder than your voice	

Reprinted with permission from D'Souza, D. C., Perry, E., MacDougall, L., Ammerman, Y., Cooper, T., Wu, Y. T., et al. (2004). The psychotomimetic effects of intravenous delta-9-tetrahydrocannabinol in healthy individuals: Implications for psychosis. *Neuropsychopharmacology*, *29*, 1558–1572.

such as depersonalization, derealization, altered body perception, feelings of unreality, and extreme slowing of time (D'Souza et al., 2004; Kleinloog et al., 2012b; Liem-Moolenaar et al., 2010; Morrison & Stone, 2011; Morrison et al., 2009). Table 1, adapted from D'Souza 2004, illustrates the quality of a range of psychotomimetic symptoms expressed on a retrospective report of peak THC effects (D'Souza et al., 2004).

More than 50 HLS have been published since the 1970s featuring THC as the predominant interventional agent, as summarized in Table 2.

The types of psychotomimetic effects of cannabinoids have been shown to be consistent across a range of doses and routes of administration, such as smoked cannabis, oral cannabis extract/THC (5–20 mg), IV THC (~0.015–0.06 mg/kg), and inhaled vaporized THC (Bhattacharyya, Crippa, et al., 2012c; D'Souza et al., 2004; Englund et al., 2013a; Kaufmann et al., 2010; Martin-Santos et al., 2012a;

TABLE 2 A Summary of HLS Featuring THC as the Predominant Interventional Agent—Psychosis-Relevant, Cognitive, and Physiological Effects

Author(s)	Year	Subjects	THC (And Other Drugs) Dose and Route	Psychosis-Relevant Effects	Cognitive Effects	Physiological Measures
Bossong et al. (2015)	2015	19 healthy volunteers with history of previous exposure	2 studies: 10 mg of dronabinol PO and 8 mg of THC inhaled using vaporizer	N/A	N/A	PET: Significant reduction in tracer binding to D2/D3 receptors in the limbic striatum
Bhattacharyya et al. (2015)	2015	15 healthy males (same population as Bhattacharyya et al., 2009)	10 mg THC-PO 600 mg CBD-PO	N/A	N/A	fMRI: opposite effects of THC and CBD on functional connectivity between dorsal striatum, PFC and hippocampus
Bossong et al. (2013)	2013	14 healthy males	6 mg followed for 3 times by1 mg of THC—inhalation using vaporizer	SUBJ: ↑ feelings of high, internal perception, external perception. ↓ calmness, alertness and contentedness	N/A	fMRI: Differential activation of different brain regions between fearful faces and happy faces.
Atakan et al. (2013)	2013	21 healthy males	10 mg of THC-PO	PANSS: ↑ positive, negative, general. SUBJ: ↓ tranquility ↑ anxiety ↑ intoxication	go/no-go task: ↑ inhibition errors	fMRI: THC versus plac = ↓ left PHG, MTG, STG, while ↑ activation of right MTG in transiently psychotic group

Sewell et al. (2013a)	2013	44 subjects [34 nonusers (27 males) and 10 chronic users (6 males)]	0.015–0.05 mg/kg of THC-IV	SUBJ: Nonusers: temporal overestimation at medium and high doses, and underproduction at all doses Chronic users: effects were blunted at all doses	N/A	N/A
Kuepper et al. (2013b)	2013	9 healthy cannabis users, 8 patients with psychotic disorder and 7 first-degree relatives	8 mg of THC—inhalation through vaporizer	SUBJ: ↑ feeling high, external perception, internal perception	N/A	PET: No dopamine release in the control group THC induced dopamine release in both the patients and relatives, most pronounced in caudate nucleus
Englund et al. (2013b)	2013	48 (22 receiving CBD, 26 receiving placebo)	1.5 mg THC-IV 600 mg CBD-PO versus placebo	PANSS: less likely ↑ in positive score with CBD SSPS: less in CBD versus placebo	HVLT-R: poorer scores in the placebo group compared to CBD	N/A
Bhattacharyya, Atakan, et al. (2012b)	2012	36 healthy males	10 mg of THC-PO	PANSS: 9 repeat carriers of DAT1 gene is associated with more psychotic symptoms induced by THC GG of AKT1 has even higher	N/A	fMRI: GG/9 carriers have attenuated activation of striatum and midbrain, which was correlated with severity of psychotic symptoms

Continued

TABLE 2 A Summary of HLS Featuring THC as the Predominant Interventional Agent—Psychosis-Relevant, Cognitive, and Physiological Effects—cont'd

Author(s)	Year	Subjects	THC (And Other Drugs) Dose and Route	Psychosis-Relevant Effects	Cognitive Effects	Physiological Measures
Ranganathan et al. (2012) [7]	2012	30 volunteers (26 males) (7 regular users)	0.0286 mg/kg of THC-IV with 25 mg Naltrexone PO	PANSS: ↑ positive, negative and general subscores (no change with naltrexone) CADS: ↑ perceptual alterations (no change with naltrexone) SUBJ: ↑ high, anxious, tired, calm (no change with naltrexone)	HVLT: impaired immediate and delayed recall (no change with naltrexone)	N/A
Kleinloog et al. (2012a)	2012	49 healthy men with mild exposure to cannabis	2, 4, and 6 mg of THC—inhalation by vaporizer with 10 mg of olanzapine PO versus two 15 mg of diphenhydramine PO	PANSS:↑ positive (which was reduced by olanzapine, but not with diphenhydramine) ↑ general SUBJ: ↑ calmness ↓ alertness	VVLT: ↓ delayed recall Stroop test: no effect	N/A
D'Souza et al. (2012a)	2012	26 healthy volunteers (17 males)	0.015 mg/kg and 0.03 mg/kg of THC- IV	PANSS: ↑ positive, negative, and general CADSS: ↑ SUBJ: ↑ high	N/A	EEG: Reduced amplitude of novelty P300a and target P300b. No change in latency.

Martin-Santos et al. (2012b)	2012	16 healthy males	10 mg THC-PO 600 mg CBD-PO	Positive psychotic symptoms with THC, none with CBD or placebo SUBJ: ↑ anxiety, dysphoria, sedation with THC, not with CBD or placebo	N/A	N/A
Bhattacharyya, Atakan, Martin-Santos, Crippa, and McGuire (2012a)	2012	15 healthy men with minimum history of previous cannabis use	10 mg THC-PO 600 mg CBD-PO	PANSS: ↑ positive with THC, none with CBD or placebo	N/A	fMRI (visual oddball detection): THC attenuated activation of right caudate, which was inversely correlated with severity of psychotic symptoms CBD resulted in opposite effects to THC on task-related changes
Morrison et al. (2011)	2011	16 healthy volunteers (7 males)	1.25 mg of THC-IV	PANSS: ↑ positive, negative, and general	N-back task: slower response times	EEG: ↓ bifrontal coherence in theta and alpha bands
Morrison and Stone (2011)	2011	19 healthy males	2.5 mg of THC-IV	PANSS negative: ↑ CAPE negative: ↑ SUBJ: no relationship between increased CAPE negative scores and self-rated sedation	N/A	N/A
Barkus et al. (2011)	2011	9 healthy males	2.5 mg of Dronabinol-IV	PANSS: ↑ positive, general, and negative	N/A	SPECT: IV THC didn't significantly increase DA release in the caudate or putamen

Continued

TABLE 2 A Summary of HLS Featuring THC as the Predominant Interventional Agent—Psychosis-Relevant, Cognitive, and Physiological Effects—cont'd

Author(s)	Year	Subjects	THC (And Other Drugs) Dose and Route	Psychosis-Relevant Effects	Cognitive Effects	Physiological Measures
Winton-Brown et al. (2011)	2011	14 healthy males	10 mg THC-PO 600 mg CBD-PO	PANSS: ↑ positive with THC, none with CBD SUBJ: ↑ sedation, intoxication, anxiety with THC, none with CBD	N/A	fMRI: THC attenuates, while CBD activates temporal areas related to processing of information
Stadelmann et al. (2011)	2011	20 healthy volunteers (10 males)	10 mg THC-PO Cannabis extract: 10 mg of THC and 5.4 mg of CBD-PO	N/A	N/A	EEG: P300: >10/>10 genotype of CNR1 gene is associated with significant decrease of P300 amplitude and significant prolongation of P300 latency with THC but not cannabis Extract For pure THC, the higher the number of AAT repeats, the smaller amplitude of P300 and the longer the latency
Kaufmann et al. (2010)	2010	15 healthy females	20 mg of THC as PO cannabis extract versus 5 mg of diazepam as an active placebo	BPRS score: ↑ SUBJ: ↑ fatigue, drowsiness, dizziness, "feeling high"	N/A	N/A

Liem-Moolenaar et al. (2010)	2010	35 healthy male volunteers with mild cannabis use	2, 4, and 6 mg of THC—intrapulmonary by vaporizer with 3 mg Haloperidol PO versus Placebo	PANSS:↑ positive (reduced by haloperidol) ↑ general SUBJ: ↑ external, internal perception, feeling high ↓ alertness (no effect of haloperidol)	Stroop test: ↓ Immediate and delayed word recall: impaired (immediate recall corrected by haloperidol)	EEG: ↑ alpha power in Fz-Cz (no effect of haloperidol)
Stone et al. (2010)	2010	16 healthy volunteers (9 females)	1.25 mg of THC-IV	SUBJ: Impairment of time perception, delay between thinking and speaking, impaired attention, concentration	Slowed tap time (related to impaired concentration)	N/A
Bocker et al. (2010)	2010	16 participants	29.3, 49.1, or 69.4 mg of THC-smoking	N/A	↑ reaction time and number of errors	EEG: resting: ↓ theta, ↑ beta power (Memory effect and theta changes are correlated)
Bhattacharyya et al. (2010)	2010	fMRI expt: Fifteen healthy men with minimal earlier exposure to cannabis (from Bhattacharyya et al., 2009)	fMRI expt: 10 mg THC-PO fMRI expt: 600 mg CBD-PO	N/A	N/A	fMRI: opposite effects of THC and CBD on striatum, hippocampus, amygdala, superior temporal cortex, occipital cortex
Bhattacharyya et al. (2010)	2010	Behavioral expt: 6 healthy volunteers	Behavioral expt: 1.25 mg THC Behavioral expt: 5 mg CBD-PO	PANSS: ↑ positive with THC, not with CBD	N/A	N/A

Continued

TABLE 2 A Summary of HLS Featuring THC as the Predominant Interventional Agent—Psychosis-Relevant, Cognitive, and Physiological Effects—cont'd

Author(s)	Year	Subjects	THC (And Other Drugs) Dose and Route	Psychosis-Relevant Effects	Cognitive Effects	Physiological Measures
Morrison et al. (2009)	2009	22 healthy males	2.5 mg of Dronabinol-IV	PANSS: ↑ positive CAPE: ↑ SUBJ: UMACL: ↓ hedonic tone—↓ energetic arousal—↑ tense arousal	RAVLT: ↓ immediate recall no change on Verbal Fluency Baddeley Reasoning Task: ↓ performance	N/A
Bhattacharyya et al. (2009)	2009	15 healthy males	10 mg THC-PO 600 mg CBD-PO	PANSS: ↑ positive, negative, general with THC, none with CBD SUBJ: ↑ sedation, intoxication, anxiety with THC, none with CBD	N/A	fMRI: THC augments activation of PHG and attenuated ventrostriatal activation—correlated with psychotic symptoms—no changes with CBD
Fusar-Poli et al. (2009)	2009	15 healthy males	10 mg THC-PO 600 mg CBD-PO	PANSS: ↑ positive, negative, and general with THC, not with CBD SUBJ: Skin-conductance responses: fluctuations in response to fearful faces ↑ with THC and ↓ with CBD ↑ Anxiety with THC, not with CBD	N/A	fMRI (showing faces): CBD reduced amygdala and anterior and posterior cingulate in response to fearful faces (correlated with decreased skin-conductance responses). THC modulated activation in frontal and parietal areas

D'Souza, Ranganathan, et al. (2008b)	2008	22 controls (14 males); 30 frequent users (21 males)	2.5 and 5 mg of THC-IV	PANSS: ↑ total score (less with frequent users than with controls) CADSS: ↑ (smaller effect in frequent users) SUBJ: Anxiety: ↑ (smaller increase in frequent users) Calm and relaxed: ↓	HVLT: immediate and delayed recall: impaired (both showed smaller effect in frequent users) ↑ number of intrusions and false positive responses Attention: ↑ omission and commission errors	N/A
D'Souza, Braley, et al. (2008d)	2008	17 healthy subjects and 11 frequent users of cannabis	0.0286 mg/kg of THC-IV with Haloperidol 0.057 mg/kg PO versus placebo	PANSS: ↑ total and positive CADSS: ↑ subjective and objective SUBJ: ↑ high, tired	Impaired verbal recall and attention	N/A
Borgwardt et al. (2008)	2008	15 healthy males	10 mg THC-PO 600 mg CBD-PO	PANSS: ↑ positive with THC, no changes with CBD. SUBJ: ↑ sedation, intoxication and anxiety with THC, no sig. effects with CBD		fMRI: THC reduced activation in right inferior frontal and anterior cingulate cortex—CBD deactivated left temporal cortex and insula
Roser et al. (2008)	2008	20 healthy volunteers (10 males)	10 mg THC-PO Cannabis extract: 10 mg of THC, 5.4 mg of CBD-PO	N/A	N/A	EEG: auditory evoked P300: significant reduction of P300 amplitude at midline frontal, central, and parietal electrodes—not corrected by CBD

Continued

TABLE 2 A Summary of HLS Featuring THC as the Predominant Interventional Agent—Psychosis-Relevant, Cognitive, and Physiological Effects—cont'd

Author(s)	Year	Subjects	THC (And Other Drugs) Dose and Route	Psychosis-Relevant Effects	Cognitive Effects	Physiological Measures
Juckel et al. (2007)	2007	22 healthy volunteers (11 males)	10 mg THC-PO Cannabis extract: 10 mg of THC, 5.4 mg of CBD-PO	N/A	N/A	EEG: auditory evoked MMN: MMN amplitudes were larger with cannabis extract but not with pure THC (in central electrodes)
Koethe et al. (2006)	2006	16 antipsychotic naïve schizophrenia or schizophreniform patients (13 males), 16 prodromal state individuals (11 males), 16 healthy controls (9 males), 16 healthy males receiving Dronabinol	120 μg/kg of Dronabinol-PO	BPRS: ↑ among patients, prodromal state, healthy controls receiving Dronabinol than healthy controls (with patients having highest scores) BDII: impaired among patients, prodromal state individuals, healthy controls receiving Dronabinol—no difference between them	N/A	N/A

Henquet et al. (2006c)	2006	30 patients with a psychotic disorder, 12 relatives of patients with a psychotic disorder, and 32 healthy controls	300 mg THC/kg body weight in tobacco cigarettes in the exposure condition, or 0 mg THC/kg body weight in tobacco cigarettes	CAPE: Carriers of the Val allele were most sensitive to the induced psychotic experiences on the condition of higher psychosis liability	Visual Verbal learning test—Abstract Visual Pattern Learning—Continuous Performance Test—Stroop Color-Word test—Digit Symbol Substitution Test: Sensitivity to memory and attention impairments is more among carriers of the Val allele	N/A
D'Souza et al. (2005)	2005	13 stable, antipsychotic-treated patients	2.5 and 5 mg of THC-IV	PANSS: ↑ positive, negative, general CADSS: ↑ clinician and participant rated	HVLT: impaired immediate, delayed recall	N/A

Continued

TABLE 2 A Summary of HLS Featuring THC as the Predominant Interventional Agent—Psychosis-Relevant, Cognitive, and Physiological Effects—cont'd

Author(s)	Year	Subjects	THC (And Other Drugs) Dose and Route	Psychosis-Relevant Effects	Cognitive Effects	Physiological Measures
Ilan et al. (2005)	2005	23 healthy users (12 men) (22 of users used for EEG and ERP analysis)	Low (1.8%) or high (3.6%)-smoking Low (between 0.1 and 0.4%) or high (more than 1 %) CBD-smoking Cannabichromene (CBC) containing cigarettes: Low (between 0.1% and 0.2%) or high (more than 5%) Cigarettes were combination of: Low THC, low CBD, high CBC High THC, low CBD, high CBC Low THC, high CBD, low CBC Low THC, high CBD, high CBC	N/A	Word presentation, working memory and word recognition tasks: reduced performance with THC—changing doses of CBD and CBC did not change effects	EEG: THC lowered amplitude of ERP and reduced EEG power—effects were not dose dependent—changing doses of CBD and CBC didn't change effects

D'Souza et al. (2004)	2004	22 healthy volunteers	2.5 and 5 mg of THC-IV	PANSS: ↑ positive, negative, general CADSS: ↑ clinician and participant rated SUBJ: ↑ anxiety, tiredness ↓ calmness	HVLT: impaired immediate, delayed recall	N/A
McDonald, Schleifer, Richards, and de Wit (2003b)	2003	37 men and women	7.5 and 15 mg of THC-PO	SUBJ: THC ↑ estimates of the duration of short intervals while not affecting estimates of longer intervals ↑ stop reaction time No effect on go reaction time, go/no-go task	HVLT, digit span forward: no effect Digit span backward: impaired	N/A
Hart et al. (2001)	2001	18 healthy volunteers (10 males), averaging 24 marijuana cigarettes per week	1.8%, 3.9% of THC-smoking	SUBJ: ↑ confused, mellow, high	MicroCog test battery: ↑ premature responses—↑ time needed to finish tasks—did not affect accuracy on cognitive flexibility, mental calculation and reasoning Impaired immediate recall	N/A

Continued

TABLE 2 A Summary of HLS Featuring THC as the Predominant Interventional Agent—Psychosis-Relevant, Cognitive, and Physiological Effects—cont'd

Author(s)	Year	Subjects	THC (And Other Drugs) Dose and Route	Psychosis-Relevant Effects	Cognitive Effects	Physiological Measures
Leweke et al. (2000)	2000	9 healthy males	1 mg of Nabilone-PO 200 mg CBD-PO	BDII: marked impairment after nabilone alone—less impairment after combined nabilone and CBD—no effect after CBD alone	N/A	N/A
Mathew et al. (1998)	1998	46 subjects (22 males)—mean marijuana use: 147 ± 165.2 "joints" per year	0.15 mg/min versus 0.25 mg/min (for 20 min) of THC-IV	N/A	N/A	PET: individuals who had reduced cerebellar blood flow after THC experienced significant alteration in time perception
Leweke et al. (1998)	1998	19 healthy volunteers	10 mg of Dronabinol-PO	N/A	↓ accuracy of classification of emotionality of words	ERP: ↑ of amplitude of ERP in response to the positive words when they are presented for the second time
Emrich et al. (1991)	1991	7 healthy subjects	3–4 mg/kg of cannabis resin-PO	BDII: impaired	N/A	N/A
Heishman et al. (1990)	1990	3 experienced users	0, 1, or 2 marijuana cigarettes (each containing 2.57% of THC)	N/A	Serial addition/ subtraction and digit recall tasks: impaired Impairment lasted for 24 h	N/A

Marks and MacAvoy (1989)	1989	6 experienced users and 6 nonusers (equal male and female ratios)	0, 2.6, and 5.2 mg of THC-smoking versus 1.19 mL/kg and 2.38 mL/kg drink containing 42% w/v ethanol	N/A	Both THC and alcohol impaired signal detection (more pronounced for peripheral signals, with users being less impaired than nonusers)	N/A
Hooker and Jones (1987)	1987	12 males (variable marijuana use)	10.7 ± 0.6 mg of THC-smoking	N/A	Delayed free recall: more errors Immediate, sustained attention, controlled retrieval from semantic memory, speed of reading, naming colors: not affected Stroop test: ↑ interference	N/A
Hicks et al. (1984)	1984	Experiment 1: 4 healthy males Experiment 2: 6 experienced users (3 males)	1.29 or 4.61% of THC-smoking (cigarettes were ~9.23 mg) with 0.2 mg of atropine sulfate-IV Smoking cigarettes containing THC	SUBJ: Experiment 1: ↑ subjective time rate and effect not related to blocking cholinergic effect Experiment 2: Increased subjective time rate is evident as time is passing	N/A	N/A

Continued

TABLE 2 A Summary of HLS Featuring THC as the Predominant Interventional Agent—Psychosis-Relevant, Cognitive, and Physiological Effects—cont'd

Author(s)	Year	Subjects	THC (And Other Drugs) Dose and Route	Psychosis-Relevant Effects	Cognitive Effects	Physiological Measures
Miller et al. (1977)	1977	2 groups of 17 male volunteers	14 mg of THC-smoking	N/A	Immediate and final free recall: ↓ Long term retention: not affected	N/A
Melges et al. (1974)	1974	6 healthy males	20 mg of THC containing cigarettes smoked over 10 (fast) or 45 min (slow) versus 120 mL of 95% alcohol drink as active placebo [consumed over 10 min (fast) or 45 min (slow)]	Inventories: ↑ temporal disorganization, delusional-like ideations, amplification, desynchronization ↑ depersonalization	↑ tracking difficulties	N/A
Tinklenberg et al. (1972)	1972	15 healthy males users	0.35 mg/kg of THC-PO versus 0.7 mL/kg of 95% ethanol	SUBJ: Underproduction of time intervals	N/A	(reported in the study (Kopell, Tinklenberg, & Hollister, 1972))
Kopell et al. (1972)	1972	12 healthy males users	0.35 mg/kg of THC-PO versus 0.7 mL/kg of 95% ethanol	N/A	N/A	EEG: amplitude of contingent negative variation ↑ with THC, ↓ with alcohol

THC: Δ-9-tetrahydrocannabinoid; Sxs: symptoms; PHG: parahippocampal gyrus; MTG: middle temporal gyrus; STG: superior temporal gyrus; RAVLT: Rey Auditory Verbal Learning Task; UMACL: University of Wales Institute of Science and Technology Mood Adjective Checklist; BDII: Binocular depth inversion illusion test; HVLT: Hopkins Verbal Learning Test; SUBJ, subjective effects.

Morrison & Stone, 2011; Morrison et al., 2009). However, the intensity of effects has also been shown to be dose dependent and to have distinct time courses depending on the route of administration (D'Souza et al., 2004; Kleinloog et al., 2012b; Liem-Moolenaar et al., 2010; Morrison & Stone, 2011; Morrison et al., 2009).

In a first study of its kind, D'Souza et al. (2004) administered two doses of IV THC (2.5 and 5 mg) to healthy adults using a double-blind, randomized, placebo-controlled, crossover design (D'Souza et al., 2004). The study found that THC induced perceptual alterations, a wide range of psychotomimetic symptoms, mood symptoms such as euphoria and anxiety, and cognitive deficits affecting attention and working and verbal memory. These findings were first replicated by Morrison et al. (2009) using a lower dose of IV THC (2.5 mg), and then by a number of researchers using a range of doses (1.25–3.5 mg) and infusion times (10–20 min) (Bhattacharyya et al., 2009, 2010, 2015; Bhattacharyya, Crippa, et al., 2012d; D'Souza, Braley, et al., 2008; D'Souza et al., 2012a; Freeman et al., 2015).

Nabilone, a synthetic analog of THC (Leweke, Schneider, Radwan, Schmidt, & Emrich, 2000; Leweke, Schneider, Thies, Munte, & Emrich, 1999), and Dronabinol, a synthetic isomer of THC (Koethe et al., 2006), have been shown to share a similar profile of effects with THC. In addition, in healthy controls, they have been found to induce deficits in the binocular depth inversion illusion task, a potential surrogate marker for psychosis present in patients with acute paranoid schizophrenia and schizophreniform psychosis (Koethe et al., 2006; Leweke et al., 1999, 2000).

As summarized in Table 2, in healthy humans, cannabinoids have also been shown to acutely induce a range of transient, dose-related deficits in cognitive processes, such as verbal learning, short-term memory, working memory, executive function, abstract thinking, decision-making, and attention (Hart, van Gorp, Haney, Foltin, & Fischman, 2001; Heishman, Huestis, Henningfield, & Cone, 1990; Hooker & Jones, 1987; Leweke et al., 1998; Marks & MacAvoy, 1989; Miller, McFarland, Cornett, & Brightwell, 1977; Ranganathan & D'Souza, 2006a). These findings have been consistent with the acute effects of cannabinoids observed in nonhuman primates and rodents (for a review, see Lichtman, Varvel, & Martin, 2002; Wilson & Nicoll, 2002). Importantly, the profile of cognitive deficits induced by cannabinoids is similar to that observed in schizophrenia (Heinrichs & Zakzanis, 1998): in both cases, working and verbal memory are the most affected domains (Heinrichs & Zakzanis, 1998; Ranganathan & D'Souza, 2006a).

Although a number of studies have shown that THC consistently induces acute deficits in working memory, deficits are not consistent across different tasks. For instance, while THC has been shown to increase reaction time and disrupt performance on the Sternberg and N-back tasks (Bocker et al., 2010, Hunault et al., 2009; Ilan, Gevins, Coleman, ElSohly, & de Wit, 2004, 2005; Ilan, Smith, & Gevins, 2004), the effects on digit span and delayed match to sample tasks have been mixed (Ballard & de Wit, 2011; D'Souza et al., 2004; D'Souza, Ranganathan, et al., 2008c; Morrison et al., 2009) and no effects have been detected on the serial sevens task (Curran, Brignell, Fletcher, Middleton, & Henry, 2002). Regarding verbal memory, THC has been shown to induce robust dose-dependent deficits in verbal learning and recall as measured by the Hopkins Verbal Learning Test (HVLT): THC disrupted both immediate and delayed (30 min) verbal recall (Fig. 4), and increased the number of "false positive" and "intrusion" responses (D'Souza et al., 2004, 2005; Morrison et al., 2009).

Finally, abnormalities in time perception that have been reported in schizophrenia (Carroll, O'Donnell, Shekhar, & Hetrick, 2009; Davalos, Kisley, & Ross, 2003; Tysk, 1983) are reproduced in healthy people under the acute effect of cannabinoids (Hicks, Gualtieri, Mayo, & Perez-Reyes, 1984; Mathew, Wilson, Turkington, & Coleman, 1998; McDonald, Schleifer, Richards, & de Wit, 2003a; Sewell et al., 2013b; Stone et al., 2010; Tinklenberg, Kopell, Melges, & Hollister, 1972). In the largest experimental study to date, Sewell et al. (2013a, 2013b) showed that different doses of THC acutely induce time overestimation and underproduction in healthy people (Sewell et al., 2013b). This is consistent with studies in nonhuman primates and rodents, which have also found that cannabinoids acutely induce deficits in time perception (Conrad, Elsmore, & Sodetz, 1972; Han & Robinson, 2001; McClure & McMillan, 1997; Schulze et al., 1988).

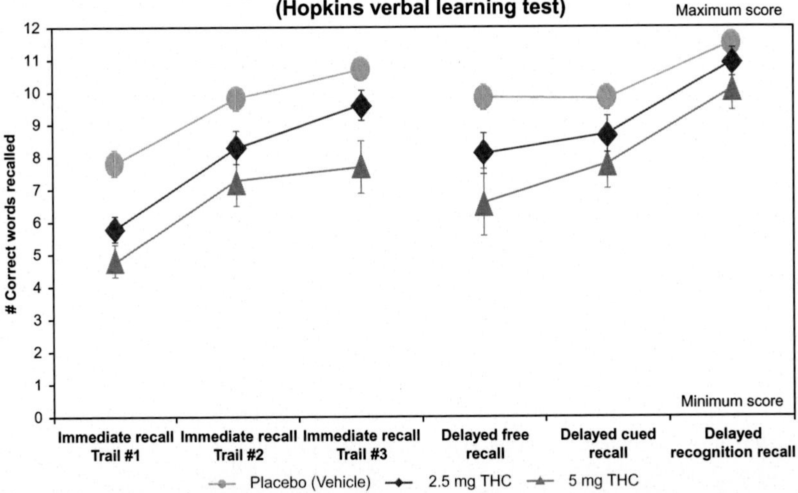

FIG. 4 Effects of THC on learning, immediate free recall, delayed free recall, delayed cued and recognition recall in healthy individuals, measured by a 12-word learning task (Hopkins Verbal Learning Test). *Reprinted with permission from D'Souza, D. C., Perry, E., MacDougall, L., Ammerman, Y., Cooper, T., Wu, Y. T., et al. (2004). The psychotomimetic effects of intravenous delta-9-tetrahydrocannabinol in healthy individuals: Implications for psychosis.* Neuropsychopharmacology, 29, *1558–1572.*

Effects in Individuals With Schizophrenia

Psychopharmacological challenge studies in clinical populations such as individuals with psychosis, cannabis use disorders, and a genetic risk for psychosis have been carried out using a variety of pharmacological stimuli including ketamine (Lahti, Koffel, LaPorte, & Tamminga, 1995), amphetamine (Laruelle et al., 1996), methylphenidate (Lieberman et al., 1984), and THC (D'Souza et al., 2005). Such studies have the potential to elucidate the complex involvement of individual neurotransmitter systems in the pathophysiology of various psychotic symptoms and substance use disorders and develop paradigms for testing treatments. However, perhaps due to theoretical safety and tolerability concerns, in addition to important ethical considerations, these studies are less numerous. Most of the HLS using cannabinoids in clinical populations are subsequently summarized in detail.

D'Souza 2005 (D'Souza et al., 2005): This was a 3-day, double-blind, randomized, placebo-controlled study, in which the behavioral and cognitive effects of 0 mg (placebo) and two active doses (2.5 and 5 mg IV THC) were characterized in 13 stable, antipsychotic-treated schizophrenia patients. Out of the 13 subjects, 10 were on typical antipsychotics (haloperidol and fluphenazine) and three were on atypical antipsychotics (risperidone and olanzapine). Behavioral outcomes included the PANSS and showed that THC transiently increased scores of the positive symptoms subscale (dose × time $F\,(4,68)=4.90$, $P<0.0016$) (Fig. 2). The positive symptoms induced by THC were not new symptoms, but symptoms previously endorsed by the subjects. On the negative subscale, THC transiently increased the scores with a significant effect of dose.

Overall this study showed that THC caused a transient increase in positive and negative symptoms, and worsened cognitive deficits in clinically stable patients with schizophrenia who were treated with antipsychotic drugs. Compared with matched controls, schizophrenia patients were more vulnerable to the psychosis-increasing and memory-impairing effects of THC. There were no serious short- or long-term adverse events associated with study participation, other than a case of exacerbation of hypertension that a subject failed to disclose at the time of screening. Furthermore, that THC did not reduce any of the core symptoms of schizophrenia raised questions about the commonly held view of a "self-medication" hypothesis of cannabis use and schizophrenia.

Henquet 2006 (Henquet et al., 2006a): Henquet et al. hypothesized that psychometric psychosis (measured by the Community Assessment of Psychotic Experiences or CAPE) and functional polymorphisms of the Catechol-O-Methyltransferase (COMT) gene (Val-Val/Val-Met/Met-Met) represented two different mechanisms impacting the same final common pathway of developing psychosis after cannabis exposure. To replicate and extend this finding, they conducted a double-blind, placebo-controlled crossover design in which patients with a psychotic disorder ($n=30$), relatives of patients with a psychotic disorder ($n=12$), and healthy controls ($n=32$) were exposed to THC or placebo cigarettes, followed by cognitive assessment and assessment of current psychotic experiences. In the 30 patients, lifetime diagnoses were: schizophrenia ($n=11$), schizoaffective disorder ($n=11$), and psychosis not otherwise specified ($n=8$). Twenty-two patients were using antipsychotic medication at the time of testing and eight were medication-free or used medication other than antipsychotic medication. Within the relatives group, three subjects had a lifetime diagnosis of bipolar disorder and one major depressive disorder. The COMT genotype distribution in the whole sample was 26% (Met/Met), 27% (Val/Val), and 47% (Val/Met).

On behavioral outcomes as measured by the PANSS, THC was not associated with a significant increase in positive symptoms, and no significant condition-by-genotype interaction was observed on the psychotic symptom outcome. However, there was a significant three-way condition-by-genotype-by-CAPE-trait interaction, which indicated that preexisting psychosis liability influenced the genetic moderation of THC-induced expression of psychosis. Thus, there was a significant condition-by-genotype interaction in the high CAPE-trait group.

On the Dutch visual verbal learning test, THC impaired verbal memory performance on immediate and delayed free recall as well as on delayed recognition. On the delayed recognition task, THC exposure caused significant impairment in subjects with the Val/Val genotype. The two-way condition-by-genotype interaction suggested that the effect in the Val/Val genotype was greater than that in the Val/Met and Met/Met genotypes.

In brief, Henquet et al. found that genetic predisposition and psychometric psychosis influenced the impact of THC on cognition and psychosis outcomes. They also concluded that the functional polymorphism of the COMT gene moderated sensitivity to the effects of THC on psychotic symptoms and that the differential sensitivity to THC associated with COMT genotype was in part conditional on additional evidence of psychosis liability (Val carriers with higher psychometric psychosis liability experienced more THC-induced transient psychotic symptoms compared with Val carriers without these additional measures of liability). This conditionality on additional psychometric psychosis liability was less evident for the cognitive measures.

Effects in Cannabis Users

Ramaekers 2009 (Ramaekers, Kauert, Theunissen, Toennes, & Moeller, 2009): This study tested whether heavy cannabis users developed tolerance to the cognition-impairing effects of THC. To test this, 24 subjects (12 occasional cannabis users and 12 heavy cannabis users) participated in a double-blind, placebo-controlled, two-way mixed model design. Both groups received single doses of placebo and 500 μg/kg THC by smoking. Performance tests included measures of perceptual motor control (critical tracking task), dual task processing (divided attention task), motor inhibition (stop signal task), and cognition (Tower of London). They found that THC-induced impairment of critical tracking performance only occurred in occasional cannabis users when compared with heavy users as indicated by a significant THC-by-cannabis use history interaction ($P=0.027$). Performance in the divided attention task was significantly affected by THC, as indicated by increased tracking error ($P=0.013$) and the number of control losses ($P=0.032$) in the primary task and a decreased number of hits in the secondary task ($P=0.024$). Again, the overall effect of THC was more prominent in occasional cannabis users compared with heavy users. In the stop signal task, stop reaction time was affected by THC-by-time after smoking and there was no influence of a previous cannabis use history. Performance on the Tower of London task was not affected by THC. Overall, these data indicated that cannabis use history strongly determined the cognitive response to single doses of THC.

D'Souza et al. (2008a)*:* In this 3-day, double-blind, randomized, placebo-controlled study, the dose-related effects of 0, 2.5, and 5 mg IV THC were studied in 30 individuals who used cannabis about once weekly and were compared to 22 healthy controls. THC transiently increased (dose × time $P<0.0001$) clinician-rated perceptual alteration scores. Weekly users showed smaller THC-induced increases in the perceptual alterations scores (group × dose × time $P<0.001$), group × dose effect ($P<0.069$), and group effect ($P<0.006$). THC transiently increased psychosis (PANSS total scores) in both groups (dose × time $P<0.0001$), but weekly users had smaller increases compared with healthy controls (group × dose × time $P<0.0025$). THC also transiently increased VAS "high" scores in both groups (dose × time $P<0.0001$), but there was no group effect on this increase.

On cognitive measures, THC impaired immediate recall in both groups in a dose-related manner ($P<0.0001$) with a significant group-by-dose interaction ($P<0.03$), with weekly users performing worse at baseline (placebo condition), yet showing smaller THC-induced recall impairments than controls. THC impaired delayed recall in both groups with a significant group-by-dose interaction ($P<0.02$), with weekly users showing smaller THC-induced recall impairments than controls. On tasks of vigilance, THC increased omission and commission errors in both groups. While there was no group effect for both omission and commission errors, there was a significant group-by-dose interaction, such that the difference between 5 mg and placebo dose was significant in weekly users but not in controls. Overall, these data suggested that weekly users of cannabis are either inherently blunted and/or develop tolerance to their response to the psychotomimetic, perceptual altering, and amnestic effects of cannabinoids.

Kuepper et al. (2013a)*:* In this study, striatal dopamine release following smoked THC was measured using a single dynamic positron emission tomography (PET) scanning session in nine healthy cannabis users (average risk of psychotic disorder), eight patients with psychotic disorder (high risk of psychotic disorder), and seven unrelated first-degree relatives (intermediate risk of psychotic disorder). Compared with the control group, both patients and relatives showed significant displacement of the ligand (F18Fallypride) in striatal subregions, indicative of dopamine release. This was most pronounced in the caudate nucleus. Behavioral measures including visual analog ratings of "feeling high," "internal perception," and "external perception" did not differ based on group (control, relatives, or patients).

Ranganathan et al. (2014): This study tested the effects of THC on individuals with a family history of alcoholism based on preclinical evidence suggesting that brain cannabinoid receptor function may mediate the effects of alcohol and risk for developing alcoholism. Thirty healthy subjects with varying degrees of family history of alcoholism were enrolled in this 3-day test study, during which they received 0.018 and 0.036 mg/kg of THC or placebo intravenously in a randomized, counterbalanced order under double-blind conditions. Greater degree of family history was correlated with greater "high" and perceptual alterations induced by THC, measured by a VAS and the CADSS, respectively. The interaction between THC dose and degree of family history on peak change in self-reported "high" was significant ($P=0.03$). On the HVLT, THC significantly reduced total immediate free recall and long delayed free recall in a dose-dependent manner, but there was no effect of family history of alcoholism.

FACTORS MODULATING THE PSYCHOTOMIMETIC AND COGNITIVE EFFECTS OF Δ^9-TETRAHYDROCANNABINOL IN HLS

Gene-Drug Interactions

As mentioned above, cannabinoids have been shown to induce a number of transient schizophrenia-like effects in healthy people. The intensity of these effects varies across individuals with similar levels of exposure to cannabinoids. The effects can range from small perceptual alterations to transient episodes of frank paranoia or psychosis in a very small percentage of people. It is likely that some of these individual differences could be explained by genetic polymorphisms implicated in schizophrenia. In one of the first gene-by-environment studies, Caspi et al. demonstrated that a common polymorphism of the COMT gene, which encodes for an enzyme that plays an important role in the breakdown of dopamine in the prefrontal cortex (Papaleo et al., 2008), moderated the risk of schizophrenia in individuals exposed to cannabis (Caspi et al., 2005). The longstanding association between schizophrenia and alterations in the dopaminergic system (Howes & Kapur, 2009) has led others to explore the relationship between the acute response to cannabinoids and genes involved in the transmission or metabolism of brain dopamine.

A common variant of the COMT gene is the Val^{158}Met polymorphism (rs4680), in which a G/A base-pair substitution results in an amino acid change from valine (Val) to methionine (Met) at codon 158 (Lachman et al., 1996). The rate of dopamine catabolism of the Val variant is up to 4 times the rate of the Met variant; thus, Val carriers have lower levels of extracellular dopamine in the prefrontal cortex (Chen et al., 2004). It has been proposed that the Val variant confers an increased risk for schizophrenia in people exposed to cannabis during adolescence; however, a number of studies examining this issue have shown mixed results (Caspi et al., 2005; Kantrowitz et al., 2009; Pelayo-Teran et al., 2010; Zammit et al., 2007). In a study assessing the interaction between the COMT polymorphism and acute effects of cannabinoids, Henquet et al. (2006a, 2006b, 2006c) found that in individuals having both one copy of the Val allele and high scores on a measure of psychosis liability, THC induced greater schizophrenia-like effects (Henquet et al., 2006b). Furthermore, individuals with two copies of the Val allele had increased sensitivity to the acute effects of THC on cognition.

The **Dopamine Transporter 1** gene (**DAT1/SLC6A3**) encodes the dopamine transporter (DAT), a transmembrane transport protein that regulates the reuptake of dopamine from the synapse into the presynaptic terminal. In the striatum, the clearance of dopamine from the synapses is highly dependent on the DAT; thus, it has been argued that polymorphisms of the DAT1 gene may be involved in the high levels of striatal dopaminergic activity observed in some patients with schizophrenia. Concordantly, a polymorphism of the DAT1 gene associated with higher striatal dopamine has been linked to schizophrenia (Talkowski et al., 2008).

The **AKT1** gene encodes the **protein kinase B**, an integral component of the dopamine signaling cascade (Beaulieu, Gainetdinov, & Caron, 2007) that has been associated with schizophrenia (Emamian, Hall, Birnbaum, Karayiorgou, & Gogos, 2004; Thiselton et al., 2008). Regarding the association between DAT1 variations and the acute effect of cannabinoids, Bhattacharyya et al. showed that carriers of the 9-repeat allele of the DAT1 gene which were also G homozygotes of the rs1130233 SNP of the AKT1 gene had increased schizophrenia-like symptoms in response to THC (Bhattacharyya, Atakan, et al., 2012b). Cannabinoids activate the AKT1 pathway via CB_1 and CB_2 receptors, representing a potential mechanism for psychosis. However, Bhattacharya et al. found that the GG genotype (a single nucleotide polymorphism) of the AKT1 gene reduced sensitivity to the acute psychosis-inducing effects of THC. More recent large epidemiological studies have further implicated other polymorphisms of this gene in increased susceptibility for THC-induced psychosis, supporting the HLS.

Other genetic variation in AKT1 has been hypothesized to increase sensitivity to the psychotomimetic effects of cannabis as well as confer increased risk for cannabis-associated psychotic disorders. This has been supported by case–control studies showing that a polymorphism of the AKT1 rs2494732 locus (C/C versus T/T) interacts with cannabis use to infer a greater likelihood of later developing a psychotic disorder (Di Forti et al., 2012) as well as worse performance in a continuous performance task in individuals with psychosis (van Winkel, van Beveren, Simons, & Genetic Risk and Outcome of Psychosis (GROUP) Investigators, 2011). Recently, using a hybrid naturalistic challenge study design, Morgan, Freeman, Powell, and Curran (2016) studied the acute response to smoked cannabis in 442 healthy young cannabis users while intoxicated with their own cannabis. They found that variation at the rs2494732 locus of the AKT1 gene predicted acute psychotomimetic response (as measured by total score on the PSI, (Mason, Morgan, Stefanovic, & Curran, 2008)), dependence on the drug, and baseline schizotypy (Morgan et al., 2016). Another study indicates that a variant in the D2 receptor gene may also increase psychosis risk, and that the risk may be greater in carriers of both this variant and the AKT1 risk allele (Colizzi et al., 2015).

Interactions With Other Drugs

A further significant finding of HLS using THC is the dissociation between serum levels and peak clinical effects. Even following IV administration, subjective effects can be delayed by 15 min. This might suggest that the psychotomimetic effects of THC are downstream to its primary site of action—the CB_1 receptor. The ubiquitous nature and known neuromodulatory role of the endocannabinoid system suggests that THC induces its psychotomimetic effects via interactions with other neurotransmitter systems with more established links to schizophrenia.

Preclinical studies demonstrate that THC administration increases striatal dopamine, but human data are mixed. D2 receptor blockade (via single doses of olanzapine or haloperidol) attenuated the acute psychotomimetic effects of THC in healthy subjects in two (Kleinloog et al., 2012a) of the three HLS using this paradigm (D'Souza, Braley, et al., 2008). In individuals with schizophrenia, long-term antipsychotic use failed to protect them from the acute psychosis-exacerbating effects of THC (D'Souza et al., 2005). Results of preclinical and human studies exploring the ability of THC to increase striatal dopamine (Barkus et al., 2011; Bossong et al., 2009; Kuepper et al., 2013b; Stokes, Mehta, Curran, Breen, & Grasby, 2009), as well as direct interactions between CB_1 and D2 receptors (Glass & Felder, 1997; Munoz-Arenas et al., 2015), are similarly mixed. Pooling of two cohorts, in order to achieve higher power, a small but significant displacement of [^{11}C]raclopride was observed in the ventral striatum after THC administration (Bossong et al., 2009), indicating increased endogenous dopamine activity at D2 receptors, but whether this small change can explain the psychotomimetic effects of THC is not clear.

Confirming a wealth of preclinical data indicating extensive interactions between the endocannabinoid and GABAergic systems, Radhakrishnan et al. showed that a pharmacologically induced GABA deficit (produced by Iomazenil—a GABA-A negative allosteric modulator) enhanced THC-induced psychotomimetic effects in healthy humans (Radhakrishnan et al., 2015). Particularly prevalent in the cortex and hippocampus, the activation of CB_1 receptors inhibits the release of GABA by cholecystokinin basket cells, leading to disruption of the cortical inhibitory/excitatory balance (Farkas et al., 2010).

This mechanism may have particular relevance to the connection between cannabis and schizophrenia (Sherif et al., 2016).

As the second most pharmacologically significant cannabinoid in cannabis, CBD has garnered increasing attention in both the scientific literature and popular press. Although its effects on the endocannabinoid system have not been fully elucidated, contrary to THC, it exhibits CB_1 receptor antagonism/inverse agonism and fatty acid amide hydrolase (FAAH) inhibition. This pattern of effects, alongside limited clinical evidence, has identified CBD as the potential natural modulator of effects of THC in cannabis (Leweke, 2007), prompting several HLS. Pretreatment with CBD has been shown to attenuate THC-induced psychotomimetic effects, paranoia, and verbal memory impairment (Bhattacharyya et al., 2015; Englund et al., 2013b; Ilan et al., 2005; Leweke et al., 2000; Martin-Santos et al., 2012b). Head to head, THC produced increased psychotic symptoms and skin conductance responses during processing of fearful faces, whereas CBD produced a reduction in anxiety and decreased skin conductance (Fusar-Poli et al., 2009). Second, during functional magnetic resonance imaging (fMRI) tasks testing verbal recall, response inhibition, processing fearful faces, and auditory and visual processing, THC and CBD had divergent effects on blood-oxygen-level dependent (BOLD) response (Bhattacharyya et al., 2010; Borgwardt et al., 2008; Fusar-Poli et al., 2009).

PSYCHOPHYSIOLOGICAL MARKERS OF PSYCHOTOMIMETIC AND COGNITIVE EFFECTS OF Δ^9-TETRAHYDROCANNABINOL IN HUMAN LABORATORY STUDIES

Typical techniques of measuring the psychotomimetic and cognitive effects of THC in HLS involve rating scales and hence are limited by subjectivity on both the part of the participant and the rater. Biomarkers—neural correlates of these clinical effects—may provide a complementary set of objective measures. In HLS, the neural effects of cannabinoids have been primarily studied using electroencephalography (EEG), which is one of the only noninvasive techniques capable of capturing neural activity in real time. In general terms, at the cellular level, neural activity consists of controlled exchanges of ions between the intra- and extracellular spaces of neurons. These exchanges cause fluctuations in the extracellular electric potential fields, that, when summed together across large populations (hundreds of millions) of similarly oriented groups of neurons (particularly pyramidal cells) (Niedermeyer & da Silva, 2005), produce electric potential fluctuations that can be measured at the scalp with an EEG. The strength of the resulting signals would depend on both the number of active pyramidal cells and the level of synchronization of the active neurons, given that desynchronized neurons would tend to cancel out by destructive interference their individual contributions to the summed electric potential (Lachaux, Axmacher, Mormann, Halgren, & Crone, 2012).

Brain functions such as attention, working memory, emotions, and language, rely on the integrity of both short-scale (within areas) and long-scale (between areas) neural networks (Bullmore & Sporns, 2009; Varela, Lachaux, Rodriguez, & Martinerie, 2001). By engaging their nodes in synchronous patterns of oscillatory electrical activity, neural networks integrate and process information (Singer, 1999). Thus, deficits affecting the generation of neural oscillations, the signal-to-noise ratio of the circuit, or the communication between nodes such as the ones observed in schizophrenia and under the acute effect of cannabinoids (Cortes-Briones, Cahill, et al., 2015b; Diez, Suazo, Casado, Martin-Loeches, & Molina, 2013; Díez et al., 2014; Higashima et al., 2007; Hinkley et al., 2011; Kubicki et al., 2003, 2005; Lawrie et al., 2002; Seok et al., 2007; Uhlhaas, Haenschel, Nikolić, & Singer, 2008; Whitfield-Gabrieli et al., 2009, Winterer & Weinberger, 2003; Winterer et al., 2000, 2004; Yang et al., 2014), likely underlie some of the disruptions in brain function (e.g., abnormal perceptual experiences) observed in both cases.

It has been proposed that CB_1 receptors located on the axon terminals of cholecystokinin (CCK)-expressing GABAergic interneurons in the cerebral cortex and hippocampus are part of a noise filter-like mechanism that enhances the signal-to-noise ratio of neural networks (Bartos & Elgueta, 2012; Csicsvari, Jamieson, Wise, & Buzsáki, 2003; Tukker, Fuentealba, Hartwich, Somogyi, & Klausberger, 2007). This mechanism relies on the physiologic, space-localized, on demand, brief activation of CB_1 receptors by endocannabinoids released by active pyramidal cells. Thus, the nonphysiologic, global, long-lasting activation of CB_1 receptors by exogenous cannabinoids such as THC could disrupt the CB_1 receptor-mediated noise filter mechanism, reducing the signal-to-noise ratio of the system. This transient change in the dynamics of neural networks induced by exogenous cannabinoids has been proposed to mimic some important aspects of the long-lasting abnormalities in the dynamics of neural networks present in schizophrenia (Cortes-Briones, Skosnik, et al., 2015; Cortes-Briones, Cahill, et al., 2015b).

Event-Related Potentials

Event-related potentials (ERPs) are transient changes in the electric potential (voltages) recorded at the scalp with EEG in response and time-locked to the presentation of stimuli. Numerous ERPs have been studied, with each thought to correspond to varying aspects of neural processing (e.g., early sensory processing, attention, and language processing) (Luck & Kappenman, 2012). Depending on the specific kind of stimuli used to elicit them, ERPs have either a positive (denoted with a "P") or a negative (denoted with an "N") polarity that peaks at a specific poststimulus latency (e.g., 300ms). This is reflected on the names of ERPs (e.g., N100 indicates a negative response peaking 100ms after the onset of the stimuli).

The ***P300*** is a positive voltage response peaking around 300ms after stimulus presentation, elicited by novel (P3a) and target deviant (P3b) stimuli (visual or auditory) presented with a low probability (~10%–20% of all stimuli) within a sequence of repetitive, highly probable stimuli (~80%–90% of all stimuli). P300 is thought to be related to directed attention, contextual updating of working memory, and the attribution of salience to novel and deviant stimuli (Polich & Criado, 2006). P300, particularly the P3b component, is thought to result from the activity of a distributed neural network including the thalamus, hippocampus, inferior parietal lobe, superior temporal gyrus, and frontal cortex (Kiehl, Laurens, Duty, Forster, & Liddle, 2001). Deficits in P300 amplitude and latency have been demonstrated in patients with schizophrenia (Bramon, Rabe-Hesketh, Sham, Murray, & Frangou, 2004; Bramon et al., 2005; Jeon & Polich, 2003; Solowij & Michie, 2007; Turetsky et al., 2007). Interestingly, studies have shown that both oral and smoked THC acutely reduces P300 amplitude (Ilan et al., 2005; Roser et al., 2008). P300 amplitude reductions have even been observed in heavy cannabis users after smoking THC, indicating that the acute deficits in P300 amplitude induced by THC may be resistant to the effects of tolerance (Theunissen et al., 2012). In a double-blind, placebo-controlled, counterbalanced, and crossover study using IV administration of THC, D'Souza et al. (2012a, 2012b) demonstrated that both P3a and P3b amplitudes were disrupted by THC in a dose-dependent manner (D'Souza et al., 2012b). These deficits were not observed during early sensory-related ERPs (N100), suggesting that THC-induced deficits are greater in late processes involving novelty, saliency, and working memory context updating. Furthermore, the reductions in P3b amplitude were correlated with the psychotomimetic effects of THC.

Mismatch negativity (MMN) is a negative voltage response peaking around 100–200 ms after stimulus presentation, elicited by auditory stimuli that deviate in frequency or duration from a sequence of standard auditory stimuli. It is thought to be an index of basic auditory processing and sensory memory that is relatively independent of attention. Its neural sources have been localized in the superior temporal and prefrontal cortex (Naatanen & Alho, 1995; Rinne, Alho, Ilmoniemi, Virtanen, & Naatanen, 2000), which is consistent with an index of early auditory processing. Regarding the acute effect of cannabinoids on MMN, Juckel, Roser, Nadulski, Stadelmann, and Gallinat (2007) examined the effects of oral THC and THC plus CBD. They found that oral THC had no effect on MMN amplitude compared to placebo (Juckel et al., 2007). Interestingly, THC + CBD was found to increase the amplitude of the MMN, which the authors interpreted as a result of the antipsychotic effects of CBD. The lack of effect on MMN amplitude with THC alone could be related to the low dose of THC chosen (10 mg orally), or due to the inter- and intraindividual variability associated with oral routes of administration.

Neural Oscillations: Steady-State Response

Reduced gamma-band (~40 Hz) auditory steady-state response (ASSR) has been observed in first episode and chronic schizophrenia (Krishnan et al.,

2009; Kwon et al., 1999; Light et al., 2006; Spencer, Salisbury, Shenton, & McCarley, 2008a). Furthermore, this deficit has been associated with some characteristic abnormalities of the disorder (e.g., positive symptoms and working memory deficits) (Light et al., 2006; Spencer, Salisbury, Shenton, & McCarley, 2008b). Cortical θ and γ oscillations are dependent on perisomatic inhibition of pyramidal neurons from basket cells expressing cholecystokinin (CCK(b) cells) and parvalbumin [PV(b) cells], respectively. Alterations in basket cells may underlie the cortical oscillation deficits in schizophrenia. CB_1 receptors are located at the terminals of CCK-containing GABAergic interneurons. In preclinical studies, cannabinoids have been shown to reduce γ oscillations in hippocampal slice preparations. Despite these findings, so far there has been only one study examining the acute effects of cannabinoids on the ASSR and the relationship between these effects and the psychotomimetic effects of THC. In a double-blind, randomized, crossover, and counterbalanced study in which subjects received IV THC (placebo, 0.015 and 0.03 mg/kg), Cortes-Briones, Skosnik, et al. (2015), Cortes-Briones, Cahill, et al. (2015a), and Cortes-Briones, Cahill, et al. (2015b) showed, for the first time in humans, that THC acutely disrupts gamma-band (40 Hz) oscillations using an ASSR paradigm (Cortes-Briones, Skosnik, et al., 2015). Specifically, this study showed a dose-dependent reduction of intertrial coherence (ITC) (a measure of the consistency of the brain's response across trials) for 40 Hz stimulation. Furthermore, the authors showed that the higher dose of THC (0.03 mg/kg) reduced evoked power during 40 Hz stimulation at a trend level, and that the effects of THC on ITC and evoked power were reduced in subjects with a recent use of cannabis. No significant effects were found for the 20 or 30 Hz ASSRs, suggesting a selective effect of cannabinoids on evoked gamma-band activity. Interestingly, an inverse relationship between ITC and the schizophrenia-like effects of THC was found, suggesting that some of the psychotomimetic effects of cannabinoids may be related to their capacity to disrupt gamma-band oscillations. Considering that ITC measures reflect the consistency (nonrandomness) of the brain's response to different occurrences (trials) of a stimulus/event, the authors hypothesized that these findings suggest that the schizophrenia-like effects of THC may be related to an increased variability (randomness) in the brain's response to stimulation.

Neural Noise

There is growing evidence supporting the idea that neural noise, that is, the randomness of neural activity, is increased in patients with schizophrenia (Diez et al., 2013; Díez et al., 2014; Winterer & Weinberger, 2003, Winterer et al., 2000, 2004). Several studies have shown increased trial-to-trial (random) variability of evoked EEG responses in schizophrenia (Diez et al., 2013; Díez et al., 2014; Ford, White, Lim, & Pfefferbaum, 1994,

Winterer et al., 2000, 2004). Furthermore, nonlinear measures of uncertainty (entropy) or randomness have shown increased *randomness* in the EEG and Magnetoencephalography (MEG) signals of patients with schizophrenia (Li et al., 2008; Takahashi et al., 2010; Fernández, Gómez, Hornero, & López-Ibor, 2013), which are higher during periods of psychotic decompensation (Takahashi et al., 2010).

So far, only one study has investigated the effect of cannabinoids on neural noise in humans, and the relationship between this effect and behavior. In a recent study, Cortes-Briones et al. used Lempel-Ziv complexity (LZC), a measure from information theory first developed to characterize the randomness of signals, to test the hypothesis that THC increases neural noise (Cortes-Briones, Cahill, et al., 2015a). Furthermore, they tested the hypothesis that there is a positive relationship between neural noise and the schizophrenia-like effects of THC that is independent of the effect of THC on signal power. Using a double-blind, randomized, crossover, and counterbalanced design in which subjects received IV THC (placebo, 0.015 and 0.03 mg/kg), the authors measured LZC in the baseline period of an oddball task. The results showed that THC increased neural noise in a dose dependent manner, and that there was a strong positive relationship between neural noise and the psychosis-like positive ($\beta=0.685$) and disorganization ($\beta=0.754$) symptoms induced by THC. Interestingly, no relationship between neural noise and negative-like symptoms was found. Importantly, these relationships were independent of the changes in signal power induced by THC. Considering the idea that random noise can interfere with and distort the information circulating between the nodes of a network, the authors hypothesized that by increasing neural noise THC may be inducing a *dysconnectivity* (aberrant connections) (Stephan, Friston, & Frith, 2009) between the nodes of the brain's networks, and that this effect could be responsible for some of the psychotomimetic effects of THC.

Neuroreceptor Imaging

In addition to EEG, other neuroimaging modalities have been utilized to identify more proximate markers of the neural changes resulting from THC administration and, in some instances, correlated with the psychotomimetic and cognitive effects (summarized in Table 2).

The results of neuroreceptor imaging to study the ability of THC to increase striatal dopamine (Barkus et al., 2011; Bossong et al., 2009; Kuepper et al., 2013b; Stokes et al., 2009) are mixed. Although there is some evidence of displacement of [^{11}C]raclopride in the ventral striatum following THC administration, the relationship with psychotomimetic effects has not been well established.

Kuepper et al. studied striatal dopamine release following vaporized THC (8 mg) and placebo and showed that compared with healthy controls, both patients and relatives of those with psychosis showed significant displacement of the ligand ([F-18]fallypride) in striatal subregions, indicative of dopamine release with the biggest effect in the caudate nucleus (Kuepper et al., 2013a).

Functional Imaging

In a large study of cannabis users using PET and O^{15} H_2O, Mathew et al. demonstrated an association between decreased cerebellar blood flow and THC-induced timing deficits (Mathew et al., 1998). The effects of THC have also been studied using fMRI. THC and CBD appear to have opposing effects on BOLD signal (Bhattacharyya et al., 2010; Winton-Brown et al., 2011). Notably, a few studies show correlations between psychotomimetic (but not yet cognitive) effects and perturbation in BOLD signal. Bhattacharya et al. reported an increased BOLD activation in the parahippocampal gyrus and attenuation in the ventral striatum, correlated with psychotomimetic symptoms (Bhattacharyya et al., 2009). In a further study, attenuated task-related activation of the right caudate was inversely correlated with severity of symptoms. In a preliminary study, Bhattacharyya et al. have also suggested that, in individuals with specific polymorphisms of both the ATK1 and DAT1 genes, THC-induced psychotomimetic effects could be mediated by attenuation of activity in the striatum and midbrain (Bhattacharyya, Atakan, et al., 2012b).

As discussed earlier, cannabinoids produce robust impairments of learning and memory. Bhattacharyya et al. reported that oral THC (10mg) altered medial temporal activation during the encoding phase of a verbal paired associate learning task in healthy males ($n=15$) (Bhattacharyya et al., 2009). THC also attenuated striatal activation during recall that was correlated with the positive symptoms of psychosis.

The effects of oral THC on response inhibition have been studied using a go/no-go task (Bhattacharyya et al., 2015; Borgwardt et al., 2008). THC attenuated activation in the inferior frontal, anterior cingulate, and precuneal cortices, and augmented activation in the medial temporal cortex and caudate. Bhattacharrya et al. further reported that the effect of THC on inferior frontal activation correlated with the severity of the positive psychotic symptoms. Of note, patients with psychosis show altered striatal and prefrontal activation during inhibition tasks (Kaladjian et al., 2007; Rubia et al., 2001).

THC has also been reported to attenuate amygdalal, hippocampal, parietal, and prefrontal activation to fearful faces, and augment activation of these regions to happy faces (Bossong et al., 2013). THC has also been reported to reduce the amygdalal response to threat-related faces (Phan et al., 2008) and may increase functional connectivity between frontal regions and amygdalal subnuclei (Gorka et al., 2015).

CONCLUSIONS

HLS using THC have reliably demonstrated an array of dose-dependent psychotomimetic and cognitive effects in healthy humans, as well as in individuals with cannabis use disorder, increased risk of psychosis, and schizophrenia. Broadly, deficits are observed in domains most relevant to schizophrenia: positive, negative, dissociative, and disorganization cluster symptoms and impairment in working and verbal memory. Laboratory models of psychosis may enrich our understanding of psychotic disorders such as schizophrenia, offering methodological advantages. The ability to extrapolate the results of HLS in healthy THC users to clinical psychosis is supported by the fact that these effects are intensified in individuals who are diagnosed with or at risk for schizophrenia. A range of more proximate markers of neuronal function have been discovered, some highly associated with clinical effects. Genetic variations and the presence of other drugs may modulate the effects.

HLS have demonstrated value in the immediate study of acute, intoxication-related, cannabis-induced psychosis, in addition to providing insights into mechanisms underlying the shared phenomenology observed with schizophrenia. Already, for example, HLS comparing the interactive effects of THC and CBD have lent support for completed and ongoing clinical trials of the latter for schizophrenia. Furthermore, novel biomarkers of psychosis, designed and refined within laboratory settings, are being tested in clinical populations with great potential relevance for screening, prediction of relapse, and monitoring of treatment response. A further novel use of laboratory-based cannabinoid challenges as a neuropsychiatric "stress test" has been proposed (Gupta, Ranganathan, & D'Souza, 2016a). As a screening test, akin to cardiac stress testing for coronary ischemia, THC may be administered within safe, controlled parameters to those suspected to be at risk of schizophrenia in order to ascertain clinical or subclinical (e.g., electrophysiological) changes, further quantifying risk and informing the utility of new or established interventions for primary or secondary prevention.

HLS provide a vital complement to other study designs within the field by allowing for more controlled confirmation and elucidation of mechanisms and more direct inference of causality. The ecological relevance of the numerous HLS using oral THC is increasing as "edibles" grow in popularity. Responding to changing content of the cannabis available in society, HLS may need to further explore higher doses of THC and the interactive effects of CBD. Similarly, HLS utilizing vaporized THC may become more prevalent. Capitalizing on established genetic, electrophysiological, biochemical, and neuroimaging-based correlates of psychosis, multimodal HLS in healthy cannabis users (perhaps in concert with other psychotomimetic drugs and individuals with schizophrenia) may further illuminate the degree of overlap between laboratory models and clinical psychotic states. The cross pollination resulting from the continued closure of the gap between the laboratory and the clinic can only serve to propel the field forward.

Key Chapter Points

- HLS using THC have reliably demonstrated a range of transient, dose-dependent psychotomimetic and deleterious cognitive effects in healthy individuals, as well as in those with cannabis use disorder, increased risk of psychosis, and schizophrenia.
- This type of study design complements more naturalistic or observational approaches by allowing for more controlled confirmation and elucidation of mechanisms as well as more direct inference of causality.
- Psychophysiological correlates of THC-induced psychosis-like states have been characterized, most notably in the EEG. These may have wider clinical potential as biomarkers of psychosis.
- Genetic variations (in COMT, DAT1, and ATK1 genes) and the presence of other drugs (such as haloperidol, iomazenil, and CBD) may modulate the effects.
- Future priorities for HLS include the further exploration of higher doses of THC, vaporized THC, and the interactive effects of CBD.

ACKNOWLEDGMENTS

John Cahill acknowledges grant funding from the Brain and Behavior Research Foundation and The Patrick and Catherine Weldon Donaghue Medical Research Foundation. Swapnil Gupta and Jose Cortes have received grant funding from the Brain and Behavior Research Foundation. Rajiv Radhakrishnan was supported by Thomas P. Detre Fellowship in Translational Neuroscience Research. Mohamed Sherif was supported by the Department of Veterans Affairs schizophrenia research fellowship. Deepak Cyril D'Souza received grant funding from NIDA, NIMH, NIAAA, VA R&D.

REFERENCES

Abrams, D. I., Vizoso, H. P., Shade, S. B., Jay, C., Kelly, M. E., & Benowitz, N. L. (2007). Vaporization as a smokeless cannabis delivery system: A pilot study. *Clinical Pharmacology and Therapeutics*, *82*, 572–578.

Agurell, S., Halldin, M., Lindgren, J. E., Ohlsson, A., Widman, M., Gillespie, H., et al. (1986). Pharmacokinetics and metabolism of delta 1-tetrahydrocannabinol and other cannabinoids with emphasis on man. *Pharmacological Reviews*, *38*, 21–43.

Atakan, Z., Bhattacharyya, S., Allen, P., Martin-Santos, R., Crippa, J. A., Borgwardt, S. J., et al. (2013). Cannabis affects people differently: Inter-subject variation in the psychotogenic effects of Delta9-tetrahydrocannabinol: A functional magnetic resonance imaging study with healthy volunteers. *Psychological Medicine*, *43*, 1255–1267.

Ballard, M. E., & de Wit, H. (2011). Combined effects of acute, very-low-dose ethanol and delta(9)-tetrahydrocannabinol in healthy human volunteers. *Pharmacology Biochemistry and Behavior*, *97*, 627–631.

Barkus, E., Morrison, P. D., Vuletic, D., Dickson, J. C., Ell, P. J., Pilowsky, L. S., et al. (2011). Does intravenous Delta9-tetrahydrocannabinol increase dopamine release? A SPET study. *Journal of Psychopharmacology*, *25*, 1462–1468.

Bartos, M., & Elgueta, C. (2012). Functional characteristics of parvalbumin-and cholecystokinin-expressing basket cells. *Journal of Physiology*, *590*, 669–681.

Beaulieu, J. M., Gainetdinov, R. R., & Caron, M. G. (2007). The Akt-GSK-3 signaling cascade in the actions of dopamine. *Trends in Pharmacological Sciences, 28*, 166–172.

Bhattacharyya, S., Atakan, Z., Martin-Santos, R., Crippa, J. A., & McGuire, P. K. (2012a). Neural mechanisms for the cannabinoid modulation of cognition and affect in man: A critical review of neuroimaging studies. *Current Pharmaceutical Design, 18*, 5045–5054.

Bhattacharyya, S., Atakan, Z., Martin-Santos, R., Crippa, J. A., Kambeitz, J., Prata, D., et al. (2012b). Preliminary report of biological basis of sensitivity to the effects of cannabis on psychosis: AKT1 and DAT1 genotype modulates the effects of delta-9-tetrahydrocannabinol on midbrain and striatal function. *Molecular Psychiatry, 17*, 1152–1155.

Bhattacharyya, S., Atakan, Z., Martin-Santos, R., Crippa, J. A., Kambeitz, J., Malhi, S., et al. (2015). Impairment of inhibitory control processing related to acute psychotomimetic effects of cannabis. *European Neuropsychopharmacology: The Journal of the European College of Neuropsychopharmacology, 25*, 26–37.

Bhattacharyya, S., Fusar-Poli, P., Borgwardt, S., Martin-Santos, R., Nosarti, C., O'Carroll, C., et al. (2009). Modulation of mediotemporal and ventrostriatal function in humans by Delta9-tetrahydrocannabinol: A neural basis for the effects of *Cannabis sativa* on learning and psychosis. *Archives of General Psychiatry, 66*, 442–451.

Bhattacharyya, S., Crippa, J. A., Allen, P., Martin-Santos, R., Borgwardt, S., Fusar-Poli, P., et al. (2012a). Induction of psychosis by Delta9-tetrahydrocannabinol reflects modulation of prefrontal and striatal function during attentional salience processing. *Archives of General Psychiatry, 69*, 27–36.

Bhattacharyya, S., Crippa, J. A., Allen, P., Martin-Santos, R., Borgwardt, S., Fusar-Poli, P., et al. (2012b). Induction of psychosis by Delta9-tetrahydrocannabinol reflects modulation of prefrontal and striatal function during attentional salience processing. *Archives of General Psychiatry, 69*, 27–36.

Bhattacharyya, S., Morrison, P. D., Fusar-Poli, P., Martin-Santos, R., Borgwardt, S., Winton-Brown, T., et al. (2010). Opposite effects of delta-9-tetrahydrocannabinol and cannabidiol on human brain function and psychopathology. *Neuropsychopharmacology, 35*, 764–774.

Bocker, K. B., Hunault, C. C., Gerritsen, J., Kruidenier, M., Mensinga, T. T., & Kenemans, J. L. (2010). Cannabinoid modulations of resting state EEG theta power and working memory are correlated in humans. *Journal of Cognitive Neuroscience, 22*, 1906–1916.

Borg, J., Gershon, S., & Alpert, M. (1975). Dose effects of smoked marihuana on human cognitive and motor functions. *Psychopharmacologia, 42*, 211–218.

Borgwardt, S. J., Allen, P., Bhattacharyya, S., Fusar-Poli, P., Crippa, J. A., Seal, M. L., et al. (2008). Neural basis of Delta-9-tetrahydrocannabinol and cannabidiol: Effects during response inhibition. *Biological Psychiatry, 64*, 966–973.

Bossong, M. G., Jansma, J. M., Bhattacharyya, S., & Ramsey, N. F. (2014). Role of the endocannabinoid system in brain functions relevant for schizophrenia: An overview of human challenge studies with cannabis or 9-tetrahydrocannabinol (THC). *Progress in Neuro-Psychopharmacology & Biological Psychiatry, 52*, 53–69.

Bossong, M. G., van Hell, H. H., Jager, G., Kahn, R. S., Ramsey, N. F., & Jansma, J. M. (2013). The endocannabinoid system and emotional processing: A pharmacological fMRI study with 9-tetrahydrocannabinol. *European Neuropsychopharmacology: The Journal of the European College of Neuropsychopharmacology, 23*, 1687–1697.

Bossong, M. G., Mehta, M. A., van Berckel, B. N., Howes, O. D., Kahn, R. S., & Stokes, P. R. (2015). Further human evidence for striatal dopamine release induced by administration of 9-tetrahydrocannabinol (THC): Selectivity to limbic striatum. *Psychopharmacology, 232*, 2723–2729.

Bossong, M. G., van Berckel, B. N., Boellaard, R., Zuurman, L., Schuit, R. C., Windhorst, A. D., et al. (2009). Delta 9-tetrahydrocannabinol induces dopamine release in the human striatum. *Neuropsychopharmacology, 34*, 759–766.

Bramon, E., Rabe-Hesketh, S., Sham, P., Murray, R. M., & Frangou, S. (2004). Meta-analysis of the P300 and P50 waveforms in schizophrenia. *Schizophrenia Research*, *70*, 315–329.

Bramon, E., McDonald, C., Croft, R. J., Landau, S., Filbey, F., Gruzelier, J. H., et al. (2005). Is the P300 wave an endophenotype for schizophrenia? A meta-analysis and a family study. *NeuroImage*, *27*, 960–968.

Buckley, P. F., Miller, B. J., Lehrer, D. S., & Castle, D. J. (2009). Psychiatric comorbidities and schizophrenia. *Schizophrenia Bulletin*, *35*, 383–402.

Bullmore, E., & Sporns, O. (2009). Complex brain networks: Graph theoretical analysis of structural and functional systems. *Nature Reviews Neuroscience*, *10*, 186–198.

Carbuto, M., Sewell, R. A., Williams, A., Forselius-Bielen, K., Braley, G., Elander, J., et al. (2012a). The safety of studies with intravenous Delta(9)-tetrahydrocannabinol in humans, with case histories. *Psychopharmacology*, *219*, 885–896.

Carbuto, M., Sewell, R. A., Williams, A., Forselius-Bielen, K., Braley, G., Elander, J., et al. (2012b). The safety of studies with intravenous Delta-tetrahydrocannabinol in humans, with case histories. *Psychopharmacology*, *219*, 885–896.

Carroll, C. A., O'Donnell, B. F., Shekhar, A., & Hetrick, W. P. (2009). Timing dysfunctions in schizophrenia span from millisecond to several-second durations. *Brain and Cognition*, *70*, 181–190.

Caspi, A., Moffitt, T. E., Cannon, M., McClay, J., Murray, R., Harrington, H., et al. (2005). Moderation of the effect of adolescent-onset cannabis use on adult psychosis by a functional polymorphism in the catechol-O-methyltransferase gene: Longitudinal evidence of a gene X environment interaction. *Biological Psychiatry*, *57*, 1117–1127.

Chen, J., Lipska, B. K., Halim, N., Ma, Q. D., Matsumoto, M., Melhem, S., et al. (2004). Functional analysis of genetic variation in catechol-O-methyltransferase (COMT): Effects on mRNA, protein, and enzyme activity in postmortem human brain. *American Journal of Human Genetics*, *75*, 807–821.

Chiang, C. W., & Barnett, G. (1984). Marijuana effect and delta-9-tetrahydrocannabinol plasma level. *Clinical Pharmacology and Therapeutics*, *36*, 234–238.

Cocchetto, D. M., Owens, S. M., Perez-Reyes, M., DiGuiseppi, S., & Miller, L. L. (1981). Relationship between plasma delta-9-tetrahydrocannabinol concentration and pharmacologic effects in man. *Psychopharmacology*, *75*, 158–164.

Colizzi, M., Iyegbe, C., Powell, J., Blasi, G., Bertolino, A., Murray, R. M., et al. (2015). Interaction between DRD2 and AKT1 genetic variations on risk of psychosis in cannabis users: A case-control study. *NPJ Schizophrenia*, *1*, 15025.

Conrad, D. G., Elsmore, T. F., & Sodetz, F. J. (1972). 9-Tetrahydrocannabinol: dose-related effects on timing behavior in chimpanzee. *Science*, *175*, 547–550.

Cortes-Briones, J., Skosnik, P. D., Mathalon, D., Cahill, J., Pittman, B., Williams, A., et al. (2015). Delta9-THC disrupts gamma (gamma)-band neural oscillations in humans. *Neuropsychopharmacology*, *40*, 2124–2134.

Cortes-Briones, J. A., Cahill, J. D., Skosnik, P. D., Mathalon, D. H., Williams, A., Sewell, R. A., et al. (2015a). The psychosis-like effects of delta-tetrahydrocannabinol are associated with increased cortical noise in healthy humans. *Biological Psychiatry*, *78*, 805–813.

Cortes-Briones, J. A., Cahill, J. D., Skosnik, P. D., Mathalon, D. H., Williams, A., Sewell, R. A., et al. (2015b). The psychosis-like effects of delta(9)-tetrahydrocannabinol are associated with increased cortical noise in healthy humans. *Biological Psychiatry*, *78*, 805–813.

Csicsvari, J., Jamieson, B., Wise, K. D., & Buzsáki, G. (2003). Mechanisms of gamma oscillations in the hippocampus of the behaving rat. *Neuron*, *37*, 311–322.

Curran, H. V., Brignell, C., Fletcher, S., Middleton, P., & Henry, J. (2002). Cognitive and subjective dose-response effects of acute oral Delta 9-tetrahydrocannabinol (THC) in infrequent cannabis users. *Psychopharmacology*, *164*, 61–70.

D'Souza, D. C., Ranganathan, M., Braley, G., Gueorguieva, R., Zimolo, Z., Cooper, T., et al. (2008a). Blunted psychotomimetic and amnestic effects of Δ-9-tetrahydrocannabinol in frequent users of cannabis. *Neuropsychopharmacology*, *33*, 2505–2516.

D'Souza, D. C., Ranganathan, M., Braley, G., Gueorguieva, R., Zimolo, Z., Cooper, T., et al. (2008b). Blunted psychotomimetic and amnestic effects of delta-9-tetrahydrocannabinol in frequent users of cannabis. *Neuropsychopharmacology*, *33*, 2505–2516.

D'Souza, D. C., Ranganathan, M., Braley, G., Gueorguieva, R., Zimolo, Z., Cooper, T., et al. (2008c). Blunted psychotomimetic and amnestic effects of delta-9-tetrahydrocannabinol in frequent users of cannabis. *Neuropsychopharmacology*, *33*, 2505–2516.

D'Souza, D. C., Perry, E., MacDougall, L., Ammerman, Y., Cooper, T., Wu, Y. T., et al. (2004). The psychotomimetic effects of intravenous delta-9-tetrahydrocannabinol in healthy individuals: Implications for psychosis. *Neuropsychopharmacology*, *29*, 1558–1572.

D'Souza, D. C., Abi-Saab, W. M., Madonick, S., Forselius-Bielen, K., Doersch, A., Braley, G., et al. (2005a). Delta-9-tetrahydrocannabinol effects in schizophrenia: implications for cognition, psychosis, and addiction. *Biological Psychiatry*, *57*, 594–608.

D'Souza, D. C., Braley, G., Blaise, R., Vendetti, M., Oliver, S., Pittman, B., et al. (2008). Effects of haloperidol on the behavioral, subjective, cognitive, motor, and neuroendocrine effects of Delta-9-tetrahydrocannabinol in humans. *Psychopharmacology*, *198*, 587–603.

D'Souza, D. C., Fridberg, D. J., Skosnik, P. D., Williams, A., Roach, B., Singh, N., et al. (2012a). Dose-related modulation of event-related potentials to novel and target stimuli by intravenous Delta(9)-THC in humans. *Neuropsychopharmacology: Official publication of the American College of Neuropsychopharmacology*, *37*, 1632–1646.

D'Souza, D. C., Fridberg, D. J., Skosnik, P. D., Williams, A., Roach, B., Singh, N., et al. (2012b). Dose-related modulation of event-related potentials to novel and target stimuli by intravenous Delta(9)-THC in humans. *Neuropsychopharmacology*, *37*, 1632–1646.

D'Souza, D. C., Abi-Saab, W. M., Madonick, S., Forselius-Bielen, K., Doersch, A., Braley, G., et al. (2005b). Delta-9-tetrahydrocannabinol effects in schizophrenia: Implications for cognition, psychosis, and addiction. *Biological Psychiatry*, *57*, 594–608.

Davalos, D. B., Kisley, M. A., & Ross, R. G. (2003). Effects of interval duration on temporal processing in schizophrenia. *Brain and Cognition*, *52*, 295–301.

Di Forti, M., Iyegbe, C., Sallis, H., Kolliakou, A., Falcone, M. A., Paparelli, A., et al. (2012). Confirmation that the AKT1 (rs2494732) genotype influences the risk of psychosis in cannabis users. *Biological Psychiatry*, *72*, 811–816.

Diez, A., Suazo, V., Casado, P., Martin-Loeches, M., & Molina, V. (2013). Spatial distribution and cognitive correlates of gamma noise power in schizophrenia. *Psychological Medicine*, *43*, 1175–1185.

Díez, Á., Suazo, V., Casado, P., Martín-Loeches, M., Perea, M. V., & Molina, V. (2014). Frontal gamma noise power and cognitive domains in schizophrenia. *Psychiatry Research: Neuroimaging*, *221*, 104–113.

ElSohly, M. A., Mehmedic, Z., Foster, S., Gon, C., Chandra, S., & Church, J. C. (2016). Changes in cannabis potency over the last 2 decades (1995–2014): Analysis of current data in the United States. *Biological Psychiatry*, *79*, 613–619.

Emamian, E. S., Hall, D., Birnbaum, M. J., Karayiorgou, M., & Gogos, J. A. (2004). Convergent evidence for impaired AKT1-GSK3beta signaling in schizophrenia. *Nature Genetics*, *36*, 131–137.

Emrich, H. M., Weber, M. M., Wendl, A., Zihl, J., von Meyer, L., & Hanisch, W. (1991). Reduced binocular depth inversion as an indicator of cannabis-induced censorship impairment. *Pharmacology Biochemistry and Behavior*, *40*, 689–690.

Englund, A., Morrison, P. D., Nottage, J., Hague, D., Kane, F., Bonaccorso, S., et al. (2013a). Cannabidiol inhibits THC-elicited paranoid symptoms and hippocampal-dependent memory impairment. *Journal of Psychopharmacology*, *27*, 19–27.

Englund, A., Morrison, P. D., Nottage, J., Hague, D., Kane, F., Bonaccorso, S., et al. (2013b). Cannabidiol inhibits THC-elicited paranoid symptoms and hippocampal-dependent memory impairment. *Journal of Psychopharmacology*, *27*, 19–27.

Farkas, I., Kallo, I., Deli, L., Vida, B., Hrabovszky, E., Fekete, C., et al. (2010). Retrograde endocannabinoid signaling reduces GABAergic synaptic transmission to gonadotropin-releasing hormone neurons. *Endocrinology*, *151*, 5818–5829.

Fernández, A., Gómez, C., Hornero, R., & López-Ibor, J. J. (2013). Complexity and schizophrenia. *Progress in Neuro-Psychopharmacology and Biological Psychiatry*, *45*, 267–276.

Fischedick, J., Van Der Kooy, F., & Verpoorte, R. (2010). Cannabinoid receptor 1 binding activity and quantitative analysis of *Cannabis sativa* L. smoke and vapor. *Chemical and Pharmaceutical Bulletin (Tokyo)*, *58*, 201–207.

Fischman, M. W., Foltin, R. W., & Brady, J. V. (1988). Human drug taking under controlled laboratory conditions. *NIDA Research Monograph*, *84*, 196–211.

Foltin, R. W., Fischman, M. W., & Byrne, M. F. (1988). Effects of smoked marijuana on food intake and body weight of humans living in a residential laboratory. *Appetite*, *11*, 1–14.

Ford, J. M., White, P., Lim, K. O., & Pfefferbaum, A. (1994). Schizophrenics have fewer and smaller P300s: A single-trial analysis. *Biological Psychiatry*, *35*, 96–103.

Freeman, D., Dunn, G., Murray, R. M., Evans, N., Lister, R., Antley, A., et al. (2015). How cannabis causes paranoia: Using the intravenous administration of 9-tetrahydrocannabinol (THC) to identify key cognitive mechanisms leading to paranoia. *Schizophrenia Bulletin*, *41*, 391–399.

Fusar-Poli, P., Allen, P., Bhattacharyya, S., Crippa, J. A., Mechelli, A., Borgwardt, S., et al. (2009). Modulation of effective connectivity during emotional processing by Delta 9-tetrahydrocannabinol and cannabidiol. *The International Journal of Neuropsychopharmacology/Official Scientific Journal of the Collegium Internationale Neuropsychopharmacologicum*, *13*, 421–432.

Garrett, E. R., & Hunt, C. A. (1974). Physiochemical properties, solubility, and protein binding of delta9-tetrahydrocannabinol. *Journal of Pharmaceutical Sciences*, *63*, 1056–1064.

Glass, M., & Felder, C. C. (1997). Concurrent stimulation of cannabinoid CB1 and dopamine D2 receptors augments cAMP accumulation in striatal neurons: Evidence for a Gs linkage to the CB1 receptor. *Journal of Neuroscience: The Official Journal of the Society for Neuroscience*, *17*, 5327–5333.

Gorka, S. M., Fitzgerald, D. A., de Wit, H., & Phan, K. L. (2015). Cannabinoid modulation of amygdala subregion functional connectivity to social signals of threat. *International Journal of Neuropsychopharmacology*, *18*(3).

Gupta, S., Ranganathan, M., & D'Souza, D. C. (2016a). The early identification of psychosis: Can lessons be learnt from cardiac stress testing? *Psychopharmacology*, *233*, 19–37.

Gupta, S., Ranganathan, M., & D'Souza, D. C. (2016b). The early identification of psychosis: Can lessons be learnt from cardiac stress testing? *Psychopharmacology*, *233*, 19–37.

Gupta, S., Cahill, J. D., Ranganathan, M., & Correll, C. U. (2014). The endocannabinoid system and schizophrenia: Links to the underlying pathophysiology and to novel treatment approaches. *Journal of Clinical Psychiatry*, *75*, 285–287.

Hampson, A. J., Grimaldi, M., Axelrod, J., & Wink, D. (1998). Cannabidiol and (-)Delta9-tetrahydrocannabinol are neuroprotective antioxidants. *Proceedings of the National Academy of Sciences of the United States of America*, *95*, 8268–8273.

Han, C. J., & Robinson, J. K. (2001). Cannabinoid modulation of time estimation in the rat. *Behavioral Neuroscience*, *115*, 243–246.

Harder, S., & Rietbrock, S. (1997). Concentration-effect relationship of delta-9-tetrahydrocannabiol and prediction of psychotropic effects after smoking marijuana. *International Journal of Clinical Pharmacology and Therapeutics*, *35*, 155–159.

Hart, C. L., van Gorp, W., Haney, M., Foltin, R. W., & Fischman, M. W. (2001). Effects of acute smoked marijuana on complex cognitive performance. *Neuropsychopharmacology*, *25*, 757–765.

Hazekamp, A., Ruhaak, R., Zuurman, L., van Gerven, J., & Verpoorte, R. (2006). Evaluation of a vaporizing device (Volcano) for the pulmonary administration of tetrahydrocannabinol. *Journal of Pharmaceutical Sciences*, *95*, 1308–1317.

Heinrichs, R. W., & Zakzanis, K. K. (1998). Neurocognitive deficit in schizophrenia: A quantitative review of the evidence. *Neuropsychology*, *12*, 426–445.

Heishman, S. J., Huestis, M. A., Henningfield, J. E., & Cone, E. J. (1990). Acute and residual effects of marijuana: Profiles of plasma THC levels, physiological, subjective, and performance measures. *Pharmacology Biochemistry and Behavior*, *37*, 561–565.

Henquet, C., Rosa, A., Krabbendam, L., Papiol, S., Fananás, L., Drukker, M., et al. (2006a). An experimental study of catechol-O-methyltransferase Val158Met moderation of Δ-9-tetrahydrocannabinol-induced effects on psychosis and cognition. *Neuropsychopharmacology*, *31*, 2748–2757.

Henquet, C., Rosa, A., Krabbendam, L., Papiol, S., Fananas, L., Drukker, M., et al. (2006b). An experimental study of catechol-O-methyltransferase Val(158)Met moderation of delta-9-tetrahydrocannabinol-induced effects on psychosis and cognition. *Neuropsychopharmacology*, *31*, 2748–2757.

Henquet, C., Rosa, A., Krabbendam, L., Papiol, S., Fananas, L., Drukker, M., et al. (2006c). An experimental study of catechol-o-methyltransferase Val158Met moderation of delta-9-tetrahydrocannabinol-induced effects on psychosis and cognition. *Neuropsychopharmacology*, *31*, 2748–2757.

Hicks, R. E., Gualtieri, C. T., Mayo, J. P., Jr., & Perez-Reyes, M. (1984). Cannabis, atropine, and temporal information processing. *Neuropsychobiology*, *12*, 229–237.

Higashima, M., Takeda, T., Kikuchi, M., Nagasawa, T., Hirao, N., Oka, T., et al. (2007). State-dependent changes in intrahemispheric EEG coherence for patients with acute exacerbation of schizophrenia. *Psychiatry Research*, *149*, 41–47.

Hinkley, L. B. N., Vinogradov, S., Guggisberg, A. G., Fisher, M., Findlay, A. M., & Nagarajan, S. S. (2011). Clinical symptoms and alpha band resting-state functional connectivity imaging in patients with schizophrenia: Implications for novel approaches to treatment. *Biological Psychiatry*, *70*, 1134–1142.

Ho, B. T., Fritchie, G. E., Kralik, P. M., Englert, L. F., McIsaac, W. M., & Idanpaan-Heikkila, J. (1970). Distribution of tritiated-1 delta 9tetrahydrocannabinol in rat tissues after inhalation. *Journal of Pharmacy and Pharmacology*, *22*, 538–539.

Hollister, L. E. (1986). Health aspects of cannabis. *Pharmacological Reviews*, *38*, 1–20.

Hooker, W. D., & Jones, R. T. (1987). Increased susceptibility to memory intrusions and the Stroop interference effect during acute marijuana intoxication. *Psychopharmacology*, *91*, 20–24.

Howes, O. D., & Kapur, S. (2009). The dopamine hypothesis of schizophrenia: Version III—The final common pathway. *Schizophrenia Bulletin*, *35*, 549–562.

Hunault, C. C., Mensinga, T. T., Bocker, K. B., Schipper, C. M., Kruidenier, M., Leenders, M. E., et al. (2009). Cognitive and psychomotor effects in males after smoking a combination of tobacco and cannabis containing up to 69 mg delta-9-tetrahydrocannabinol (THC). *Psychopharmacology*, *204*, 85–94.

Hunt, C. A., & Jones, R. T. (1980). Tolerance and disposition of tetrahydrocannabinol in man. *Journal of Pharmacology and Experimental Therapeutics*, *215*, 35–44.

Ilan, A. B., Smith, M. E., & Gevins, A. (2004). Effects of marijuana on neurophysiological signals of working and episodic memory. *Psychopharmacology, 176*, 214–222.

Ilan, A. B., Gevins, A., Coleman, M., ElSohly, M. A., & de Wit, H. (2005). Neurophysiological and subjective profile of marijuana with varying concentrations of cannabinoids. *Behavioral Pharmacology, 16*, 487–496.

Jeon, Y. W., & Polich, J. (2003). Meta-analysis of P300 and schizophrenia: Patients, paradigms, and practical implications. *Psychophysiology, 40*, 684–701.

Johansson, E., Noren, K., Sjovall, J., & Halldin, M. M. (1989). Determination of delta 1-tetrahydrocannabinol in human fat biopsies from marihuana users by gas chromatography-mass spectrometry. *Biomedical Chromatography, 3*, 35–38.

Juckel, G., Roser, P., Nadulski, T., Stadelmann, A. M., & Gallinat, J. (2007). Acute effects of Delta9-tetrahydrocannabinol and standardized cannabis extract on the auditory evoked mismatch negativity. *Schizophrenia Research, 97*, 109–117.

Kaladjian, A., Jeanningros, R., Azorin, J. M., Grimault, S., Anton, J. L., & Mazzola-Pomietto, P. (2007). Blunted activation in right ventrolateral prefrontal cortex during motor response inhibition in schizophrenia. *Schizophrenia Research, 97*(1), 184–193.

Kantrowitz, J. T., Nolan, K. A., Sen, S., Simen, A. A., Lachman, H. M., Bowers, M. B., Jr. (2009). Adolescent cannabis use, psychosis and catechol-O-methyltransferase genotype in African Americans and Caucasians. *Psychiatric Quarterly, 80*, 213–218.

Kaufmann, R. M., Kraft, B., Frey, R., Winkler, D., Weiszenbichler, S., Backer, C., et al. (2010). Acute psychotropic effects of oral cannabis extract with a defined content of Delta9-tetrahydrocannabinol (THC) in healthy volunteers. *Pharmacopsychiatry, 43*, 24–32.

Kiehl, K. A., Laurens, K. R., Duty, T. L., Forster, B. B., & Liddle, P. F. (2001). Neural sources involved in auditory target detection and novelty processing: An event-related fMRI study. *Psychophysiology, 38*, 133–142.

Kleinloog, D., Liem-Moolenaar, M., Jacobs, G., Klaassen, E., de Kam, M., Hijman, R., et al. (2012a). Does olanzapine inhibit the psychomimetic effects of Delta(9)-tetrahydrocannabinol? *Journal of Psychopharmacology, 26*, 1307–1316.

Kleinloog, D., Liem-Moolenaar, M., Jacobs, G., Klaassen, E., de Kam, M., Hijman, R., et al. (2012b). Does olanzapine inhibit the psychomimetic effects of {Delta}9-tetrahydrocannabinol? *Journal of Psychopharmacology, 26*, 1307–1316.

Koethe, D., Gerth, C. W., Neatby, M. A., Haensel, A., Thies, M., Schneider, U., et al. (2006). Disturbances of visual information processing in early states of psychosis and experimental delta-9-tetrahydrocannabinol altered states of consciousness. *Schizophrenia Research, 88*, 142–150.

Kopell, B. S., Tinklenberg, J. R., & Hollister, L. E. (1972). Contingent negative variation amplitudes. Marihuana and alcohol. *Archives of General Psychiatry, 27*, 809–811.

Krishnan, G. P., Hetrick, W. P., Brenner, C. A., Shekhar, A., Steffen, A. N., & O'Donnell, B. F. (2009). Steady state and induced auditory gamma deficits in schizophrenia. *NeuroImage, 47*, 1711–1719.

Kubicki, M., Westin, C.-F., Nestor, P. G., Wible, C. G., Frumin, M., Maier, S. E., et al. (2003). Cingulate fasciculus integrity disruption in schizophrenia: A magnetic resonance diffusion tensor imaging study. *Biological Psychiatry, 54*, 1171–1180.

Kubicki, M., Park, H., Westin, C., Nestor, P., Mulkern, R., Maier, S., et al. (2005). DTI and MTR abnormalities in schizophrenia: Analysis of white matter integrity. *NeuroImage, 26*, 1109–1118.

Kuepper, R., Ceccarini, J., Lataster, J., van Os, J., van Kroonenburgh, M., van Gerven, J. M., et al. (2013a). Delta-9-tetrahydrocannabinol-induced dopamine release as a function of psychosis risk: 18 F-fallypride positron emission tomography study. *PLoS One, 8*, e70378.

Kuepper, R., Ceccarini, J., Lataster, J., van Os, J., van Kroonenburgh, M., van Gerven, J. M., et al. (2013b). Delta-9-tetrahydrocannabinol-induced dopamine release as a function of psychosis risk: 18F-fallypride positron emission tomography study. *PLoS One*, *8*, e70378.

Kwon, J. S., O'Donnell, B. F., Wallenstein, G. V., Greene, R. W., Hirayasu, Y., Nestor, P. G., et al. (1999). Gamma frequency-range abnormalities to auditory stimulation in schizophrenia. *Archives of General Psychiatry*, *56*, 1001–1005.

Lachaux, J.-P., Axmacher, N., Mormann, F., Halgren, E., & Crone, N. E. (2012). High-frequency neural activity and human cognition: Past, present and possible future of intracranial EEG research. *Progress in Neurobiology*, *98*, 279–301.

Lachman, H. M., Papolos, D. F., Saito, T., Yu, Y. M., Szumlanski, C. L., & Weinshilboum, R. M. (1996). Human catechol-O-methyltransferase pharmacogenetics: Description of a functional polymorphism and its potential application to neuropsychiatric disorders. *Pharmacogenetics*, *6*, 243–250.

Lahti, A. C., Koffel, B., LaPorte, D., & Tamminga, C. A. (1995). Subanesthetic doses of ketamine stimulate psychosis in schizophrenia. *Neuropsychopharmacology*, *13*, 9–19.

Laruelle, M., Abi-Dargham, A., van Dyck, C. H., Gil, R., D'Souza, C. D., Erdos, J., et al. (1996). Single photon emission computerized tomography imaging of amphetamine-induced dopamine release in drug-free schizophrenic subjects. *Proceedings of the National Academy of Sciences of the United States of America*, *93*, 9235–9240.

Lawrie, S. M., Buechel, C., Whalley, H. C., Frith, C. D., Friston, K. J., & Johnstone, E. C. (2002). Reduced frontotemporal functional connectivity in schizophrenia associated with auditory hallucinations. *Biological Psychiatry*, *51*, 1008–1011.

Lemberger, L., & Rowe, H. (1975). Clinical pharmacology of nabilone, a cannabinol derivative. *Clinical Pharmacology and Therapeutics*, *18*, 720–726.

Lemberger, L., Tamarkin, N. R., Axelrod, J., & Kopin, I. J. (1971). Delta-9-tetrahydrocannabinol: Metabolism and disposition in long-term marihuana smokers. *Science*, *173*, 72–74.

Leuschner, J. T., Harvey, D. J., Bullingham, R. E., & Paton, W. D. (1986). Pharmacokinetics of delta 9-tetrahydrocannabinol in rabbits following single or multiple intravenous doses. *Drug Metabolism and Disposition*, *14*, 230–238.

Leweke, F. M. (2007). *Cannabidiol as an antipsychotic—New perspectives*. In: *2nd international cannabis and mental health conference, London*.

Leweke, F. M., Schneider, U., Thies, M., Munte, T. F., & Emrich, H. M. (1999). Effects of synthetic delta9-tetrahydrocannabinol on binocular depth inversion of natural and artificial objects in man. *Psychopharmacology*, *142*, 230–235.

Leweke, F. M., Schneider, U., Radwan, M., Schmidt, E., & Emrich, H. M. (2000). Different effects of nabilone and cannabidiol on binocular depth inversion in Man. *Pharmacology Biochemistry and Behavior*, *66*, 175–181.

Leweke, M., Kampmann, C., Radwan, M., Dietrich, D. E., Johannes, S., Emrich, H. M., et al. (1998). The effects of tetrahydrocannabinol on the recognition of emotionally charged words: An analysis using event-related brain potentials. *Neuropsychobiology*, *37*, 104–111.

Li, Y., Tong, S., Liu, D., Gai, Y., Wang, X., Wang, J., et al. (2008). Abnormal EEG complexity in patients with schizophrenia and depression. *Clinical Neurophysiology*, *119*, 1232–1241.

Lichtman, A. H., Varvel, S. A., & Martin, B. R. (2002). Endocannabinoids in cognition and dependence. *Prostaglandins Leukotrienes and Essential Fatty Acids*, *66*, 269–285.

Lieberman, J. A., Kane, J. M., Gadaleta, D., Brenner, R., Lesser, M. S., & Kinon, B. (1984). Methylphenidate challenge as a predictor of relapse in schizophrenia. *American Journal of Psychiatry*, *141*, 633–638.

Liem-Moolenaar, M., te Beek, E. T., de Kam, M. L., Franson, K. L., Kahn, R. S., Hijman, R., et al. (2010). Central nervous system effects of haloperidol on THC in healthy male volunteers. *Journal of Psychopharmacology*, *24*, 1697–1708.

Light, G. A., Hsu, J. L., Hsieh, M. H., Meyer-Gomes, K., Sprock, J., Swerdlow, N. R., et al. (2006). Gamma band oscillations reveal neural network cortical coherence dysfunction in schizophrenia patients. *Biological Psychiatry*, *60*, 1231–1240.

Luck, S. J., & Kappenman, E. S. (2012). *Oxford handbook of event-related potential components*. Oxford: Oxford University Press.

Marks, D. F., & MacAvoy, M. G. (1989). Divided attention performance in cannabis users and non-users following alcohol and cannabis separately and in combination. *Psychopharmacology*, *99*, 397–401.

Martin-Santos, R., Crippa, J. A., Batalla, A., Bhattacharyya, S., Atakan, Z., Borgwardt, S., et al. (2012a). Acute effects of a single, oral dose of d9-tetrahydrocannabinol (THC) and cannabidiol (CBD) administration in healthy volunteers. *Current Pharmaceutical Design*, *18*, 4966–4979.

Martin-Santos, R., Crippa, J. A., Batalla, A., Bhattacharyya, S., Atakan, Z., Borgwardt, S., et al. (2012b). Acute effects of a single, oral dose of d9-tetrahydrocannabinol (THC) and cannabidiol (CBD) administration in healthy volunteers. *Current Pharmaceutical Design*, *18*, 4966–4979.

Mason, O. J., Morgan, C. J., Stefanovic, A., & Curran, H. V. (2008). The psychotomimetic states inventory (PSI): Measuring psychotic-type experiences from ketamine and cannabis. *Schizophrenia Research*, *103*, 138–142.

Mathew, R. J., Wilson, W. H., Turkington, T. G., & Coleman, R. E. (1998). Cerebellar activity and disturbed time sense after THC. *Brain Research*, *797*, 183–189.

McClure, G. Y., & McMillan, D. E. (1997). Effects of drugs on response duration differentiation. VI. Differential effects under differential reinforcement of low rates of responding schedules. *Journal of Pharmacology and Experimental Therapeutics*, *281*, 1368–1380.

McDonald, J., Schleifer, L., Richards, J. B., & de Wit, H. (2003a). Effects of THC on behavioral measures of impulsivity in humans. *Neuropsychopharmacology*, *28*, 1356–1365.

McDonald, J., Schleifer, L., Richards, J. B., & de Wit, H. (2003b). Effects of THC on behavioral measures of impulsivity in humans. *Neuropsychopharmacology: Official Publication of the American College of Neuropsychopharmacology*, *28*, 1356–1365.

McPartland, J. M., & Pruitt, P. L. (1997). Medical marijuana and its use by the immunocompromised. *Alternative Therapies in Health and Medicine*, *3*, 39–45.

Mehmedic, Z., Chandra, S., Slade, D., Denham, H., Foster, S., Patel, A. S., et al. (2010). Potency trends of Delta9-THC and other cannabinoids in confiscated cannabis preparations from 1993 to 2008. *Journal of Forensic Science*, *55*, 1209–1217.

Melges, F. T., Tinklenberg, J. R., Deardorff, C. M., Davies, N. H., Anderson, R. E., & Owen, C. A. (1974). Temporal disorganization and delusional-like ideation. Processes induced by hashish and alcohol. *Archives of General Psychiatry*, *30*, 855–861.

Miller, L. L., McFarland, D., Cornett, T. L., & Brightwell, D. (1977). Marijuana and memory impairment: Effect on free recall and recognition memory. *Pharmacology Biochemistry and Behavior*, *7*, 99–103.

Morgan, C. J., Freeman, T. P., Powell, J., & Curran, H. V. (2016). AKT1 genotype moderates the acute psychotomimetic effects of naturalistically smoked cannabis in young cannabis smokers. *Translational Psychiatry*, *6*, e738.

Morrison, P. D., & Stone, J. M. (2011). Synthetic delta-9-tetrahydrocannabinol elicits schizophrenia-like negative symptoms which are distinct from sedation. *Human Psychopharmacology*, *26*, 77–80.

Morrison, P. D., Zois, V., McKeown, D. A., Lee, T. D., Holt, D. W., Powell, J. F., et al. (2009). The acute effects of synthetic intravenous Delta9-tetrahydrocannabinol on psychosis, mood and cognitive functioning. *Psychological Medicine*, *39*, 1607–1616.

Morrison, P. D., Nottage, J., Stone, J. M., Bhattacharyya, S., Tunstall, N., Brenneisen, R., et al. (2011). Disruption of frontal theta coherence by Delta(9)-tetrahydrocannabinol is associated with positive psychotic symptoms. *Neuropsychopharmacology*, *36*, 827–836.

Munoz-Arenas, G., Paz-Bermudez, F., Baez-Cordero, A., Caballero-Floran, R., Gonzalez-Hernandez, B., Floran, B., et al. (2015). Cannabinoid CB1 receptors activation and coactivation with D2 receptors modulate GABAergic neurotransmission in the globus pallidus and increase motor asymmetry. *Synapse*, *69*, 103–114.

Naatanen, R., & Alho, K. (1995). Generators of electrical and magnetic mismatch responses in humans. *Brain Topography*, *7*, 315–320.

Naef, M., Russmann, S., Petersen-Felix, S., & Brenneisen, R. (2004). Development and pharmacokinetic characterization of pulmonal and intravenous delta-9-tetrahydrocannabinol (THC) in humans. *Journal of Pharmaceutical Sciences*, *93*, 1176–1184.

Niedermeyer, E., & da Silva, F. L. (2005). *Electroencephalography: Basic principles, clinical applications, and related fields*. Philadelphia, PA: Lippincott Williams & Wilkins.

Ohlsson, A., Lindgren, J. E., Wahlen, A., Agurell, S., Hollister, L. E., & Gillespie, H. K. (1980). Plasma delta-9 tetrahydrocannabinol concentrations and clinical effects after oral and intravenous administration and smoking. *Clinical Pharmacology and Therapeutics*, *28*, 409–416.

Ohlsson, A., Lindgren, J. E., Wahlen, A., Agurell, S., Hollister, L. E., & Gillespie, H. K. (1982). Single dose kinetics of deuterium labelled delta 1-tetrahydrocannabinol in heavy and light cannabis users. *Biomedical Mass Spectrometry*, *9*, 6–10.

Papaleo, F., Crawley, J. N., Song, J., Lipska, B. K., Pickel, J., Weinberger, D. R., et al. (2008). Genetic dissection of the role of catechol-O-methyltransferase in cognition and stress reactivity in mice. *Journal of Neuroscience*, *28*, 8709–8723.

Pelayo-Teran, J. M., Perez-Iglesias, R., Mata, I., Carrasco-Marin, E., Vazquez-Barquero, J. L., & Crespo-Facorro, B. (2010). Catechol-O-Methyltransferase (COMT) Val158Met variations and cannabis use in first-episode non-affective psychosis: Clinical-onset implications. *Psychiatry Research*, *179*, 291–296.

Perez-Reyes, M., Timmons, M. C., Lipton, M. A., Davis, K. H., & Wall, M. E. (1972). Intravenous injection in man of 9-tetrahydrocannabinol and 11-OH-9-tetrahydrocannabinol. *Science*, *177*, 633–635.

Perez-Reyes, M., Timmons, M. C., Lipton, M. A., Christensen, H. D., Davis, K. H., & Wall, M. E. (1973). A comparison of the pharmacological activity of delta 9-tetrahydrocannabinol and its monohydroxylated metabolites in man. *Experientia*, *29*, 1009–1010.

Phan, K. L., Angstadt, M., Golden, J., Onyewuenyi, I., Popovska, A., & De Wit, H. (2008). Cannabinoid modulation of amygdala reactivity to social signals of threat in humans. *Journal of Neuroscience*, *28*(10), 2313–2319.

Plasse, T. F., Gorter, R. W., Krasnow, S. H., Lane, M., Shepard, K. V., & Wadleigh, R. G. (1991). Recent clinical experience with dronabinol. *Pharmacology, Biochemistry and Behavior*, *40*, 695–700.

Polich, J., & Criado, J. R. (2006). Neuropsychology and neuropharmacology of P3a and P3b. *International Journal of Psychophysiology*, *60*, 172–185.

Radhakrishnan, R., Skosnik, P. D., Cortes-Briones, J., Sewell, R. A., Carbuto, M., Schnakenberg, A., et al. (2015). GABA deficits enhance the psychotomimetic effects of Delta9-THC. *Neuropsychopharmacology*, *40*, 2047–2056.

Ramaekers, J. G., Kauert, G., Theunissen, E., Toennes, S. W., & Moeller, M. (2009). Neurocognitive performance during acute THC intoxication in heavy and occasional cannabis users. *Journal of Psychopharmacology*, *23*(3), 266–277.

Ranganathan, M., & D'Souza, D. C. (2006a). The acute effects of cannabinoids on memory in humans: A review. *Psychopharmacology*, *188*, 425–444.

Ranganathan, M., & D'Souza, D. (2006b). The acute effects of cannabinoids on memory in humans: A review. *Psychopharmacology*, *188*, 425–444.

Ranganathan, M., Schnakenberg, A., Skosnik, P. D., Cohen, B. M., Pittman, B., Sewell, R. A., et al. (2012). Dose-related behavioral, subjective, endocrine, and psychophysiological effects of the kappa opioid agonist salvinorin A in humans. *Biological Psychiatry*, *72*, 871–879.

Ranganathan, M., Sewell, R. A., Carbuto, M., Elander, J., Schnakenberg, A., Radhakrishnan, R., et al. (2014). Effects of Δ9-tetrahydrocannabinol in individuals with a familial vulnerability to alcoholism. *Psychopharmacology*, *231*, 2385–2393.

Rinne, T., Alho, K., Ilmoniemi, R. J., Virtanen, J., & Naatanen, R. (2000). Separate time behaviors of the temporal and frontal mismatch negativity sources. *NeuroImage*, *12*, 14–19.

Roser, P., Juckel, G., Rentzsch, J., Nadulski, T., Gallinat, J., & Stadelmann, A. M. (2008). Effects of acute oral Delta9-tetrahydrocannabinol and standardized cannabis extract on the auditory P300 event-related potential in healthy volunteers. *European Neuropsychopharmacology: The Journal of the European College of Neuropsychopharmacology*, *18*, 569–577.

Rubia, K., Russell, T., Bullmore, E. T., Soni, W., Brammer, M. J., Simmons, A., et al. (2001). An fMRI study of reduced left prefrontal activation in schizophrenia during normal inhibitory function. *Schizophrenia Research*, *52*(1), 47–55.

Schofield, D., Tennant, C., Nash, L., Degenhardt, L., Cornish, A., Hobbs, C., et al. (2006). Reasons for cannabis use in psychosis. *Australian and New Zealand Journal of Psychiatry*, *40*, 570–574.

Schulze, G. E., McMillan, D., Bailey, J., Scallet, A., Ali, S., Slikker, W., et al. (1988). Acute effects of delta-9-tetrahydrocannabinol in rhesus monkeys as measured by performance in a battery of complex operant tests. *Journal of Pharmacology and Experimental Therapeutics*, *245*, 178–186.

Schwarcz, G., Karajgi, B., & McCarthy, R. (2009). Synthetic Δ-9-tetrahydrocannabinol (dronabinol) can improve the symptoms of schizophrenia. *Journal of Clinical Psychopharmacology*, *29*, 255–258.

Seok, J.-H., Park, H.-J., Chun, J.-W., Lee, S.-K., Cho, H. S., Kwon, J. S., et al. (2007). White matter abnormalities associated with auditory hallucinations in schizophrenia: A combined study of voxel-based analyses of diffusion tensor imaging and structural magnetic resonance imaging. *Psychiatry Research: Neuroimaging*, *156*, 93–104.

Sewell, R. A., Schnakenberg, A., Elander, J., Radhakrishnan, R., Williams, A., Skosnik, P. D., et al. (2013a). Acute effects of THC on time perception in frequent and infrequent cannabis users. *Psychopharmacology*, *226*, 401–413.

Sewell, R. A., Schnakenberg, A., Elander, J., Radhakrishnan, R., Williams, A., Skosnik, P. D., et al. (2013b). Acute effects of THC on time perception in frequent and infrequent cannabis users. *Psychopharmacology*, *226*, 401–413.

Sherif, M., Radhakrishnan, R., D'Souza, D. C., & Ranganathan, M. (2016). Human laboratory studies on cannabinoids and psychosis. *Biological Psychiatry*, *79*, 526–538.

Singer, W. (1999). Neuronal synchrony: a versatile code for the definition of relations? *Neuron*, *24*, 49–65 111–125.

Solowij, N., & Michie, P. T. (2007). Cannabis and cognitive dysfunction: Parallels with endophenotypes of schizophrenia? *Journal of Psychiatry and Neuroscience*, *32*, 30–52.

Spencer, K. M., Salisbury, D. F., Shenton, M. E., & McCarley, R. W. (2008a). Gamma-band auditory steady-state responses are impaired in first episode psychosis. *Biological Psychiatry*, *64*, 369–375.

Spencer, K. M., Salisbury, D. F., Shenton, M. E., & McCarley, R. W. (2008b). γ-Band auditory steady-state responses are impaired in first episode psychosis. *Biological Psychiatry*, *64*, 369–375.

Stadelmann, A. M., Juckel, G., Arning, L., Gallinat, J., Epplen, J. T., & Roser, P. (2011). Association between a cannabinoid receptor gene (CNR1) polymorphism and cannabinoid-induced alterations of the auditory event-related P300 potential. *Neuroscience Letters*, *496*, 60–64.

Stephan, K. E., Friston, K. J., & Frith, C. D. (2009). Dysconnection in schizophrenia: From abnormal synaptic plasticity to failures of self-monitoring. *Schizophrenia Bulletin*, *35*, 509–527 sbn176.

Stokes, P. R., Mehta, M. A., Curran, H. V., Breen, G., & Grasby, P. M. (2009). Can recreational doses of THC produce significant dopamine release in the human striatum? *NeuroImage*, *48*, 186–190.

Stone, J. M., Morrison, P. D., Nottage, J., Bhattacharyya, S., Feilding, A., & McGuire, P. K. (2010). Delta-9-tetrahydrocannabinol disruption of time perception and of self-timed actions. *Pharmacopsychiatry*, *43*, 236–237.

Strougo, A., Zuurman, L., Roy, C., Pinquier, J. L., van Gerven, J. M., Cohen, A. F., et al. (2008). Modelling of the concentration-effect relationship of THC on central nervous system parameters and heart rate—Insight into its mechanisms of action and a tool for clinical research and development of cannabinoids. *Journal of Psychopharmacology*, *22*, 717–726.

Swendsen, J., Ben-Zeev, D., & Granholm, E. (2011). Real-time electronic ambulatory monitoring of substance use and symptom expression in schizophrenia. *American Journal of Psychiatry*, *168*, 202–209.

Takahashi, T., Cho, R. Y., Mizuno, T., Kikuchi, M., Murata, T., Takahashi, K., et al. (2010). Antipsychotics reverse abnormal EEG complexity in drug-naive schizophrenia: A multiscale entropy analysis. *NeuroImage*, *51*, 173–182.

Talkowski, M. E., Kirov, G., Bamne, M., Georgieva, L., Torres, G., Mansour, H., et al. (2008). A network of dopaminergic gene variations implicated as risk factors for schizophrenia. *Human Molecular Genetics*, *17*, 747–758.

Theunissen, E. L., Kauert, G. F., Toennes, S. W., Moeller, M. R., Sambeth, A., Blanchard, M. M., et al. (2012). Neurophysiological functioning of occasional and heavy cannabis users during THC intoxication. *Psychopharmacology*, *220*, 341–350.

Thiselton, D. L., Vladimirov, V. I., Kuo, P. H., McClay, J., Wormley, B., Fanous, A., et al. (2008). AKT1 is associated with schizophrenia across multiple symptom dimensions in the Irish study of high density schizophrenia families. *Biological Psychiatry*, *63*, 449–457.

Tinklenberg, J. R., Kopell, B. S., Melges, F. T., & Hollister, L. E. (1972). Marihuana and alcohol, time production and memory functions. *Archives of General Psychiatry*, *27*, 812–815.

Tukker, J. J., Fuentealba, P., Hartwich, K., Somogyi, P., & Klausberger, T. (2007). Cell type-specific tuning of hippocampal interneuron firing during gamma oscillations in vivo. *Journal of Neuroscience*, *27*, 8184–8189.

Turetsky, B. I., Calkins, M. E., Light, G. A., Olincy, A., Radant, A. D., & Swerdlow, N. R. (2007). Neurophysiological endophenotypes of schizophrenia: the viability of selected candidate measures. *Schizophrenia Bulletin*, *33*, 69–94.

Turner, C. E., Elsohly, M. A., & Boeren, E. G. (1980). Constituents of *Cannabis sativa* L. XVII. A review of the natural constituents. *Journal of Natural Products*, *43*, 169–234.

Tysk, L. (1983). Time estimation by healthy subjects and schizophrenic patients: A methodological study. *Perceptual and Motor Skills*, *56*, 983–988.

Uhlhaas, P. J., Haenschel, C., Nikolić, D., & Singer, W. (2008). The role of oscillations and synchrony in cortical networks and their putative relevance for the pathophysiology of schizophrenia. *Schizophrenia Bulletin*, *34*, 927–943.

van Winkel, R., van Beveren, N. J., Simons, C., Genetic Risk and Outcome of Psychosis (GROUP) Investigators (2011). AKT1 moderation of cannabis-induced cognitive alterations in psychotic disorder. *Neuropsychopharmacology*, *36*, 2529–2537.

Varela, F., Lachaux, J. P., Rodriguez, E., & Martinerie, J. (2001). The brainweb: Phase synchronization and large-scale integration. *Nature Reviews Neuroscience*, *2*, 229–239.

Wall, M. E., Sadler, B. M., Brine, D., Taylor, H., & Perez-Reyes, M. (1983). Metabolism, disposition, and kinetics of delta-9-tetrahydrocannabinol in men and women. *Clinical Pharmacology and Therapeutics*, *34*, 352–363.

Watanabe, K., Matsunaga, T., Yamamoto, I., Funae, Y., & Yoshimura, H. (1995). Involvement of CYP2C in the metabolism of cannabinoids by human hepatic microsomes from an old woman. *Biological and Pharmaceutical Bulletin*, *18*, 1138–1141.

Watanabe, K., Yamaori, S., Funahashi, T., Kimura, T., & Yamamoto, I. (2007). Cytochrome P450 enzymes involved in the metabolism of tetrahydrocannabinols and cannabinol by human hepatic microsomes. *Life Sciences*, *80*, 1415–1419.

Whitfield-Gabrieli, S., Thermenos, H. W., Milanovic, S., Tsuang, M. T., Faraone, S. V., McCarley, R. W., et al. (2009). Hyperactivity and hyperconnectivity of the default network in schizophrenia and in first-degree relatives of persons with schizophrenia. *Proceedings of the National Academy of Sciences*, *106*, 1279–1284.

Wilson, R. I., & Nicoll, R. A. (2002). Endocannabinoid signaling in the brain. *Science*, *296*, 678–682.

Winterer, G., & Weinberger, D. R. (2003). Cortical signal-to-noise ratio: Insight into the pathophysiology and genetics of schizophrenia. *Clinical Neuroscience Research*, *3*, 55–66.

Winterer, G., Coppola, R., Goldberg, T. E., Egan, M. F., Jones, D. W., Sanchez, C. E., et al. (2004). Prefrontal broadband noise, working memory, and genetic risk for schizophrenia. *American Journal of Psychiatry*, *161*, 490–500.

Winterer, G., Ziller, M., Dorn, H., Frick, K., Mulert, C., Wuebben, Y., et al. (2000). Schizophrenia: Reduced signal-to-noise ratio and impaired phase-locking during information processing. *Clinical Neurophysiology*, *111*, 837–849.

Winton-Brown, T. T., Allen, P., Bhattacharyya, S., Borgwardt, S. J., Fusar-Poli, P., Crippa, J. A., et al. (2011). Modulation of auditory and visual processing by delta-9-tetrahydrocannabinol and cannabidiol: An FMRI study. *Neuropsychopharmacology*, *36*, 1340–1348.

Yang, G. J., Murray, J. D., Repovs, G., Cole, M. W., Savic, A., Glasser, M. F., et al. (2014). Altered global brain signal in schizophrenia. *Proceedings of the National Academy of Sciences of the United States of America*, *111*, 7438–7443.

Zammit, S., Spurlock, G., Williams, H., Norton, N., Williams, N., O'Donovan, M. C., et al. (2007). Genotype effects of CHRNA7, CNR1 and COMT in schizophrenia: Interactions with tobacco and cannabis use. *British Journal of Psychiatry*, *191*, 402–407.

Zammit, S., Moore, T. H. M., Lingford-Hughes, A., Barnes, T. R. E., Jones, P. B., Burke, M., et al. (2008). Effects of cannabis use on outcomes of psychotic disorders: Systematic review. *British Journal of Psychiatry*, *193*, 357–363.

Chapter 5

Psychotomimetic and Cognitive Effects of Cannabis Use in the General Population

Nadia Solowij
University of Wollongong, Wollongong, NSW, Australia

INTRODUCTION

The induction of an overt schizophrenia-spectrum disorder is one of the most serious potential outcomes of cannabis exposure and one that rarely manifests among the millions of cannabis users worldwide. Less extreme but related, and also potentially debilitating outcomes that are experienced by a greater portion of cannabis users are psychotomimetic effects (defined here as subclinical psychotic-like symptoms and experiences) and cognitive impairment. Psychotic experiences that are considered diagnostically subthreshold are similar to those that define a psychotic disorder, but are experienced in an attenuated form; transiently or selectively, and as such fail to meet criteria for a diagnosis. These are in fact relatively common in the general population, particularly among adolescents, and are experienced along a continuum (Linscott & van Os, 2013). There is substantial evidence that the likelihood of experiencing subclinical psychotic-like symptoms is significantly increased in cannabis users.

Δ^9-Tetrahydrocannabinol (THC), as the primary psychoactive constituent of the cannabis plant, has "psychotomimetic" properties; that is, it induces effects that mimic aspects of psychosis. Multiple clinical and preclinical studies have demonstrated that administration of THC leads to a range of effects that resemble psychotic-like symptoms and phenotypes. Cognitive impairment is one of the most debilitating domains of dysfunction in schizophrenia, and is predictive of functional outcomes. This chapter outlines the evidence for psychotic-like symptoms and experiences and cognitive impairment in cannabis users in the general population from experimental, naturalistic, observational, and epidemiologic studies. The development of psychotic symptoms and cognitive deficits is evident following both acute exposure to cannabis, particularly at high doses, as well as with regular use in long-term or heavy users. Laboratory

The Complex Connection between Cannabis and Schizophrenia. http://dx.doi.org/10.1016/B978-0-12-804791-0.00005-7

studies of acute administration are the topic of a separate chapter in this volume (Chapter 4), as is the association between cannabis use and frank psychotic disorders (Chapter 10). The focus here is on subclinical psychotic-like symptoms and experiences, and cognitive effects.

Deliberations pertaining to causality in the association between cannabis use and schizophrenia, addressed elsewhere, are pertinent also to the development of psychotic-like symptoms and cognitive deficits, though studies are increasingly excluding the possibility of reverse causation. Self-medication hypotheses posit that individuals with psychosis or prodromal symptoms start using cannabis as a means of coping with or relieving aspects of their symptoms (or medication side effects). Recent studies have tended to examine cannabis use that specifically preceded any observable symptom development or treatment initiation to dispel the possible explanation of reverse causation. Studies of cognitive effects in cannabis users require similar attention with regard to temporal sequencing.

PSYCHOTIC-LIKE SYMPTOMS AND EXPERIENCES IN CANNABIS USERS IN THE GENERAL POPULATION

That cannabis use can induce transient psychotic-like symptoms has long been recognized (Moreau, 1845; Baudelaire, 1860). A prescient early study gave healthy volunteers (medical staff) a single acute exposure to cannabis and reported the induction of psychotic-like symptoms that included perceptual alterations/hallucinations, thought disorder, paranoid ideation, and delusions (Ames, 1958). Following the escalation of recreational cannabis use that occurred in the 1960s and 1970s, a series of clinical observations, case reports, and case–control studies were published from samples in India, Egypt, the Caribbean, Greece, South Africa, the United States, and the United Kingdom (Solowij, 1998). The notion of a specific cannabis psychosis was posited (Tunving, 1985) with ensuing debate over whether this constituted a short-lasting toxic psychosis versus a precipitation of a longer-lasting functional psychosis—a debate that continues to this day with mixed evidence (Hall & Degenhardt, 2004; Fiorentini et al., 2011; Baldacchino et al., 2012).

The first modern epidemiological study linking cannabis use with the development of psychotic disorders, specifically schizophrenia (Andreasson, Allebeck, Engstrom, & Rydberg, 1987), received some skepticism from the scientific and medical communities. However, it was subsequently substantiated by the authors with follow-up and reanalysis of the sample (Zammit, Allebeck, Andreasson, Lundberg, & Lewis, 2002), as well as significant supportive evidence from a succession of further epidemiological studies demonstrating associations with both overt disorders and psychotic-like experiences. Prospective epidemiologic studies provide the strongest evidence of association between cannabis use and the experience of subclinical psychotic-like symptoms, and these studies are highlighted below. In addition, a range of cross-sectional epidemiologic studies and small naturalistic, observational, or empirical studies of cannabis users from the general population have provided evidence of elevated symptoms and associated phenomena (e.g., impaired cognition) and are also reviewed here.

Schizotypal Traits

Studies on the association between cannabis use and schizotypy are reviewed in more detail in Chapter 6. A brief overview is provided here given that some schizotypal traits can be seen as psychotic-like experiences. At one end of a potential continuum of psychotic-like experiences in the community are schizotypal traits. Such traits (often measured using the Schizotypal Personality Questionnaire; SPQ) are similar to those of schizophrenia but in an attenuated form, and trait schizotypy is thought to interact with environmental and genetic factors to confer risk for schizophrenia (Nelson, Seal, Pantelis, & Phillips, 2013). Schizotypal traits have been found to increase following acute exposure to cannabis (Mass, Bardong, Kindl, & Dahme, 2001), while in the unintoxicated state, schizotypal traits are conceptualized as reflecting psychosis proneness. High baseline schizotypy increases acute psychotomimetic responses to cannabis and the aftereffects of intoxication (Barkus, Stirling, Hopkins, & Lewis, 2006; Barkus & Lewis, 2008; Stirling et al., 2008; Mason et al., 2009). In studies of psychosis-relevant cognitive deficits and electroencephalogram measures of neural synchronization in cannabis users, Skosnik and colleagues (2001, 2006, 2008) found that frequent cannabis use, greater duration of use, and neuroticism (Fridberg, Vollmer, O'Donnell, & Skosnik, 2011) were associated with high positive schizotypy subscale scores on the SPQ. Social anxiety and depression have also been shown to moderate the relationship between schizotypy and frequent cannabis use, with high social anxiety, in particular, predictive of infrequent cannabis use in low schizotypes, but in high schizotypes high social anxiety was associated with frequent cannabis use and more cannabis-related problems (Najolia, Buckner, & Cohen, 2012). Cannabis users had higher SPQ and Magical Ideation Scale scores from the Chapman Psychosis Proneness Scale in one large undergraduate sample (Dumas et al., 2002), and in another, disorganized schizotypy (odd thinking and behavior) was associated with greater use of cannabis, alcohol, and tobacco, though only cannabis use was associated with elevated cognitive-perceptual (positive) schizotypy (Esterberg, Goulding, McClure-Tone, & Compton, 2009). Nunn and colleagues (2001) reported higher positive schizotypal traits, as well as delusional ideation from the Peters et al. Delusion Inventory (PDI), in cannabis users compared to nonusers or alcohol-only users. Greater interpersonal (negative) schizotypy was associated with younger age of onset of cannabis use in a community sample (Compton, Chien, & Bollini, 2009), whereas some other studies have reported lower negative schizotypal traits in cannabis users (Cohen, Buckner, Najolia, & Stewart, 2011; Skosnik et al., 2008) or no difference related to cannabis use on this dimension (Fridberg et al., 2011).

Szoke et al. (2014) conducted a meta-analysis of cross-sectional studies of schizotypal features among cannabis users and found that use of cannabis at least once was associated with higher schizotypy scores across positive, negative, and disorganized dimensions, with small to medium effect sizes. A dose-response relationship was found for total and positive schizotypy scores in the majority of studies, and positive and disorganized schizotypy scores were higher in current users than in ever users. Most recently, Spriggens and Hides (2015) reported that both lifetime cannabis use and schizotypy scores predicted psychotic-like experiences measured by the positive scale of the Community Assessment of Psychic Experiences (CAPE), and trait hopelessness was a moderator of the association between recent cannabis use and psychotic experiences; those with high trait hopelessness who had used within days experienced a significantly greater number of psychotic-like experiences than those with average and low levels of trait hopelessness.

In a population study, Davis and coworkers (2013) assessed the prevalence of individual features of schizotypal personality disorder and found that these increased with the severity of cannabis use, from use through abuse to dependence, with a dose-response relationship for all disorder features. The authors highlighted features most strongly associated with the intensity of cannabis use—paranoid ideation or suspiciousness, inappropriate or constricted affect, and odd or eccentric behavior or appearance—though high adjusted odds ratios (ORs), particularly in association with cannabis dependence, were also reported for ideas of reference; odd beliefs, thinking, and speech; and excessive social anxiety. Participants were instructed not to report symptoms they had experienced only when using drugs. Like lifetime use, the same pattern of results was observed among individuals with past-year cannabis use. A previous study (Anglin et al., 2012) used a prospective cohort and showed that cannabis use prior to age 14 strongly predicted schizotypal personality disorder features in adulthood after controlling for a range of confounds, including preexisting symptoms.

van Winkel et al. (2015) showed that cannabis use in unaffected siblings of probands with schizophrenia increased latent psychosis vulnerability expression in terms of positive schizotypal similarity between the sibling and the patient, and the sibling and a parent, but parental schizotypy did not predict cannabis use in offspring. This supports a gene-by-environment interaction in that cannabis use moderates familial liability by shifting the expression of psychotic-like experiences toward psychosis. Previous work from this group, among others, had shown a greater sensitivity to the effects of cannabis among unaffected siblings of patients with schizophrenia compared with controls (GROUP, 2011), as well as mediation by AKT1 genetic variation of the relationship between short- or long-term effects of cannabis use and psychosis expression (van Winkel et al., 2011).

Psychotomimetic Experiences

Another measure of psychotic-like experiences that has been applied to samples of cannabis users in several studies is the Cannabis Experiences Questionnaire

(CEQ; Barkus et al., 2006; Barkus & Lewis, 2008). The CEQ retrospectively quantifies psychotic and paranoid/dysphoric experiences following use of cannabis. This scale has been used to measure baseline psychosis proneness, alongside the SPQ and the Green Paranoia Scale in acute administration studies (Englund et al., 2013), and in association with striatal dopamine synthesis capacity (with which no relationship was found) (Bloomfield, Morgan, Egerton, et al., 2014). Higher CEQ paranoid/dysphoric and/or psychosis-like experiences in shorter-term users (5 years) compared to longer-term (26 years) users were reported in a small sample of regular users and were associated with more recent use and a younger age of onset of use (Greenwood et al., 2014). Short-term users, but not long-term users, also exhibited higher SPQ total scores and CAPE total frequency scores compared to age-matched nonuser controls. This pattern of results could potentially be explained by exposure to more potent forms of cannabis at a younger age in the short-term users than the long-term users, due to the increasing THC but decreasing cannabidiol (CBD) content of cannabis preparations in recent decades (e.g., ElSohly et al., 2016; see more on this below). However, the CEQ psychosis-like symptoms were found in the *longer*-term users to correlate with reduced mismatch negativity amplitude (a candidate schizophrenia endophenotype) (Greenwood et al., 2014), though not with P50 sensory gating deficits (Broyd et al., 2013).

Since the CEQ is a retrospective assessment of experiences after cannabis use, Mason and colleagues (2008) developed the Psychotomimetic States Inventory (PSI) to capture these experiences during intoxication in exposure studies. The PSI measures delusory thinking, perceptual distortions, cognitive disorganization, anhedonia, mania, and paranoia. All of these subscales were shown to be sensitive to the acute effects of cannabis in a naturalistic setting (Mason et al., 2008, 2009; Morgan, Schafer, Freeman, & Curran, 2010; Morgan et al., 2012). Highly psychosis-prone individuals (based on SPQ scores) showed enhanced psychotomimetic symptoms on the PSI after acute cannabis use (Mason et al., 2009).

Psychotic-Like Symptoms and Experiences Assessed in Epidemiological Studies

Table 1 provides a listing of key prospective cohort and case-control studies that have assessed the association between cannabis use and psychotic-like symptoms or experiences in the general population, along with reported ORs and 95% confidence intervals (CIs). Generalized psychosis-related symptoms have been assessed by means of select items from the Symptom Checklist 90 (SCL-90), showing an association with cannabis use after controlling for confounders and discounting reverse causation (Fergusson et al., 2003, 2005), as well as an association with adolescent-onset use in a large, community-based, longitudinal study (Rössler, Hengartner, Angst, Ajdacic-Gross, 2012). The CAPE was developed to assess the continuum of psychotic-like symptoms in the general

TABLE 1 Observational Prospective Cohort and Case-Control Studies Predicting Risk of Developing Psychotic Symptoms or Experiences in Cannabis Users

	Sample Size and Follow-up Period[a]	Cannabis Use Measure	Psychotic Symptom or Experiences Measures	Adjusted Odds Ratio (95% CI)
Tien and Anthony (1990)	2295; 1 year	Daily use (yes/no)	Interview-based experiences measure (DIS)	2.0 (1.25–3.12)[b]
van Os et al. (2002)	4045; 3 years	Any use and cumulative frequency (5 levels)	Symptom severity (BPRS)	2.76 (1.18–6.47) for any use; 6.81 (1.79–25.92) for heaviest use
Fergusson, Horwood, and Swain-Campbell (2003), Fergusson, Horwood, and Ridder (2005)	1055; 7 years	Frequency (5 levels) and dependence (CIDI)	Psychotic-like symptoms (SCL-90)	1.8 (1.2–2.6)
Stefanis et al. (2004)	3500	Lifetime frequency and age at first use	Positive and negative symptoms (CAPE)	4.3 (1.0–17.9)[c]
Ferdinand et al. (2005)	1580; 14 years	Lifetime use (at least 5 times) and age at onset	Symptoms and age at onset (CIDI)	2.81 (1.79–4.43)[d]
Henquet et al. (2005)	2437; 3.5 years	Five or more times at baseline; heaviest frequency (6 levels)	Symptoms 4 years later (CIDI)	1.67 (1.13–2.46) for any symptom; 2.23 (1.52–3.29) for at least two symptoms; 2.23 (1.30–3.84) for daily use, any symptom

Wiles et al. (2006)	1795; 18 months	Dependence (3 levels)	Self-reported symptoms (PSQ)	1.47 (0.55–3.94) for dependence
Miettunen et al., 2008	6298 adolescents	Ever use (4 levels)	Prodromal symptoms (PROD-screen)	2.23 (1.70–2.94) for ≥3 symptoms
McGrath et al. (2010)[e]	3801 (228 sibling pairs)	Duration since first use	Hallucinations (CIDI), delusions (PDI)	2.8 (1.9–4.1) for hallucinations; 4.2 (4.2–5.8) for delusions
Kuepper, van Os, Lieb, et al. (2011)	1923; 10 years	Five or more times at baseline (yes/no) and incident use over follow-up	Psychotic symptoms (CIDI)	1.9 (1.1–3.1) for incident use; 2.2 (1.2–4.2) for continued use
Schubart, van Gastel, et al. (2011)	17,698 young adults	Amount spent on cannabis per week and age at initial use	Positive, negative, depressive symptoms—top 10% (CAPE)	Use ≤12 years of age: 3.1 (2.1–4.3) for positive symptoms; 1.7 (1.1–2.5) for negative symptoms; Heavy use: 3.0 (2.4–3.6) for positive symptoms; 3.4 (2.9–4.1) for negative symptoms; 2.8 (2.3–3.3) for depressive symptoms
Zammit, Owen, Evans, Heron, and Lewis (2011)	2630 adolescents; 2 years	Use >20 times by age 14	Severity of experiences (PLIKSi) at age 16	2.5 (1.62–3.77) (unadjusted)
Rössler et al. (2012)	591 (66% high and 33% lower SCL-90 scorers at baseline); 30 years	Frequency (3 levels) in adolescence and in adulthood	Schizotypal symptoms and schizophrenia nuclear symptoms constructed from SCL-90-R	2.60 (1.59–4.23) for adolescent use and schizotypal symptom development over 30 years

Continued

TABLE 1 Observational Prospective Cohort and Case-Control Studies Predicting Risk of Developing Psychotic Symptoms or Experiences in Cannabis Users—cont'd

	Sample Size and Follow-up Period[a]	Cannabis Use Measure	Psychotic Symptom or Experiences Measures	Adjusted Odds Ratio (95% CI)
Davis et al. (2013)[e]	34,653	Lifetime use, abuse, dependence	Schizotypal symptoms (AUDADIS-IV)	Dose-response relationship for all symptoms; OR range for dependence from 3.66 (2.81–5.02) to 9.32 (5.99–14.49)
Gage et al. (2014)	1756; 2 years	Cumulative exposure by age 16 (4 levels)	Severity of experiences (4 levels) (PLIKSi) at age 18	1.12 (0.76–1.65)
Morgan et al. (2014)	1680	Ever use; past year use (yes/no)	Experiences (PSQ)	2.47 (1.73–3.53) for past-year use
Bechtold, Hipwell, Lewis, Loeber, and Pardini (2016)	1009 boys; followed from ages 13 to 18	Current weekly use and years of weekly use	Symptoms (Youth Self Report psychosis-like items)	Paranoia: 2.04 (1.29–3.23) for current weekly use; 4.96 (2.14–11.50) for ≥2 years weekly use; Hallucinations: 3.64 (1.07–12.38) for ≥2 years weekly use

AUDADIS-IV, Alcohol Use Disorder and Associated Disabilities Interview Schedule, DSM-IV Version; BPRS, Brief Psychiatric Rating Scale; CIDI, Composite International Diagnostic Interview; DIS, Diagnostic Interview Schedule; PDI, Peters et al. Delusions Inventory; PLIKSi, Psychosis-like Symptoms interview; PSQ, Psychosis Screening Questionnaire; SCL-90, Symptom Checklist 90; SCL-90-R, Symptom Checklist 90–Revised.

[a] *Follow-up period for longitudinal studies.*

[b] *Relative risk.*

[c] *Reported in Henquet et al. (2005) Schizophr Bull, 31(3): 608–612.*

[d] *Hazard ratio (unadjusted).*

[e] *Also assessed formal diagnoses (e.g., non-affective psychosis, schizophreniform disorder, schizotypal personality disorder), not reported here.*

population. Significant elevations in scores have been shown with cannabis use across positive, negative, and/or depressive dimensions in two population studies (Stefanis et al., 2004; Schubart, van Gastel, et al., 2011), as well as in several human laboratory studies (e.g., Broyd et al., 2013; Greenwood et al., 2014). In a large, cross sectional, community sample of adolescents, lifetime use and recent frequency of use were associated with perceptual abnormalities (auditory and visual hallucinations) measured by the CAPE positive symptom dimension (Hides et al., 2009).

Symptoms typical of the prodromal phase of a psychotic disorder were assessed by Miettunen et al. (2008) in adolescents. These included feelings that something strange or inexplicable is taking place in oneself or in the environment, feelings that one is being followed or influenced in some special way, experiences of thoughts running wild, or difficulty in controlling the speed of thoughts. Lifetime use of cannabis was associated with an approximately twofold increased risk of experiencing three or more such symptoms; the association remained after controlling for a range of potential confounders (including tobacco and other substance use, parental substance use disorder, childhood behavioral/emotional problems, and social class), and a dose-response relationship was observed (from once to ≥5 times using cannabis).

More overt psychotic-like symptoms have been measured using a range of clinical structured interviews and remote survey question methods. These include the Diagnostic Interview Schedule (DIS; Tien & Anthony, 1990), the Brief Psychiatric Rating Scale (BPRS; van Os et al., 2002), the Psychosis Screening Questionnaire (PSQ; Wiles et al., 2006), the Psychosis-like Symptoms interview (PLIKSi; Zammit et al., 2011; Gage et al., 2014), the Alcohol Use Disorder and Associated Disabilities Interview Schedule, DSM-IV Version (AUDADIS-IV; Davis et al., 2013), and the PDI (McGrath et al., 2010). A number of studies have used items from the Composite International Diagnostic Interview (CIDI; Ferdinand et al., 2005; Henquet et al., 2005; McGrath et al., 2010; Kuepper, van Os, Lieb, et al., 2011). Items assessed or endorsed span both positive and negative psychotic domains, including visual and auditory hallucinations, delusions (of being spied on, persecution, thoughts being read, reference, control, grandiosity), thought interference (broadcasting, insertion, withdrawal), unusual thought content, conceptual disorganization, hypomania, paranoia, and strange experiences. The majority of studies have focused on positive psychotic symptoms.

Most recently, Bechtold and colleagues (2016) used five items from the Youth Self-Report to assess feelings in the past year of paranoia, hallucinations, and bizarre thinking, rated as sometimes or very true, in a large sample of adolescent boys in each year from ages 13 to 18, along with the number of days that they used cannabis. Current weekly use and prior years of weekly use were significantly associated with paranoia and subclinical hallucinations, but not bizarre thinking, which was nevertheless prevalent in the sample. With modeling, the authors showed that for each additional year of weekly cannabis use, the odds of experiencing paranoia rose by 133%, and hallucinations by 92%; furthermore, total symptoms, paranoia, and hallucinations persisted after a year of abstinence from cannabis. There was no evidence for reverse causation. Follow-up assessments in young adulthood revealed that 2.3% of the sample had transitioned to a psychotic disorder (diagnosed using the DIS between ages 20 and 36) (Bechtold et al., 2016).

Overall Evidence for Associations, Potential Confounders, and Interactions

The cumulative evidence suggests that cannabis use increases the risk of developing psychotic symptoms, may at least double the risk of developing schizophrenia, and that essentially, the greater the extent of use, the greater the risk. There is little evidence for reverse causation from studies that have addressed the issue (e.g., Fergusson et al., 2003, 2005; Kuepper, van Os, & Henquet, 2011, Kuepper, van Os, Lieb, et al., 2011; Bechtold et al., 2016; but not Ferdinand et al., 2005). A meta-analysis conducted in 2007 (Moore et al., 2007) reported a 40% increase in risk of any psychotic outcome in those who had ever used cannabis, with psychotic outcome defined as symptoms/experiences or diagnosed disorder (adjusted OR: 1.41, 95% CI: 1.20–1.65). A dose-response relationship was found, with increased risk (50%–200% across the studies included) in the most frequent users (adjusted OR: 2.09, 95% CI: 1.54–2.84). The most recent meta-analysis conducted by Marconi, Di Forti, Lewis, Murray, and Vassos (2016) examined cannabis use as a continuum prior to psychosis onset (to exclude reverse causation), also demonstrating a dose-response relationship with an OR of 3.90 (95% CI: 2.84–5.34) for risk of schizophrenia or psychosis-related outcomes among the heaviest cannabis users. Differentiating psychotic symptoms versus disorder diagnosis as outcomes, the OR was 3.59 (2.42–5.32) for the former and 5.07 (3.62–7.09) for the latter. The fact that disorder diagnosis showed a stronger association with degree of cannabis use than simply experiencing psychotic symptoms indicates that the latter association is unlikely to be contaminated by experiences during intoxication (see further below). Outcomes were similar across cross-sectional and prospective cohort studies, and no increase in risk was found by date of study, suggesting that previous arguments of increased risk being attributable to increasing cannabis potency (see discussion of Di Forti et al., 2014, Di Forti, Marconi, Carra, et al., 2015, below) were not substantiated in this meta-analysis.

Many epidemiological studies have provided evidence that an association between cannabis use and psychotic experiences remains after adjustment for numerous potential confounders (e.g., Fergusson et al., 2003, 2005; Gage et al., 2014; Schubart, van Gastel, et al., 2011; Bechtold et al., 2016), although adjustment for confounders has attenuated (e.g., Wiles et al., 2006) or even eliminated the association in some instances (e.g., Rössler et al., 2012). Tobacco use and other illicit drug use are examples of significant confounders that weaken the association between cannabis use and psychosis, although Schubart, van Gastel, and colleagues (2011) found that adjusting for other drug use only strengthened the odds for the association between top 10% total CAPE score and heavy cannabis use (OR: 14.35, 95% CI: 3.3–61.6) and only slightly diminished the association with early age at initiation of cannabis use and positive symptoms (≤12 years: OR: 2.3, 95% CI: 0.6–8.7; compare to the OR in Table 1). Gage et al. (2014) raised concerns about adjusting for these highly correlated measures and the inherent difficulty in teasing out confounding versus mediating effects. Specificity of cannabis use and not substance use in general, or alcohol use, was demonstrated by Verdoux, Sorbara, Gindre, Swendsen, and van Os (2002) in reporting an association between frequency of cannabis use and CAPE positive and negative dimension symptoms in a large sample of females, and was also shown in a meta-analysis of studies examining the association between cannabis use and an earlier age at onset of psychotic disorders (Large, Sharma, Compton, Slade, & Nielssen, 2011). In a population study of broader mental health problems associated with cannabis use during secondary school years, van Gastel, Tempelaar, et al. (2013) reported a past-month cannabis use association with clinically relevant scores on the Strengths and Difficulties Questionnaire, with an OR of 4.46 (95% CI: 3.46–5.76) that reduced to trend level after adjusting for other risk factors for poor psychosocial functioning. A number of these (e.g., alcohol, tobacco, and other drug use; truancy and frequent illness absences; poor performance at school) were associated with cannabis use. In another study, van Gastel, MacCabe, et al. (2013) showed that tobacco use and cannabis use were equally strongly associated with the frequency of CAPE psychotic-like experiences, and suggested that the association with cannabis was confounded by tobacco use. Gage, Hickman, and Zammit (2016) discussed the likely effect of ongoing or unmeasured residual confounding leading to an overestimation of the association between cannabis use and psychosis, since regular cannabis users and those at risk of developing a mental illness share many characteristics, which some argue is sufficient to explain the association without resorting to a causal interpretation (Ksir & Hart, 2016). However, neither genetic nor shared environment effects could explain the association found between the duration of cannabis use and psychotic symptoms in a cross-sectional sibling pair study (McGrath et al., 2010).

There is evidence that cannabis use interacts with childhood adversity to increase the likelihood of psychotic experiences. A recent study demonstrated a synergistic relationship between childhood adversity (physical or sexual abuse) and cannabis use in the past year, increasing the odds of psychotic experiences from 2.47 (95% CI: 1.73–3.53) to 5.54 (3.33–9.21) (Morgan et al., 2014). Prior studies had reported ORs for the combination of childhood adversity and cannabis use ranging from 6.93 to 30.5, demonstrating a significantly greater degree of risk than for either risk factor alone (Houston, Murphy, Adamson, Sringer, & Shevlin, 2008; Harley et al., 2010; Konings et al., 2012).

A number of factors may influence underestimation of the association between cannabis use and psychotic disorder risk (Gage et al., 2016), which apply equally to the association between cannabis use and psychotic symptoms or experiences. These include misclassification, selection bias, and study attrition. Gage et al. (2016) note that cognitive impairment in cannabis users (see further below) or in those prone to psychotic outcomes could affect estimates of the extent of their use of cannabis, and the accuracy of self-report together with the lack of standardized measures of cannabis use (Solowij, Lorenzetti, & Yücel, 2016) is an ongoing problem for research in this area. Due to the high lipid solubility of THC, residual cannabinoids are often present for extended periods of time in body fat stores of heavy users, which may result in a state of subacute "chronic intoxication." As such, the psychotic-like symptoms experienced by cannabis users could be conceptualized to potentially reflect low-level intoxication or subacute effects (Gage et al., 2016). As indicated above, there is substantial heterogeneity in the means by which psychotic symptoms or experiences have been assessed, and some studies concede that the methods do not allow for symptoms of cannabis intoxication to be identified separately from nonintoxication experiences (e.g., Zammit et al., 2011), while other studies specifically instruct participants not to report symptoms they had experienced only when using drugs, and these tended to find a dose-response relationship with symptoms (Davis et al., 2013). That the association between cannabis use and symptom experience is unlikely to be contaminated by intoxication effects is supported by the degree of cannabis use showing a stronger association with psychotic disorder diagnosis (Marconi et al., 2016).

Assessment of former users, abstinent for a prolonged period, provides another means of dissociating subacute intoxication from the experience of psychotic symptoms. Kuepper, van Os, Lieb, et al. (2011) found only weak evidence for persistence of symptoms in the substantially reduced subsample of former users from their epidemiologic cohort, whereas, worryingly, Bechtold et al. (2016) showed that symptoms persisted despite a year of abstinence from cannabis. The Kuepper, van Os, Lieb, et al., 2011 study demonstrates that emergent cannabis use over a 10-year period increased the risk of later incident psychotic symptoms, and continued use of cannabis increased the persistence of symptoms, which is important given that the persistence of symptoms increases the risk of later psychotic disorder onset (Kuepper, van Os, & Henquet, 2011; Shevlin, McElroy, Bentall, Reininghaus, & Murphy, 2017). That only 2.3% of the Bechtold et al. (2016) sample transitioned to a psychotic disorder indicates that we still have much to understand with regard to vulnerability.

Vulnerabilities and Potential Mechanisms

Identifying vulnerability factors that increase the risk of psychotic outcomes after cannabis use is a high priority. Verdoux, Gindre, Sorbara, Tournier, and Swendsen (2003), in an experience sampling study of daily life, showed that cannabis-induced psychotic experiences were much greater in those who showed psychosis liability as measured by the CAPE. Henquet et al. (2005)

reported that the 3.5-year risk of developing psychotic-like symptoms in young cannabis users increased from 21% to 51% among those with a predisposition for psychosis measured by SCL-90-R psychoticism and paranoid ideation subscales.

Familial liability is a clear candidate for vulnerability research; estimates of up to a 20% risk elevation have been posited for developing a psychotic disorder in regular cannabis users with a first-degree relative with a psychotic disorder (Gage et al., 2016). There is evidence for a shared genetic etiology for psychosis and for cannabis use, but this only explains a small proportion of the variance in the association between them (Power et al., 2014). In a community-based twin sample of 4830 16-year-old pairs, Shakoor and colleagues (2015) determined that cannabis use explains 2%–5% of the variance in positive, cognitive, and negative psychotic-like experiences measured using the Specific Psychotic Experiences Questionnaire, and that their co-occurrence is due to environmental factors and correlates common to both cannabis use and psychotic experiences, and only in part heritable factors explaining each individually. The latter explained all of the covariation between cannabis use and paranoia, cognitive disorganization, and parent-rated negative symptoms, but the relationship between cannabis use and hallucinations was associated with familial influences. The polygenic risk score for schizophrenia was reported to be unrelated to cannabis use or the potency of cannabis used (Di Forti, Vassos, Lynskey, Morgan, & Murray, 2015). Variation in a range of individual candidate genes has been implicated in potentially modulating the psychotic-like effects of cannabis—largely genes involved in dopamine neurotransmission (e.g., COMT, DRD2, AKT1, Morgan, Freeman, Powell, & Curran, 2016)—but findings have either failed or remain to be replicated. Reduced striatal dopamine synthesis capacity in cannabis users has been associated with apathy (Bloomfield, Morgan, Kapur, Curran, & Howes, 2014) but not with cannabis-induced positive psychotic symptoms (Bloomfield, Morgan, Egerton, et al., 2014). Alterations in the endogenous cannabinoid system—CB_1 receptors and anandamide, in particular—are also implicated in psychotogenic mechanisms (Leweke et al., 2007; Morgan et al., 2013; see also Curran et al., 2016).

O'Tuathaigh and colleagues (2014) reviewed preclinical mechanistic studies identifying genetic and biological underpinnings of the emergence of psychotic phenomena with cannabis exposure (as best as can be modeled in animals), but highlight the lack of molecular, cellular, and physiological identifiers of neuronal alterations. A neurotransmitter systems functionality study by this group (Behan et al., 2012) showed a COMT genotype interaction for THC effects in adolescent mice on dopaminergic cell size in the ventral tegmental area and parvalbumin cell size in prefrontal cortex, but not cell density or CB_1 protein expression in the hippocampus.

Suggestions for future research in reports of epidemiological studies have continued to highlight the need for specific studies to investigate the mechanisms by which cannabis use causes psychotic phenomena, including identification of risk modification by genetic or other factors, cannabis use parameters, and particularly vulnerable groups, as well as further studies of the long-term effects of cannabis use on neuropsychological domains relevant to psychotic states (Arseneault, Cannon, Witton, & Murray, 2004; Moore et al., 2007; Marconi et al., 2016). Furthermore, understanding the biological underpinnings of the relationship between cannabis use and the development of psychotic-like symptoms may inform the development of more effective and novel interventions (Anglin et al., 2012). A range of different methods are being pursued to achieve these research goals, including ongoing preclinical studies, human studies of acute cannabinoid administration, brain structural and functional imaging, and cognitive studies. Associations between smaller hippocampal volumes and greater cumulative exposure to cannabis in long-term, heavy cannabis users, and higher subclinical positive psychotic-like symptoms assessed by means of the Scale for the Assessment of Positive Symptoms, have been demonstrated (Yücel et al., 2008). Confusing anomalous perceptual experiences and negative affect have been shown to explain THC-induced paranoia in vulnerable individuals (selected for high paranoid ideation) (Freeman et al., 2015).

Effects of Proportional Exposure to THC Versus CBD

CBD has antipsychotic properties (Leweke et al., 2012) and has been shown to modulate the psychotomimetic effects of THC in laboratory administration studies (see Chapter 4). Morgan and Curran (2008) analyzed the association between CBD content in the hair of a community sample of cannabis users and psychotic symptoms measured by the Oxford Liverpool Inventory of Life Experiences (OLIFE) and the PDI. Compared to those who had CBD detected in hair and those with no cannabinoids detected, those with only THC detected in hair showed higher scores on the PDI and on the OLIFE unusual experiences factor, which encompasses positive symptoms, including hallucinations and delusions. Schubart, Sommer, et al. (2011) have estimated relative exposure to CBD from cannabis preparation preferences for typical use of 1877 participants in a web-based, cross-sectional study. They reported an inverse relationship between CBD content and CAPE positive symptoms specifically, but the effect size was small and total CAPE score as well as positive, negative, and depressive symptom scores increased with frequency of cannabis use regardless of CBD content. The internet provides opportunities to assess large samples and is increasingly being used in a range of studies. Moritz, van Quaquebeke, Lincoln, Köther, and Andreou (2013) warn, however, of potential overestimation of psychotic-like experiences in internet-based studies; they have demonstrated that simulators (clinical experts and students asked to simulate a person with schizophrenia) have reported inflated positive symptom scores (relative to schizophrenia patients). Di Forti et al. (2014) showed that the risk of psychosis associated with use of high-THC-potency cannabis is much higher than the risk associated with the use of hashish, which contains similar proportions of THC to CBD. THC potency was a critical factor in the earlier induction of psychosis by 6 years relative to never users, and Di Forti, Marconi, Carra, and colleagues (2015) went on to show that potency may be more important than frequency of exposure.

Neuroprotective properties have also been purported for CBD, and Yücel et al. (2016) provided evidence in a small sample of cannabis users of protection from hippocampal volume loss by exposure to CBD; similar findings were also reported by Demirakca et al. (2011). There remains much to be understood with regard to the potential protective effects of CBD in terms of psychotic phenomena (see Chapter 14). Not all experimental studies have shown that CBD ameliorates the psychotomimetic properties of THC acutely (e.g., Morgan et al., 2010), and effects may be dose-dependent. Currently, chronic exposure to low-level CBD in recreational cannabis users in the general community is thought to possibly mitigate psychotic-like consequences in the longer-term (Morgan et al., 2012).

COGNITIVE EFFECTS OF CANNABIS USE ON USERS IN THE GENERAL POPULATION

Although psychotic-like experiences in cannabis users have tended to predominate research in recent years, interest in the cognitive effects of cannabis use has also been increasing, in part as a means of further elucidating related mechanisms by which cannabis may induce psychosis, and also in order to attempt to understand vulnerability, symptom persistence, and recovery. Several reviews of acute and chronic effects of cannabis use on cognition have recently been published (Volkow, Baler, Compton, & Weiss, 2014; Volkow et al., 2016), and two were particularly thorough (Broyd, van Hell, Beale, Yücel, & Solowij, 2016; Curran et al., 2016). This section will therefore not elaborate extensively on the evidence for or nature of cognitive effects and the reader is referred to those reviews. Instead, a brief summary of what is known about acute and chronic effects, and recovery from those effects, is provided, followed by a focus on their similarity to the cognitive deficits associated with psychosis.

There is fairly consistent evidence across studies that cannabis intoxication impairs verbal learning and episodic memory (largely the encoding of new information), working memory, attention (task- and dose-dependent), psychomotor function, and inhibition; several executive functions (such as decision-making and planning ability) have also been shown to be consistently impaired following acute exposure (Broyd et al., 2016). Tolerance to some of these effects is evident with blunted—but not absent—deficits in frequent users (Broyd et al., 2016). With long-term and generally heavy use of cannabis, impaired verbal learning and memory—and impaired attention and attentional bias—are the most consistently observed cognitive effects (Broyd et al., 2016). Mixed evidence has been found for executive function and decision-making, as well as psychomotor function. This evidence comes from studies that aimed to assess cannabis users in the unintoxicated state, but the length of abstinence imposed or requested of users ranged from several hours to several weeks, so impaired cognitive function in cannabis users in these studies could at least partially reflect chronic intoxication or the effect of drug residues. However, many studies report associations between a range of cannabis use parameters—including frequency of use, duration of use, and age at initiation of use—aiding interpretation of shorter- versus longer-lasting effects. Mostly, the cognition-impairing effects of chronic cannabis use have held after controlling for other substance use, but not in all studies (see Broyd et al., 2016). The facts that the cognitive deficits observed are associated with both acute and chronic exposure to cannabis, and that the findings are supported by preclinical research (see Curran et al., 2016), adds weight to arguments for causation.

Recovery of cognitive function remains underinvestigated and hence contentious. Cross-sectional studies that imposed several weeks of abstinence from cannabis, as well as a few longitudinal studies, have provided mixed evidence for the persistence of deficits versus recovery in different cognitive domains and in varying samples comprising adolescents and adults. The evidence suggests likely persistent effects of cannabis on attention and psychomotor function, and possibly on learning and memory, although the latter have appeared to recover in some studies (see Broyd et al., 2016; Curran et al., 2016). Downregulation of cannabinoid receptors following chronic exposure has been shown to reverse after two days to four weeks of abstinence (Sim-Selley, 2003; Hirvonen et al., 2012; D'Souza et al., 2016), and preliminary evidence for recovery of hippocampal volume and integrity was provided by a recent small-sample, cross-sectional study (Yücel et al., 2016). Downstream effects may take substantial time to recover but ultimately are expected to, given the plasticity of the human brain.

Impaired cognition during acute intoxication and for days to weeks after use is likely to impact educational outcomes, other functioning in daily life, and cumulatively, life goals achievement (Volkow et al., 2014). An earlier age at onset of cannabis use during adolescence is often reported to be associated with worse cognitive outcomes, findings supported by neuropsychological and functional imaging studies, as well as by preclinical research (Broyd et al., 2016; Curran et al., 2016). However, evidence is confounded by other substance use and mental health issues that co-occur with adolescent cannabis use (see Lorenzetti, Alonso-Lana, et al., 2016). In a prospective study of a population cohort followed to age 38, persistent cannabis use or dependence was associated with an intelligence quotient (IQ) decline of about 6 points (or approximately 8 points when dependence commenced in adolescence), reflecting a broad decline in functioning not specific to any particular cognitive domain, and cessation of use did not restore IQ in adolescent-onset users (Meier et al., 2012). However, Curran et al. (2016) discuss evidence from recent studies that did not replicate these findings and raise questions regarding confounding and causality.

The mechanisms underlying cognitive impairment in cannabis users remain to be elucidated. There is preclinical and growing human evidence that THC is responsible for much of the harm to the brain, while CBD may potentially protect against cognitive impairment induced by THC (particularly memory impairment) in both acute administration and naturalistic studies, similar to the findings described above for psychotic-like symptoms (e.g., Morgan et al., 2010; Englund et al., 2013). Greater cumulative dose of cannabis exposure, as well as an earlier age at onset of cannabis use, has been found to be associated with brain structural alterations in regions dense in cannabinoid receptors (e.g., hippocampus, prefrontal cortex, amygdala, cerebellum) (Lorenzetti, Solowij, and Yücel, 2016)—regions that are involved in higher cognitive functioning, and also implicated in psychosis. There is preliminary evidence that CBD may protect against these neuroanatomical alterations (Demirakca et al., 2011; Yücel et al., 2016). Changes in neurotransmitter signaling, synaptic plasticity, and disrupted functional circuitry resulting from both acute and chronic exposure to THC are known to occur (Castillo, Younts, Chavez, & Hashimotodani, 2012; Fratta & Fattore, 2013; Mechoulam & Parker, 2013; Curran et al., 2016), and are posited to explain cognitive impairment and potentially psychotic-like outcomes. However, much further research is required to understand the latter, as well as substrates of recovery after cessation of cannabis use.

There is similarity between the cognitive impairment in cannabis users and that observed in schizophrenia, albeit the impairment is to a lesser degree in cannabis users (Solowij & Michie, 2007). Fig. 1 shows proportions of MATRICS (Measurement and Treatment Research to Improve Cognition in Schizophrenia) Consensus Cognitive Battery impaired cognitive domains in current and abstinent cannabis users in studies conducted since 2004 (as reviewed by Broyd et al., 2016), weighted by the number of studies that assessed each domain and the total sample size across those studies. Verbal learning and memory are clearly the most consistently impaired domain. Differences in the relative weight of evidence for working memory, attention, and processing speed between this figure and summaries of systematic reviews (e.g., Broyd et al., 2016) reflects the nature of cognitive tasks assigned to the MATRICS domains. For example, multiple tasks assigned to the processing speed domain within MATRICS were considered as attention (e.g., rapid visual information processing, tracking, trail making) or psychomotor (e.g., digit-symbol substitution, grooved pegboard, finger tapping) tasks by Broyd et al. (2016). A recent study by Mollon and colleagues (2016) has revealed that the neuropsychological profile of a sample of individuals with psychotic experiences differed from that typical of schizophrenia, and highlighted deficits in processing speed as being the most predictive of transition to a disorder. They reported that past-year cannabis use was significantly correlated with both psychotic experiences and poorer neuropsychological functioning, and that cannabis use could be a mediator or moderator of poorer neurocognitive functioning in those who experience subthreshold psychotic-like symptoms. Further studies of processing speed in cannabis users may be informative in this regard. Social cognition is another MATRICS domain that has been underinvestigated in cannabis users, despite some evidence for affect recognition deficits in heavy users (Platt, Kamboj, Morgan, & Curran, 2010). Social cognitive

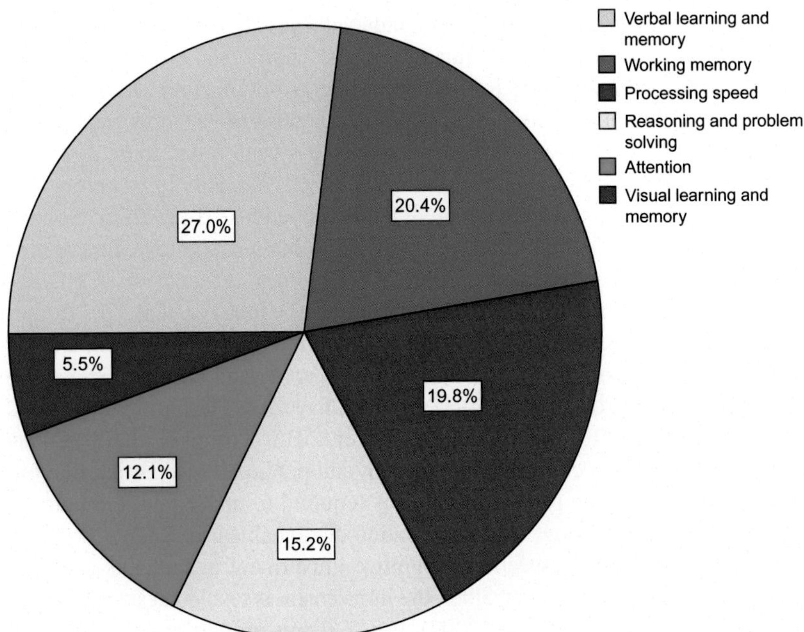

FIG. 1 Proportion of MATRICS Consensus Cognitive Battery cognitive domains in which cannabis users show impaired performance (adjusted by number of studies assessing each domain and sample size).

deficits are established in schizophrenia, and there is increasing evidence of their prevalence in association with psychotic experiences, with resulting effects on functioning and quality of life. However, Arnold, Allott, Farhall, Killackey, and Cotton (2015) have reported that faster processing speed and better social cognition predicted recency and frequency of cannabis use in a sample of people with first-episode psychosis. Counterintuitively, better cognitive functioning in people with schizophrenia who use cannabis than in nonusers may reflect a need for better social skills in order to procure drugs, or a neurocognitively less vulnerable sample who only developed psychosis after early exposure to cannabis. Further discussion of this issue is beyond the scope of this chapter (see the meta-analysis by Yücel et al., 2012).

NEEDED RESEARCH

Given rapid movement toward legalization of cannabis for medicinal and recreational purposes, understanding the level of use that may result in adverse consequences and the implications of the ratio of THC-to-CBD content for both immediate and long-term effects is imperative in order to minimize harm. Sex differences in the effects of cannabis remain largely underinvestigated with regard to both cognitive and psychotomimetic outcomes. Evidence for sex differences in cannabinoid metabolism, action, and brain morphology in cannabis users (Medina et al., 2009; Fattore & Fratta, 2010; McQueeny et al., 2011) suggests that sex effects may be important in understanding the outcomes. Differential pubertal changes and brain maturation in males and females during adolescence through young adulthood may impact understandings of differential effects on age at onset (Curran et al., 2016), and these require further elucidation. Cannabis effects interacting with normal cognitive decline that occurs with aging, as well as in a range of vulnerable populations including those with neurological or other medical conditions, are also prime areas for further research. Of note, additional research may identify and differentiate therapeutic and beneficial properties of cannabinoids from the harms.

Greater standardization of cannabis use metrics is a necessary step for further research in this area (Solowij et al., 2016), along with biomarkers such as blood, urine, and hair cannabinoid levels to support self-report (Curran et al., 2016; Lorenzetti, Solowij, et al., 2016), and to clarify the specific parameters of cannabis exposure that confer the greatest risk for the development of adverse consequences such as psychotic-like phenomena and cognitive deficits (Volkow et al., 2016). An interesting potential line of research may be to examine the interrelatedness of psychotomimetic and cognitive outcomes in cannabis users; that is, whether psychotic-like phenomena and cognitive impairment co-occur, or whether their dissociation may further inform mechanisms and vulnerability or resilience. Further examination of additive, interactive, or synergistic effects of cannabinoids with other substances, and a range of other psychological and environmental vulnerability factors, is required. The effects of cannabis use disorders are increasingly recognized as being important to understand in addition to the effects of exposure (Lorenzetti, Cousijn, et al., 2016); dependence was the criterion for some studies of psychotomimetic effects (e.g., Fergusson et al., 2003, 2005; Wiles et al., 2006; Davis et al., 2013) or cognitive effects (e.g., Field, 2005; Gonzalez et al., 2012; Cousijn et al., 2013), but most have focused on exposure parameters.

Cognitive changes associated with exposure, including altered reward, salience, impulsivity, and learning and memory, are relevant to the maintenance of addictive behaviors, may interact with genetic vulnerability to increase the risk of developing a disorder (Curran et al., 2016), and this transition may be underpinned by specific neural adaptations (Lorenzetti, Cousijn, et al., 2016). The characterization of neural, behavioral, genetic, and epigenetic factors that confer risk to developing psychotic-like symptoms or cognitive deficits following exposure to cannabis, and individual differences in vulnerability or resilience, is key to furthering understanding in this area. As Volkow et al. (2016) highlighted, the effects of cannabis exposure may be modest overall in the general population and the extent of their expression depends upon the presence of predispositional, perhaps genetic, and other environmental risk factors. Data sharing within large consortia (e.g., ENIGMA; Hibar et al., 2015; Mackey et al., 2016) may go a long way toward achieving some of these goals.

Importantly, additional well-controlled prospective studies are required that monitor the development of cognitive impairment and restoration of cognition/brain function and structure from before cannabis use through current use (e.g., US National Institutes of Health-funded Adolescent Brain Cognition Development (ABCD) Study), and ultimately through cessation of use and prolonged abstinence. A combination of both epidemiological studies and smaller laboratory-based observational studies, together with preclinical research, may inform mechanisms, which would hold the most promise for understanding the psychotomimetic and cognitive effects of cannabis use.

Key Chapter Points

- Subclinical psychotic-like symptoms and experiences are prevalent among cannabis users in the general population. Positive symptoms are most frequently described; negative and disorganized symptoms are reported in some studies.
- Psychotic-like symptoms emerge following acute exposure to cannabis and are also evident in chronic cannabis users and associated most often with frequent, high-THC-potency exposure, and in some studies more often with recent use than lifetime use, as well as with early-onset use during adolescence.
- Both acute exposure and chronic cannabis use induce cognitive impairments. The extent to which these recover following cessation of chronic use remains underinvestigated and hence uncertain.
- Regarding cannabis-associated cognitive effects, impairments in the following domains has been reported: learning and memory; working memory; attention, processing speed, and psychomotor functioning; and executive functioning.
- Elucidating vulnerability factors for the development of psychotic-like phenomena and cognitive impairment, and the mechanisms by which cannabis exposure results in these outcomes (and their neural underpinnings), will require further research.

REFERENCES

Ames, F. (1958). A clinical and metabolic study of acute intoxication with Cannabis sativa and its role in the model psychoses. *The Journal of Mental Science*, *104*, 972–999.

Andreasson, S., Allebeck, P., Engstrom, A., & Rydberg, U. (1987). Cannabis and schizophrenia. *Lancet*, *2*, 1483–1486.

Anglin, D. M., Corcoran, C. M., Brown, A. S., Chen, H., Lighty, Q., Brook, J. S., et al. (2012). Early cannabis use and schizotypal personality disorder symptoms from adolescence to middle adulthood. *Schizophrenia Research*, *137*, 45–49.

Arnold, C., Allott, K., Farhall, J., Killackey, E., & Cotton, S. (2015). Neurocognitive and social cognitive predictors of cannabis use in first-episode psychosis. *Schizophrenia Research*, *168*, 231–237.

Arseneault, L., Cannon, M., Witton, J., & Murray, R. M. (2004). Causal association between cannabis and psychosis: Examination of the evidence. *The British Journal of Psychiatry*, *184*, 110–117.

Baldacchino, A., Hughes, Z., Kehoe, M., Blair, H., Teh, Y., Windeatt, S., et al. (2012). Cannabis psychosis: Examining the evidence for a distinctive psychopathology in a systematic and narrative review. *The American Journal on Addictions*, *21*, S88–S98.

Barkus, E., & Lewis, S. (2008). Schizotypy and psychosis-like experiences from recreational cannabis in a non-clinical sample. *Psychological Medicine*, *38*, 1267–1276.

Barkus, E. J., Stirling, J., Hopkins, R. S., & Lewis, S. (2006). Cannabis-induced psychosis-like experiences are associated with high schizotypy. *Psychopathology*, *39*, 175–178.

Baudelaire, C. (1860). *Les Paradis Artificiel*. Paris: Poulet-Malassis.

Bechtold, J., Hipwell, A., Lewis, D. A., Loeber, R., & Pardini, D. (2016). Concurrent and sustained cumulative effects of adolescent marijuana use on subclinical psychotic symptoms. *The American Journal of Psychiatry*, *173*, 781–789.

Behan, A. T., Hryniewiecka, M., O'Tuathaigh, C. M. P., Kinsella, A., Cannon, M., Karayiorgou, M., et al. (2012). Chronic adolescent exposure to delta-9-tetrahydrocannabinol in COMT mutant mice: Impact on indices of dopaminergic, endocannabinoid and GABAergic pathways. *Neuropsychopharmacology*, *37*, 1773–1783.

Bloomfield, M. A. P., Morgan, C. J. A., Egerton, A., Kapur, S., Curran, H. V., & Howes, O. D. (2014). Dopaminergic function in cannabis users and its relationship to cannabis-induced psychotic symptoms. *Biological Psychiatry*, *75*, 470–478.

Bloomfield, M. A. P., Morgan, C. J. A., Kapur, S., Curran, H. V., & Howes, O. D. (2014). The link between dopamine function and apathy in cannabis users: An [^{18}F]-DOPA PET imaging study. *Psychopharmacology*, *231*, 2251–2259.

Broyd, S. J., Greenwood, L.-m., Croft, R. J., Dalecki, A., Todd, J., Michie, P. T., et al. (2013). Chronic effects of cannabis on sensory gating. *International Journal of Psychophysiology*, *89*, 381–389.

Broyd, S. J., van Hell, H. H., Beale, C., Yücel, M., & Solowij, N. (2016). Acute and chronic effects of cannabinoids on cognition—A systematic review. *Biological Psychiatry*, *79*, 557–567.

Castillo, P. E., Younts, T. J., Chavez, A. E., & Hashimotodani, Y. (2012). Endocannabinoid signaling and synaptic function. *Neuron*, *76*, 487–492.

Cohen, A. S., Buckner, J. D., Najolia, G. M., & Stewart, D. W. (2011). Cannabis and psychometrically-defined schizotypy: Use, problems and treatment considerations. *Journal of Psychiatric Research*, *45*, 548–554.

Compton, M. T., Chien, V. H., & Bollini, A. M. (2009). Associations between past alcohol, cannabis, and cocaine use and current schizotypy among first-degree relatives of patients with schizophrenia and non-psychiatric controls. *The Psychiatric Quarterly*, *80*, 143–154.

Cousijn, J., Watson, P., Koenders, L., Vingerhoets, W. A., Goudriaan, A. E., & Wiers, R. W. (2013). Cannabis dependence, cognitive control and attentional bias for cannabis words. *Addictive Behaviors*, *38*, 2825–2832.

Curran, H. V., Freeman, T. P., Mokrysz, C., Lewis, D. A., Morgan, C. J. A., & Parsons, L. H. (2016). Keep off the grass? Cannabis, cognition and addiction. *Nature Reviews Neuroscience*, *17*, 293–306.

D'Souza, D. C., Cortes-Briones, J. A., Ranganathan, M., Thurnauer, H., Creatura, G., Surti, T., et al. (2016). Rapid changes in CB1 receptor availability in cannabis dependent males after abstinence from cannabis. *Biological Psychiatry: Cognitive Neuroscience and Neuroimaging*, *1*, 60–67.

Davis, G. P., Compton, M. T., Wang, S., Levin, F. R., & Blanco, C. (2013). Association between cannabis use, psychosis, and schizotypal personality disorder: Findings from the national epidemiologic survey on alcohol and related conditions. *Schizophrenia Research*, *151*, 197–202.

Demirakca, T., Sartorius, A., Ende, G., Meyer, N., Welzel, H., Skopp, G., et al. (2011). Diminished gray matter in the hippocampus of cannabis users: Possible protective effects of cannabidiol. *Drug and Alcohol Dependence*, *114*, 242–245.

Di Forti, M., Marconi, A., Carra, E., Fraietta, S., Trotta, A., Bonomo, M., et al. (2015). Proportion of patients with first-episode psychosis attributable to use of high potency cannabis: The example of south London. *The Lancet Psychiatry*, *2*, 233–238.

Di Forti, M., Sallis, H., Allegri, F., Trotta, A., Ferraro, L., Stilo, S. A., et al. (2014). Daily use, especially of high-potency cannabis, drives the earlier onset of psychosis in cannabis users. *Schizophrenia Bulletin*, *40*, 1509–1517.

Di Forti, M., Vassos, E., Lynskey, M., Morgan, C., & Murray, R. M. (2015). Cannabis and psychosis: Authors' reply. *The Lancet Psychiatry*, *2*, 382.

Dumas, P., Saoud, M., Bouafia, S., Gutknecht, C., Ecochard, R., Daléry, J., et al. (2002). Cannabis use correlates with schizotypal personality traits in healthy students. *Psychiatry Research*, *19*, 27–35.

ElSohly, M. A., Mehmedic, Z., Foster, S., Gon, C., Chandra, S., & Church, J. C. (2016). Changes in cannabis potency over the last two decades (1995-2014)—Analysis of current data in the United States. *Biological Psychiatry*, *79*, 613–619.

Englund, A., Morrison, P. D., Nottage, J., Hague, D., Kane, F., Bonaccorso, S., et al. (2013). Cannabidiol inhibits THC-elicited paranoid symptoms and hippocampal-dependent memory impairment. *Journal of Psychopharmacology*, *27*, 19–27.

Esterberg, M. L., Goulding, S. M., McClure-Tone, E. B., & Compton, M. T. (2009). Schizotypy and nicotine, alcohol, and cannabis use in a non-psychiatric sample. *Addictive Behaviors*, *34*, 374–379.

Fattore, L., & Fratta, W. (2010). How important are sex differences in cannabinoid action? *British Journal of Pharmacology*, *160*, 544–548.

Ferdinand, R. F., Sondeijker, F., van der Ende, J., Selten, J. P., Huizink, A., & Verhulst, F. C. (2005). Cannabis use predicts future psychotic symptoms, and vice versa. *Addiction*, *100*, 612–618.

Fergusson, D. M., Horwood, L. J., & Ridder, E. M. (2005). Tests of causal linkages between cannabis use and psychotic symptoms. *Addiction*, *100*, 354–366.

Fergusson, D. M., Horwood, L. J., & Swain-Campbell, N. R. (2003). Cannabis dependence and psychotic symptoms in young people. *Psychological Medicine*, *33*, 15–21.

Field, M. (2005). Cannabis 'dependence' and attentional bias for cannabis-related words. *Behavioural Pharmacology*, *16*, 473–476.

Fiorentini, A., Volonteri, L. S., Dragogna, F., Rovera, C., Maffini, M., Mauri, M. C., et al. (2011). Substance-induced psychoses: A critical review of the literature. *Current Drug Abuse Reviews*, *4*, 228–240.

Fratta, W., & Fattore, L. (2013). Molecular mechanisms of cannabinoid addiction. *Current Opinion in Neurobiology*, *23*, 70–81.

Freeman, D., Dunn, G., Murray, R. M., Evans, N., Lister, R., Antley, A., et al. (2015). How cannabis causes paranoia: Using the intravenous administration of Δ^9-tetrahydrocannabinol (THC) to identify key cognitive mechanisms leading to paranoia. *Schizophrenia Bulletin*, *41*, 391–399.

Fridberg, D. J., Vollmer, J. M., O'Donnell, B. F., & Skosnik, P. D. (2011). Cannabis users differ from non-users on measures of personality and schizotypy. *Psychiatry Research*, *186*, 46–52.

Gage, S. H., Hickman, M., Heron, J., Munafo, M., Lewis, G., Macleod, J., et al. (2014). Associations of cannabis and cigarette use with psychotic experiences at age 18: Findings from the Avon longitudinal study of parents and children. *Psychological Medicine*, *44*, 3435–3444.

Gage, S. H., Hickman, M., & Zammit, S. (2016). Association between cannabis and psychosis: Epidemiologic evidence. *Biological Psychiatry*, *79*, 549–556.

Genetic Risk and Outcome of Psychosis (GROUP) Investigators. (2011). Evidence that familial liability for psychosis is expressed as differential sensitivity to cannabis: An analysis of patient-sibling and sibling-control pairs. *Archives of General Psychiatry, 68*, 138–147.

Gonzalez, R., Schuster, R. M., Mermelstein, R. J., Vassileva, J., Martin, E. M., & Diviak, K. R. (2012). Performance of young adult cannabis users on neurocognitive measures of impulsive behavior and their relationship to symptoms of cannabis use disorders. *Journal of Clinical and Experimental Neuropsychology, 34*, 962–976.

Greenwood, L.-m., Broyd, S. J., Croft, R., Todd, J., Michie, P. T., Johnstone, S., et al. (2014). Chronic effects of cannabis use on the auditory mismatch negativity. *Biological Psychiatry, 75*, 449–458.

Hall, W., & Degenhardt, L. (2004). Is there a specific 'cannabis psychosis'? In D. Castle & R. Murray (Eds.), *Marijuana and madness* (pp. 89–100). Cambridge: Cambridge University Press.

Harley, M., Kelleher, I., Clarke, M., Lynch, F., Arseneault, L., Connor, D., et al. (2010). Cannabis use and childhood trauma interact additively to increase the risk of psychotic symptoms in adolescence. *Psychological Medicine, 40*, 1627–1634.

Henquet, C., Krabbendam, L., Spauwen, J., Kaplan, C., Lieb, R., Wittchen, H. U., et al. (2005). Prospective cohort study of cannabis use, predisposition for psychosis, and psychotic symptoms in young people. *British Medical Journal, 330*, 11.

Hibar, D. P., Stein, J. L., Renteria, M. E., Arias-Vasquez, A., Desrivieres, S., Jahanshad, N., et al. (2015). Common genetic variants influence human subcortical brain structures. *Nature, 520*, 224–229.

Hides, L., Lubman, D. I., Buckby, J., Yuen, H. P., Cosgrave, E., Baker, K., et al. (2009). The association between early cannabis use and psychotic-like experiences in a community adolescent sample. *Schizophrenia Research, 112*, 130–135.

Hirvonen, J., Goodwin, R. S., Li, C.-T., Terry, G. E., Zoghbi, S. S., Morse, C., et al. (2012). Reversible and regionally selective downregulation of brain cannabinoid CB1 receptors in chronic daily cannabis smokers. *Molecular Psychiatry, 17*, 642–649.

Houston, J. E., Murphy, J., Adamson, G., Sringer, M., & Shevlin, M. (2008). Childhood sexual abuse, early cannabis use, and psychosis: Testing an interaction model based on the National Comorbidity Survey. *Schizophrenia Bulletin, 34*, 580–585.

Konings, M., Stefanis, N., Kuepper, R., de Graaf, R., ten Have, M., van Os, J., et al. (2012). Replication in two independent population-based samples that childhood maltreatment and cannabis use synergistically impact on psychosis risk. *Psychological Medicine, 42*, 149–159.

Ksir, C., & Hart, C. L. (2016). Cannabis and psychosis: A critical overview of the relationship. *Current Psychiatry Reports, 18*, 12.

Kuepper, R., van Os, J., & Henquet, C. (2011). Cannabis use and psychosis: Author's reply. *British Medical Journal, 342*, d1973.

Kuepper, R., van Os, J., Lieb, R., Wittchen, H. U., Hofler, M., & Henquet, C. (2011). Continued cannabis use and risk of incidence and persistence of psychotic symptoms: 10 year follow-up cohort study. *British Medical Journal, 342*, d738.

Large, M., Sharma, S., Compton, M. T., Slade, T., & Nielssen, O. (2011). Cannabis use and earlier onset of psychosis: A systematic meta-analysis. *Archives of General Psychiatry, 68*, 555–561.

Leweke, F. M., Giuffrida, A., Koethe, D., Schreiber, D., Nolden, B. M., Kranaster, L., et al. (2007). Anandamide levels in cerebrospinal fluid of first episode schizophrenic patients: Impact of cannabis use. *Schizophrenia Research, 94*, 29–36.

Leweke, F. M., Piomelli, D., Pahlisch, F., Muhl, D., Gerth, C. W., Hoyer, C., et al. (2012). Cannabidiol enhances anandamide signaling and alleviates psychotic symptoms of schizophrenia. *Translational Psychiatry, 2*, e94.

Linscott, R. J., & van Os, J. (2013). An updated and conservative systematic review and meta-analysis of epidemiological evidence on psychotic experiences in children and adults: On the pathway from proneness to persistence to dimensional expression across mental disorders. *Psychological Medicine, 43*, 1133–1149.

Lorenzetti, V., Alonso-Lana, S., Youssef, G. J., Verdejo-Garcia, A., Suo, C., Cousijn, J., et al. (2016). Adolescent cannabis use: What is the evidence for functional brain alteration? *Current Pharmaceutical Design, 22*, 6353–6365.

Lorenzetti, V., Cousijn, J., Solowij, N., Garavan, H., Suo, C., Yücel, M., et al. (2016). Cannabis use disorders: A call for evidence. *Frontiers in Behavioral Neuroscience, 10*, 86.

Lorenzetti, V., Solowij, N., & Yücel, M. (2016). The role of cannabinoids in neuroanatomic alterations in cannabis users. *Biological Psychiatry, 79*, e17–e31.

Mackey, S., Algaier, N., Chaarani, B., Spechler, P., Orr, C., Bunn, J., et al. (2016). Genetic imaging consortium for addiction medicine: From neuroimaging to genes. *Progress in Brain Research, 224*, 203–223.

Marconi, A., Di Forti, M., Lewis, C. M., Murray, R. M., & Vassos, E. (2016). Meta-analysis of the association between the level of cannabis use and risk of psychosis. *Schizophrenia Bulletin, 42*, 1262–1269.

Mason, O., Morgan, C. J., Dhiman, S. K., Patel, A., Parti, N., Patel, A., et al. (2009). Acute cannabis use causes increased psychotomimetic experiences in individuals prone to psychosis. *Psychological Medicine, 39*, 951–956.

Mason, O. J., Morgan, C. J. M., Stefanovic, A., & Curran, H. V. (2008). The psychotomimetic states inventory (PSI): Measuring psychotic-type experiences from ketamine and cannabis. *Schizophrenia Research, 103*, 138–142.

Mass, R., Bardong, C., Kindl, K., & Dahme, B. (2001). Relationship between cannabis use, schizotypal traits, and cognitive function in healthy subjects. *Psychopathology, 34*, 209–214.

McGrath, J., Welham, J., Scott, J., Varghese, D., Degenhardt, L., Hayatbakhsh, M. R., et al. (2010). Association between cannabis use and psychosis-related outcomes using sibling pair analysis in a cohort of young adults. *Archives of General Psychiatry, 67*, 440–447.

McQueeny, T., Padula, C. B., Price, J., Medina, K. L., Logan, P., & Tapert, S. F. (2011). Gender effects on amygdala morphometry in adolescent marijuana users. *Behavioural Brain Research, 224*, 128–134.

Mechoulam, R., & Parker, L. A. (2013). The endocannabinoid system and the brain. *Annual Review of Psychology, 64*, 21–47.

Medina, K. L., McQueeny, T., Nagel, B. J., Hanson, K. L., Yang, T. T., & Tapert, S. F. (2009). Prefrontal cortex morphometry in abstinent adolescent marijuana users: Subtle gender effects. *Addiction Biology, 14*, 457–468.

Meier, M. H., Caspi, A., Ambler, A., Harrington, H., Houts, R., Keefe, R. S. E., et al. (2012). Persistent cannabis users show neuropsychological decline from childhood to midlife. *Proceedings of the National Academy of Sciences of the United States of America, 109*, E2657–E2664.

Miettunen, J., Tormanen, S., Murray, G. K., Jones, P. B., Maki, P., Ebeling, H., et al. (2008). Association of cannabis use with prodromal symptoms of psychosis in adolescence. *The British Journal of Psychiatry, 192*, 470–471.

Mollon, J., David, A. S., Morgan, C., Frissa, S., Glahn, D., Pilecka, I., et al. (2016). Psychotic experiences and neuropsychological functioning in a population-based sample. *JAMA Psychiatry, 73*, 129–138.

Moore, T. H., Zammit, S., Lingford-Hughes, A., Barnes, T. R. E., Jones, P. B., Burke, M., et al. (2007). Cannabis use and risk of psychotic or affective mental health outcomes: A systematic review. *Lancet, 370*, 319–328.

Moreau, J.-J. (1845). *Du Haschisch et de l'Alienation Mentale*. Paris: Masson.

Morgan, C. J. A., & Curran, H. V. (2008). Effects of cannabidiol on schizophrenia-like symptoms in people who use cannabis. *The British Journal of Psychiatry, 192*, 306–307.

Morgan, C. J. A., Freeman, T. P., Powell, J., & Curran, H. V. (2016). AKT1 genotype moderates the acute psychotomimetic effects of naturalistically smoked cannabis in young cannabis smokers. *Translational Psychiatry, 6*, e738.

Morgan, C. J. A., Gardener, C., Schafer, G., Swan, S., Demarchi, C., Freeman, T. P., et al. (2012). Subchronic impact of cannabinoids in street cannabis on cognition, psychotic-like symptoms and psychological well-being. *Psychological Medicine, 42*, 391–400.

Morgan, C. J. A., Page, E., Schaefer, C., Chatten, K., Manocha, A., Gulati, S., et al. (2013). Cerebrospinal fluid anandamide levels, cannabis use and psychotic-like symptoms. *The British Journal of Psychiatry, 202*, 381–382.

Morgan, C., Reininghaus, U., Reichenberg, A., Frissa, S., SELCoH study team, Hotopf, M., et al. (2014). Adversity, cannabis use and psychotic experiences: evidence of cumulative and synergistic effects. *The British Journal of Psychiatry, 204*, 346–353.

Morgan, C. J. A., Schafer, G., Freeman, T. P., & Curran, H. V. (2010). Impact of cannabidiol on the acute memory and psychotomimetic effects of smoked cannabis: Naturalistic study. *The British Journal of Psychiatry, 197*, 285–290.

Moritz, S., van Quaquebeke, N., Lincoln, T. M., Köther, U., & Andreou, C. (2013). Can we trust the internet to measure psychotic symptoms? *Schizophrenia Research and Treatment*, 457010.

Najolia, G. M., Buckner, J. D., & Cohen, A. S. (2012). Cannabis use and schizotypy: The role of social anxiety and other negative affective states. *Psychiatry Research, 200*, 660–668.

Nelson, M. T., Seal, M. L., Pantelis, C., & Phillips, L. J. (2013). Evidence of a dimensional relationship between schizotypy and schizophrenia: A systematic review. *Neuroscience & Biobehavioral Reviews, 37*, 317–327.

Nunn, J. A., Rizza, F., & Peters, E. R. (2001). The incidence of schizotypy among cannabis and alcohol users. *The Journal of Nervous and Mental Disease, 189*, 741–748.

O'Tuathaigh, C. M. P., Gantois, I., & Waddington, J. L. (2014). Genetic dissection of the psychotomimetic effects of cannabinoid exposure. *Progress in Neuro-Psychopharmacology & Biological Psychiatry, 52*, 33–40.

Platt, B., Kamboj, S., Morgan, C. J., & Curran, H. V. (2010). Processing dynamic facial affect in frequent cannabis-users: Evidence of deficits in the speed of identifying emotional expressions. *Drug and Alcohol Dependence, 112*, 27–32.

Power, R., Verweij, K., Zuhair, M., Montgomery, G. W., Henders, A. K., Heath, A. C., et al. (2014). Genetic predisposition to schizophrenia associated with increased use of cannabis. *Molecular Psychiatry, 19*, 1201–1204.

Rössler, W., Hengartner, M. P., Angst, J., & Ajdacic-Gross, V. (2012). Linking substance use with symptoms of subclinical psychosis in a community cohort over 30 years. *Addiction, 107*, 1174–1184.

Schubart, C. D., Sommer, I. E., van Gastel, W. A., Goetgebuer, R. L., Kahn, R. S., & Boks, M. P. (2011). Cannabis with high cannabidiol content is associated with fewer psychotic experiences. *Schizophrenia Research, 130*, 216–221.

Schubart, C. D., van Gastel, W. A., Breetvelt, E. J., Beetze, S. L., Ophoff, R. A., Sommer, I. E. C., et al. (2011). Cannabis use at a young age is associated with psychotic experiences. *Psychological Medicine, 41*, 1301–1310.

Shakoor, S., Zavos, H. M. S., McGuire, P., Cardno, A. G., Freeman, D., & Ronald, A. (2015). Psychotic experiences are linked to cannabis use in adolescents in the community because of common underlying environmental risk factors. *Psychiatry Research, 227*, 144–151.

Shevlin, M., McElroy, E., Bentall, R. P., Reininghaus, U., & Murphy, J. (2017). The psychosis continuum: Testing a bifactor model of psychosis in a general population sample. *Schizophrenia Bulletin*, *43*, 133–141.

Sim-Selley, L. J. (2003). Regulation of cannabinoid CB1 receptors in the central nervous system by chronic cannabinoids. *Critical Reviews in Neurobiology*, *15*, 91–119.

Skosnik, P. D., Krishnan, G. P., Aydt, E. E., Kuhlenshmidt, H. A., & O'Donnell, B. F. (2006). Psychophysiological evidence of altered neural synchronization in cannabis use: Relationship to schizotypy. *The American Journal of Psychiatry*, *163*, 1798–1805.

Skosnik, P. D., Park, S., Dobbs, L., & Gardner, W. L. (2008). Affect processing and positive syndrome schizotypy in cannabis users. *Psychiatry Research*, *157*, 279–282.

Skosnik, P. D., Spatz-Glenn, L., & Park, S. (2001). Cannabis use is associated with schizotypy and attentional disinhibition. *Schizophrenia Research*, *48*, 83–92.

Solowij, N. (1998). *Cannabis and cognitive functioning*. Cambridge: Cambridge University Press.

Solowij, N., Lorenzetti, V., & Yücel, M. (2016). The effects of cannabis use on human behavior: A call for standardization of cannabis use metrics. *JAMA Psychiatry*, *73*, 995–996.

Solowij, N., & Michie, P. T. (2007). Cannabis and cognitive dysfunction: Parallels with endophenotypes of schizophrenia? *Journal of Psychiatry & Neuroscience*, *32*, 30–52.

Spriggens, L., & Hides, L. (2015). Patterns of cannabis use, psychotic-like experiences and personality styles in young cannabis users. *Schizophrenia Research*, *165*, 3–8.

Stefanis, N. C., Delespaul, P., Henquet, C., Bakoula, C., Stefanis, C. N., & van Os, J. (2004). Early adolescent cannabis exposure and positive and negative dimensions of psychosis. *Addiction*, *99*, 1333–1341.

Stirling, J., Barkus, E. J., Nabosi, L., Irshad, S., Roemer, G., Schreudergoidheijt, B., et al. (2008). Cannabis-induced psychotic-like experiences are predicted by high schizotypy. Confirmation of preliminary results in a large cohort. *Psychopathology*, *41*, 371–378.

Szoke, A., Galliot, A.-M., Richard, J.-R., Ferchiou, A., Baudin, G., Leboyer, M., et al. (2014). Association between cannabis use and schizotypal dimensions—A meta-analysis of cross-sectional studies. *Psychiatry Research*, *219*, 58–66.

Tien, A. Y., & Anthony, J. C. (1990). Epidemiological analysis of alcohol and drug use as risk factors for psychotic experiences. *The Journal of Nervous and Mental Disease*, *178*, 473–480.

Tunving, K. (1985). Psychiatric effects of cannabis use. *Acta Psychiatrica Scandinavica*, *72*, 209–217.

van Gastel, W. A., MacCabe, J. H., Schubart, C. D., Vreeker, A., Tempelaar, W., Kahn, R. S., et al. (2013). Cigarette smoking and cannabis use are equally strongly associated with psychotic-like experiences: A cross-sectional study in 1929 young adults. *Psychological Medicine*, *43*, 2393–2401.

van Gastel, W. A., Tempelaar, W., Bun, C., Schubart, C. D., Kahn, R. S., Plevier, C., et al. (2013). Cannabis use as an indicator of risk for mental health problems in adolescents: A population-based study at secondary schools. *Psychological Medicine*, *43*, 1849–1856.

van Os, J., Bak, M., Hanssen, M., Bijl, R. V., de Graaf, R., & Verdoux, H. (2002). Cannabis use and psychosis: A longitudinal population-based study. *American Journal of Epidemiology*, *156*, 319–327.

van Winkel, R., & Genetic Risk and Outcome of Psychosis (GROUP) Investigators. (2011). Family-based analysis of genetic variation underlying psychosis-inducing effects of cannabis: Sibling analysis and proband follow-up. *Archives of General Psychiatry*, *68*, 148–157.

van Winkel, R., & GROUP Investigators. (2015). Further evidence that cannabis moderates familial correlation of psychosis-related experiences. *PLoS ONE*, *10*, e0137625.

Verdoux, H., Gindre, C., Sorbara, F., Tournier, M., & Swendsen, J. D. (2003). Effects of cannabis and psychosis vulnerability in daily life: An experience sampling test study. *Psychological Medicine*, *33*, 23–32.

Verdoux, H., Sorbara, F., Gindre, C., Swendsen, J., & van Os, J. (2002). Cannabis use and dimensions of psychosis in a non-clinical population of female subjects. *Schizophrenia Research, 59*, 77–84.

Volkow, N. D., Baler, R. D., Compton, W. M., & Weiss, S. R. B. (2014). Adverse health effects of marijuana use. *The New England Journal of Medicine, 370*, 2219–2227.

Volkow, N. D., Swanson, J. M., Evins, E., DeLisi, L. E., Meier, M. H., Gonzalez, R., et al. (2016). Effects of cannabis use on human behavior, including cognition, motivation, and psychosis: A review. *JAMA Psychiatry, 73*, 292–297.

Wiles, N. J., Zammit, S., Bebbington, P., Singleton, N., Meltzer, H., & Lewis, G. (2006). Self-reported psychotic symptoms in the general population: Results from the longitudinal study of the British National Psychiatric Morbidity Survey. *The British Journal of Psychiatry, 188*, 519–526.

Yücel, M., Bora, E., Lubman, D. I., Solowij, N., Brewer, W. J., Cotton, S., et al. (2012). The impact of cannabis use on cognitive functioning in patients with schizophrenia: A meta-analysis of existing findings and new data in a first-episode sample. *Schizophrenia Bulletin, 38*, 316–330.

Yücel, M., Lorenzetti, V., Suo, C., Zalesky, A., Fornito, A., Takagi, M. J., et al. (2016). Hippocampal harms, protection and recovery following regular cannabis use. *Translational Psychiatry, 6*, e710.

Yücel, M., Solowij, N., Respondek, C., Whittle, S., Fornito, A., Pantelis, C., et al. (2008). Regional brain abnormalities associated with long-term heavy cannabis use. *Archives of General Psychiatry, 65*, 694–701.

Zammit, S., Allebeck, P., Andreasson, S., Lundberg, I., & Lewis, G. (2002). Self reported cannabis use as a risk factor for schizophrenia in Swedish conscripts of 1969: Historical cohort study. *British Medical Journal, 325*, 1199–1201.

Zammit, S., Owen, M., Evans, J., Heron, J., & Lewis, G. (2011). Cannabis, COMT and psychotic experiences. *The British Journal of Psychiatry, 199*, 6.

Chapter 6

The Association Between Cannabis Use and Schizotypy

Angelo B. Cedeño
Lenox Hill Hospital, New York, NY, United States

CHARACTERISTICS OF SCHIZOTYPY

As a clinical and research phenomenon, the concept of schizotypy has received much interest from researchers and clinicians hoping to elucidate the specifics of this multidimensional concept that is linked to psychosis-spectrum disorders generally and to schizophrenia in particular. Schizotypy generally refers to a set of behavioral, cognitive, perceptual, and affective traits that are found in the general population with a base rate of approximately 10% (Meehl, 1990). Paul Meehl, whose work strongly influenced the modern conceptualization of schizotypy and its relation to schizophrenia, put forth a theory of schizotaxia, schizotypy, and schizophrenia. Meehl suggested that a necessary condition for schizotypy is the concept of schizotaxia (Meehl, 1962). According to this theory, schizotaxia is an integrative neural defect that is produced by genetic mutations. Meehl indicated that this neural defect is inherited and that the interaction of this neural defect with a social learning history produces a personality organization, which he called the schizotype. This theory also suggests that the most severe schizotypes develop clinical schizophrenia, possibly due to very poor parenting, traumatic events, or other factors. Although this theory does not argue that all schizotypes develop clinical schizophrenia, it does suggest that the presence of a general schizotaxic vulnerability is necessary to develop schizotypy and, potentially, schizophrenia. Family studies show that positive and negative traits of individuals diagnosed with schizophrenia are associated with positive and negative traits in their relatives with schizotypy (Arendt, Mortense, Rosenberg, Pedersen, & Waltoft, 2008). Generally, then, schizotypy can be conceptualized as psychosis-proneness, or as a liability to schizophrenia.

Modern researchers have found support, through studies conducted in the United States and abroad, for organizing schizotypal traits according to three general factors: positive, negative, and disorganized schizotypal traits (Vollema & van den Bosch, 1995). Positive schizotypal traits include unusual perceptual

The Complex Connection between Cannabis and Schizophrenia. http://dx.doi.org/10.1016/B978-0-12-804791-0.00006-9

experiences, odd beliefs and magical thinking, odd or eccentric behavior, suspiciousness, and ideas of reference. Negative schizotypal traits include excessive social anxiety, lack of close friends, and social isolation. Disorganized schizotypal traits include odd/eccentric thinking, odd/eccentric speech, and allusive thinking (i.e., loose associative processing). Some earlier studies of schizotypy have suggested a fourth factor, impulse nonconformity, but follow-up factor analytic studies have not found strong support for this factor and the three-factor model of schizotypy has been the most commonly investigated (Claridge et al., 1996; Ettinger, Meyhöfer, Steffens, Wagner, & Koutsouleris, 2014).

Given the nature of the traits theorized to comprise the construct of schizotypy, one major clinical phenomenon that is related to schizotypy is Schizotypal Personality Disorder (STPD). As defined by the Diagnostic and Statistical Manual of Mental Disorders, Fifth Edition (DSM-5), STPD is a "pervasive pattern of social and interpersonal deficits marked by acute discomfort with, and reduced capacity for, close relationships as well as by cognitive or perceptual distortions and eccentricities of behavior, beginning by early adulthood and present in a variety of contexts" (American Psychiatric Association, 2013). In order to be diagnosed with DSM-5 STPD, at least five of the following signs/symptoms must be present: ideas of reference, strange beliefs or magical thinking, abnormal perceptual experiences, strange thinking and speech, paranoia, inappropriate or constricted affect, strange behavior or appearance, lack of close friends, and excessive social anxiety that does not abate and stems from paranoia rather than from negative judgments about oneself. In addition, the symptoms must not occur during the course of a disorder with similar symptoms, such as schizophrenia or an autism-spectrum disorder. Although the construct of schizotypy is closely related to STPD, it has been suggested that schizotypy and STPD are not isomorphic. Indeed, if schizotypy can be conceptualized as psychosis-proneness, then this proneness can express itself in a variety of ways, of which STPD and schizophrenia are just a few examples (Lenzenweger, 2015).

COMMONLY USED MEASURES OF SCHIZOTYPY

Numerous measures of schizotypy exist in the literature and many of these measures assess different aspects of this construct. Although an exhaustive list of schizotypy measures is beyond the scope of this chapter, the most commonly used measures will be presented and briefly described.

Perhaps the most often used measure of schizotypy is the Schizotypal Personality Questionnaire (SPQ; Raine, 1991), which is a 74-item questionnaire comprised of nine subscales that correspond to the DSM criteria for STPD. The SPQ's subscales generally fall along the three schizotypy dimensions mentioned earlier: cognitive-perceptual (positive) traits, interpersonal (negative) traits, and disorganized traits. A modified version of this measure, the Schizotypal Personality Questionnaire-Brief (SPQ-B; Raine & Benishay, 1995) is comprised of 22 items covering the same three dimensions.

The Perceptual Aberration Scale (PerAb; Chapman, Chapman, & Raulin, 1978) is comprised of 35 true/false items and assesses unusual perceptual or body-image experiences often reported by individuals with psychotic disorders. The main focus of the PerAb is unusual body sensations. The Physical Anhedonia Scale (PAS; Chapman, Chapman, & Raulin, 1976) is a 40-item true/false questionnaire that measures the lowered ability to experience physical pleasure (e.g., eating, sex, touching, etc.). The Social Anhedonia Scale (SAS; Chapman et al., 1976) is a similar 48-item true/false questionnaire that measures the lowered ability to experience interpersonal pleasure (talking, being with people, competing, etc.). The Magical Ideation Scale (MIS; Eckblad & Chapman, 1983) is a 30-item true/false questionnaire that assesses magical ideation, or beliefs in events that do not have a causal relationship despite evidence to the contrary.

The Oxford-Liverpool Inventory of Feelings (O-LIFE; Mason, Claridge, & Jackson, 1995) assesses four aspects of schizotypy: Unusual Experiences, Introvertive Anhedonia, Cognitive Disorganization, and Impulsive Nonconformity. The first three factors correspond to the aforementioned positive, negative, and disorganized dimensions, respectively, from the SPQ. The Impulsive Nonconformity factor is purported to measure the more antisocial aspects of schizotypy. The Peters et al. Delusions Inventory (PDI; Peters, Joseph, & Garety, 1999) measures the aspects of schizotypy that relate to delusional ideation. The questionnaire assesses distress associated with delusions, preoccupation with delusions, and conviction. The Community Assessment of Psychic Experiences (CAPE; Stefanis et al., 2002) is comprised of 42 items covering positive, negative, and depressive dimensions using 4-point Likert scales for the frequency of symptoms and distress of symptoms. The CAPE was modified based on the PDI and measures the lifetime prevalence of psychotic-like experiences in the general population.

SIGNIFICANT ASSOCIATIONS BETWEEN SCHIZOTYPY AND CANNABIS USE

Many studies have found that schizotypy and cannabis use are significantly associated (Williams, Wellman, & Rawlins, 1996). Specifically, many researchers propose that the positive and disorganized dimensions of schizotypy are strongly related to cannabis use. In addition, these studies have also examined some aspects of neurocognitive functioning.

In a study by Mass, Bardong, Kindl, and Dahme (2001) examining college students, for example, 20 current cannabis users (defined as those who reported using cannabis within the past month) were matched to 20 controls (those who reported having never used cannabis). Participants completed the PerAb, the SPQ, and neuropsychological tests of executive functioning. Results of the study indicated that current cannabis users scored higher on the PerAb and the disorganized subscale of the SPQ. Results also showed that current cannabis users had poorer executive functioning than nonusers and that poorer executive functioning was associated with increased schizotypy scores.

A similar study by Skosnik and colleagues (2001) examined associations among cannabis use, schizotypal characteristics, and some of the neurocognitive deficits that are observed in patients with schizophrenia. In this study, the SPQ, measures of spatial working memory, and measures of attentional disinhibition were administered to a sample of 40 college students and members of the community who were placed into the following groups: current cannabis use group, defined as regular use at least once weekly; past cannabis use group, defined as cannabis consumption at least once in the past; and a control group of participants with no history of cannabis use. Participants in the current cannabis use group showed increased scores on the positive schizotypal traits measured by the SPQ and did not show increased scores on the negative traits. The results also suggested that current cannabis use was associated with increased attentional disinhibition.

In a study by Nunn, Rizza, and Peters (2001) examining the relationship between schizotypal traits and cannabis use, researchers administered the O-LIFE, the PDI, and the Hospital Anxiety and Depression Scale (HADS) to a sample of 196 undergraduates. Participants were divided into four groups: a cannabis/alcohol group, defined as users who smoke two or more cannabis cigarettes weekly and who drink two or more units of alcohol weekly; a cannabis group, defined as those who smoke two or more cannabis cigarettes weekly and who drink less than two units of alcohol weekly; an alcohol group, defined as those who had not smoked cannabis before and who drink two or more units of alcohol weekly; and a no drinking/no drugs group. Results revealed that cannabis use, but not alcohol use, was associated with higher scores on delusional ideation and the unusual experiences facet of schizotypy. Cannabis and alcohol use was associated with lower scores on the introvertive anhedonia domain as compared with those who did not use alcohol or cannabis.

Dumas et al. (2002) found similar results. In this study, the SPQ, MIS, PerAb, a revised version of the PAS, and a revised version of the SAS were administered to a sample of 232 college students. Based on their answers to questions regarding cannabis use, students were placed into the following groups: never-users, defined as those who had never used cannabis; past or occasional users, defined as those who had used cannabis at least once in the past or who currently used cannabis once weekly or less; and regular users, defined as those who used cannabis twice weekly or more. Cannabis use and positive schizotypal traits were significantly associated, but cannabis use and only one negative schizotypal trait (SPQ constricted affect) were significantly associated.

Other researchers have contributed to this literature and have found support for the aforementioned findings. Schiffman, Nakamura, Earleywine, and Labrie (2005), for example, conducted a study in which the SPQ-B and a brief questionnaire assessing cannabis use were completed by 189 undergraduate students. Participants were categorized as recent cannabis users if they reported using cannabis in the past 90 days. Participants who used cannabis beyond the 90-day period or who had never used cannabis were grouped together as nonusers/occasional users. Cannabis users displayed higher scores on positive and disorganized schizotypal traits as compared with nonusers/occasional users. In addition, cannabis users displayed decreased scores on negative schizotypal traits that approached significance. A similar study by Fridberg, Vollmer, O'Donnell, and Skosnik (2011) found that cannabis users scored significantly higher on the PerAb and on the positive and disorganized scales of the SPQ as compared with nonusers; yet, there were no significant differences between groups on the negative dimension of the SPQ.

The aforementioned studies relied almost exclusively on relatively small, mostly undergraduate samples. These findings, however, have also been replicated in some larger undergraduate samples, community and clinical samples, and in some longitudinal studies. Esterberg, Goulding, McClure-Tone, and Compton (2009), for example, collected information about substance use and administered the SPQ to a nonclinical sample of 825 undergraduates. Cannabis use was significantly associated with the positive dimension of the SPQ, although alcohol and nicotine use were not.

In a study by the Genetic Risk and Outcome in Psychosis (GROUP) investigators (2011), researchers assessed cannabis use and administered the CAPE to a sample of patients with a psychotic disorder, their siblings, and community controls. Scores on the positive dimension of schizotypy were more closely associated between siblings exposed to cannabis and their patient relative as compared with siblings who had not been exposed to cannabis. There were no significant findings for the negative dimension of schizotypy. The associations between cannabis use and schizotypy were also examined in Wave 2 of the National Epidemiologic Survey on Alcohol and Related Conditions (NESARC), which is a nationally representative survey of individuals 18 years and older living in all 50 states of the US and the District of Columbia. Cannabis users were significantly more at risk of STPD than non-cannabis users and the risk of STPD increased with greater use of cannabis in a dose-dependent manner (Davis, Compton, Wang, Levin, & Blanco, 2013).

In a longitudinal study by Stefanis and colleagues (2004), the CAPE and a questionnaire assessing lifetime cannabis use were administered to a sample of 3500 19-year-olds, selected from a large, nationwide longitudinal survey. Cannabis use was significantly associated with both positive and negative dimensions of psychosis and this association was not influenced by depression. Another longitudinal study examined a cohort of children from childhood into adulthood and assessed cannabis use and the presence of STPD traits. Cannabis use before age 14 predicted symptoms of STPD in adulthood (Anglin et al., 2012); this association remained significant after controlling for adolescent STPD symptoms, anxiety, depression, cigarette use, and other drug use.

BROADENING THE SCHIZOTYPY-CANNABIS ASSOCIATION

The above-mentioned studies generally suggest that cannabis users score higher on measures of schizotypy as compared with nonusers. These studies also generally suggest that the positive and disorganized dimensions of schizotypy tend to be those most associated with cannabis use. Other studies, however, have found alternative results to challenge this view.

In a study by Verdoux and colleagues (2003), researchers administered the CAPE and a questionnaire assessing demographic information and frequency of alcohol and cannabis use to a sample of 632 undergraduate students. Greater cannabis use was associated with both positive and negative dimensions of psychotic experiences. In addition, alcohol was not significantly associated with psychotic experiences. In a similar study by Bailey and Swallow (2004), a sample of 60 undergraduate students completed the SPQ-B and a questionnaire assessing cannabis use. Participants were categorized as cannabis users if they responded "yes" to the question, "Have you smoked cannabis?" Participants could choose from the following options: daily, two/three times a week, weekly, monthly, and less than monthly. Total SPQ-B scores were significantly higher for cannabis users as compared with nonusers. Cannabis users also scored significantly higher than nonusers on all three SPQ-B subscales (i.e., positive, negative, disorganized). Similarly, Compton, Chien, and Bollini (2009) examined a sample of 60 participants who completed the SPQ and a measure of substance use. These participants were either first-degree relatives of patients with schizophrenia or schizoaffective disorder or participants who did not have a history of psychosis in first- or second-degree relatives. Individuals who had ever used cannabis displayed significantly higher total SPQ scores, as well as higher scores on the positive and negative SPQ dimensions as compared with nonusers. In addition, for individuals in the 25–29-year age range, the frequency and amount of cannabis use was significantly correlated with all four SPQ scores (i.e., total, positive, negative, and disorganized).

Barkus, Stirling, Hopkins, and Lewis (2006) conducted a study that challenged many previous findings. The SPQ-B, a questionnaire assessing frequency of cannabis use, and the Cannabis Experiences Questionnaire (CEQ) were completed by a sample of 137 undergraduate students. The CEQ was developed by the authors to explore the effects of cannabis and is comprised of three subscales: pleasurable experiences (e.g., feeling happy, feeling laid-back), psychotic-like experiences (e.g., feeling paranoid, experiencing auditory hallucinations), and after-effects (e.g., loss of drive, reduced attention). Schizotypy scores did not significantly differ between cannabis users and nonusers. Higher schizotypy scores in cannabis users were associated with a higher likelihood of experiencing psychotic-like symptoms during cannabis use and experiencing unpleasant after-effects. Results were replicated by Barkus and Lewis (2008), who found that schizotypy scores were not significantly related to frequency of cannabis use. Significantly higher scores were observed, however, on the disorganized dimension of the SPQ for those who had smoked cannabis at least once as compared with individuals who had not smoked cannabis. In another study by Stirling et al. (2008), researchers once again did not find significantly higher schizotypy scores in cannabis users as compared with nonusers. This study did reveal, however, that a greater frequency of cannabis use was significantly associated with higher schizotypy scores among cannabis users, and that high schizotypy for current cannabis users was associated with negative psychosis-like experiences (e.g., agitation, paranoia).

Another study examined the specific items in a commonly used measure of schizotypy. Earleywine (2006) compared individual items from the SPQ-B between current cannabis users (defined as those who used cannabis once weekly or more) and former users (defined as those who had used cannabis in their lifetimes but not in the past year) in a community sample of adults. SPQ-B scores were significantly higher for current cannabis users as compared with former users. Results also showed that two SPQ-B items: "I sometimes use words in unusual ways" and "Have you ever noticed a common event or object that seemed to be a special sign for you?," accounted for the difference. Removing those items eliminated the significant differences in SPQ-B scores between the two groups.

Finally, Cohen, Buckner, Najolia, and Stewart (2011) assessed frequency of cannabis use and administered the SPQ-B to a sample of 1665 undergraduate students. Users were categorized into the following schizotypy groups: schizotypy (top 5% of scores), non-schizotypy (bottom 50% of scores), and unconventional (scores within the 50th to 85th percentiles). There was a significant difference in cannabis use in the schizotypy group as compared with the other groups, with the schizotypy group endorsing cannabis use more frequently. Furthermore, when compared with the other groups, cannabis use in the schizotypy group was not associated with higher positive or disorganized traits. Results also indicated that cannabis use was significantly associated with lower scores on the negative dimension of the SPQ-B, consistent with some prior findings.

DIRECTIONS FOR FUTURE STUDY

The literature examining the associations between schizotypy and cannabis use raises some interesting questions and provides researchers in this area with many considerations to elucidate further the nature of this relationship. Taken together, there is some evidence suggesting that when compared with cannabis nonusers, cannabis users tend to display higher scores on measures of schizotypy. Specifically, cannabis users tend to score higher on dimensions assessing positive schizotypal traits and disorganized schizotypal traits. Some studies, however, have found significant differences not only in the positive and disorganized dimensions of schizotypy, but also in the negative dimension. Although there is support for the view of significant differences between users and nonusers, other studies have not found significant differences between the two groups in schizotypy scores. In addition, these studies suggest that higher disorganized traits in cannabis users might reflect aspects of cannabis use and not schizotypy per se.

Given these mixed results, certain areas of the literature are worth expanding upon. First, it is crucial for researchers to continue to further refine the definition of schizotypy and how to measure this construct. As stated previously, schizotypy is associated with a variety of behavioral, affective, cognitive, and perceptual traits that have been examined and replicated in many studies. As this construct relates to cannabis use, it is important to consider which aspects of schizotypy will be studied (i.e., positive, negative, disorganized) and the rationale for this. Studies suggesting that some of the associations previously found between schizotypy and cannabis use might be better explained by cannabis use than by schizotypy per se (Barkus & Lewis, 2008; Barkus et al., 2006) are crucial to understanding the nature of this relationship. In addition, an overwhelming number of studies examining schizotypy measure the construct by self-report measures. As Lenzenweger (2015) suggests, schizotypy can be expressed in a variety of ways and laboratory measures would add to existing self-report measures.

Second, many studies examining the relationship between schizotypy and cannabis use depend on college samples. Although such samples provide an adequate age range to examine psychosis-proneness, these samples might not be representative of the general population in regards to demographic and clinical characteristics. In keeping with the model of schizotaxia (Meehl, 1962), more studies examining schizotypy and substance use in family members of individuals diagnosed with schizophrenia and other psychosis-spectrum disorders would provide key information.

Finally, most studies examining schizotypy and cannabis use are cross-sectional in nature. One of the key questions in this literature is whether schizotypy precedes cannabis use or vice versa. Indeed, the findings are mixed in this regard and the cross-sectional nature of many of the studies makes it difficult to make causal statements about this relationship. Longitudinal studies will be instrumental in better understanding the nuances of the relationship between schizotypy and cannabis use.

REFERENCES

American Psychiatric Association. (2013). *Diagnostic and statistical manual of mental disorders* (5th ed.). Washington, DC: APA.

Anglin, D. M., Corcoran, C. M., Brown, A. S., Chen, H., Lighty, Q., Brook, J. S., & Cohen, P. R. (2012). Early cannabis use and schizotypal personality disorder symptoms from adolescence to middle adulthood. *Schizophrenia Research*, *137*, 45–49.

Arendt, M., Mortensen, P. B., Rosenberg, R., Pedersen, C. B., & Waltoft, B. L. (2008). Familial predisposition for psychiatric disorder. *Archives of General Psychiatry*, *65*, 1269–1274.

Bailey, E. L., & Swallow, B. L. (2004). The relationship between cannabis use and schizotypal symptoms. *European Psychiatry*, *19*, 113–114.

Barkus, E., & Lewis, S. (2008). Schizotypy and psychosis-like experiences from recreational cannabis in a non-clinical sample. *Psychological Medicine*, *38*, 1267–1276.

Barkus, E. J., Stirling, J., Hopkins, R. S., & Lewis, S. (2006). Cannabis-induced psychotic-like experiences are associated with high schizotypy. *Psychopathology*, *39*, 175–178.

Chapman, L. J., Chapman, J. P., & Raulin, M. L. (1976). Scales for physical and social anhedonia. *Journal of Abnormal Psychology*, *85*, 374–382.

Chapman, L. J., Chapman, J. P., & Raulin, M. L. (1978). Body-image aberration in schizophrenia. *Journal of Abnormal Psychology*, *87*, 399–407.

Claridge, G., McCreery, C., Mason, O., Bentall, R., Boyle, G., Slade, P., & Popplewell, D. (1996). The factor structure of 'schizotypal' traits: A large replication study. *British Journal of Clinical Psychology*, *35*, 103–115.

Cohen, A. S., Buckner, J. D., Najolia, G. M., & Stewart, D. W. (2011). Cannabis and psychometrically-defined schizotypy: Use, problems and treatment considerations. *Journal of Psychiatric Research*, *45*, 548–554.

Compton, M. T., Chien, V. H., & Bollini, A. M. (2009). Associations between past alcohol, cannabis, and cocaine use and current schizotypy among first-degree relatives of patients with schizophrenia and non-psychiatric controls. *Psychiatric Quarterly*, *80*, 143–154.

Davis, G. P., Compton, M. T., Wang, S., Levin, F. R., & Blanco, C. (2013). Association between cannabis use, psychosis, and schizotypal personality disorder: Findings from the National Epidemiologic Survey on Alcohol and Related Conditions. *Schizophrenia Research*, *151*, 197–202.

Dumas, P., Saoud, M., Bouafia, S., Gutknecht, C., Ecochard, R., Daléry, J., Rochet, T., & d'Amato, T. (2002). Cannabis use correlates with schizotypal personality traits in healthy students. *Psychiatry Research*, *109*, 27–35.

Earleywine, M. (2006). Schizotypy, marijuana, and differential item functioning. *Journal of Clinical Psychiatry*, *55*, 391–393.

Eckblad, M., & Chapman, L. J. (1983). Magical ideation as an indicator of schizotypy. *Journal of Consulting and Clinical Psychology*, *51*, 215–225.

Esterberg, M. L., Goulding, S. M., McClure-Tone, E. B., & Compton, M. T. (2009). Schizotypy and nicotine, alcohol, and cannabis use in a non-psychiatric sample. *Addictive Behaviors*, *34*, 374–379.

Ettinger, U., Meyhöfer, I., Steffens, M., Wagner, M., & Koutsouleris, N. (2014). Genetics, cognition and neurobiology of schizotypal personality: A review of the overlap with schizophrenia. *Frontiers in Psychiatry*, *18*, 2–16.

Fridberg, D. J., Vollmer, J. M., O'Donnell, B. F., & Skosnik, P. D. (2011). Cannabis users differ from non-users on measures of personality and schizotypy. *Psychiatry Research*, *186*, 46–52.

Investigators GROUP. (2011). Evidence that familial liability for psychosis is expressed as differential sensitivity to cannabis. *Archives of General Psychiatry*, *68*, 138–147.

Lenzenweger, M. (2015). Thinking clearly about schizotypy: Hewing to the schizophrenia liability core, considering interesting tangents, and avoiding conceptual quicksand. *Schizophrenia Bulletin*, *41*, 483–491.

Mason, O., Claridge, G., & Jackson, M. (1995). New scales for the assessment of schizotypy. *Personality and Individual Differences*, *53*, 727–730.

Mass, R., Bardong, C., Kindl, K., & Dahme, B. (2001). Relationship between cannabis use, schizotypal traits, and cognitive function in healthy subjects. *Psychopathology*, *34*, 209–214.

Meehl, P. E. (1962). Schizotaxia, schizotypy, schizophrenia. *American Psychologist*, *17*, 827–838.

Meehl, P. E. (1990). Toward an integrated theory of schizotaxia, schizotypy, and schizophrenia. *Journal of Personality Disorders*, *4*, 1–99.

Nunn, J. A., Rizza, F., & Peters, E. R. (2001). The incidence of schizotypy among cannabis and alcohol users. *Journal of Nervous and Mental Disease*, *189*, 741–748.

Peters, E., Joseph, S. R., & Garety, P. A. (1999). The measurement of delusional ideation in the normal population: Introducing the PDI (Peters et al., Delusions Inventory). *Schizophrenia Bulletin*, *25*, 553–556.

Raine, A. (1991). The SPQ: A scale for the assessment of schizotypal personality based on DSM-III-R criteria. *Schizophrenia Bulletin*, *17*, 555–564.

Raine, A., & Benishay, D. (1995). The SPQ-B: A brief screening instrument for schizotypal personality disorder. *Journal of Personality Disorders*, *9*, 346–355.

Schiffman, J., Nakamura, B., Earleywine, M., & Labrie, J. (2005). Symptoms of schizotypy precede cannabis use. *Psychiatry Research*, *134*, 37–42.

Skosnik, P. D., Spatz-Glenn, L., & Park, S. (2001). Cannabis use is associated with schizotypy and attentional disinhibition. *Schizophrenia Research*, *48*, 83–92.

Stefanis, N. C., Delespaul, P., Henquet, C., Bakoula, C., Stefanis, C. N., & van Os, J. (2004). Early adolescent cannabis exposure and positive and negative dimensions of psychosis. *Addiction*, *99*, 1333–1341.

Stefanis, N. C., Hanssen, M., Smirnis, N. K., Avramopoulos, D. A., Evdokimidis, I. K., Stefanis, C. N., Verdoux, H., & van Os, J. (2002). Evidence that three dimensions of psychosis have a distribution in the general population. *Psychological Medicine*, *32*, 347–358.

Stirling, J., Barkus, E. J., Nabosi, L., Irshad, S., Roemer, G., Schreudergoidheijt, B., & Lewis, S. (2008). Cannabis-induced psychotic-like experiences are predicted by high schizotypy: confirmation of preliminary results in a large cohort. *Psychopathology*, *41*, 371–378.

Verdoux, H., Sorbara, F., Gindre, C., Swendsen, J. D., & van Os, J. (2003). Cannabis use and dimensions of psychosis in a nonclinical population of female subjects. *Schizophrenia Research*, *59*, 77–84.

Vollema, M. G., & van den Bosch, R. J. (1995). The multidimensionality of schizotypy. *Schizophrenia Bulletin*, *21*, 19–31.

Williams, J. H., Wellman, N. A., & Rawlins, J. N. P. (1996). Cannabis use correlates with schizotypy in healthy people. *Addiction*, *91*, 869–877.

Chapter 7

Effects of Cannabis Use in Those at Ultra-High Risk for Psychosis

Brian O'Donoghue, Meredith McHugh, Barnaby Nelson, Patrick McGorry
Orygen, The National Centre of Excellence in Youth Mental Health, Melbourne, VIC, Australia
University of Melbourne, Melbourne, VIC, Australia

OVERVIEW OF THE CONCEPT OF ULTRA-HIGH RISK FOR PSYCHOSIS

Among help-seeking young people, it is now possible to identify those at increased risk for developing a psychotic disorder. However, a major challenge to prospectively identifying the prodrome remains in that the initial, earliest psychiatric symptoms of a psychotic disorder are often nonspecific (Yung & Nelson, 2013). The prodrome technically can only be diagnosed retrospectively among the proportion of individuals who go on to develop a full-threshold psychotic disorder. The need for prospective research led to the development of the concept of and criteria for identifying young people who are at elevated risk of developing psychosis. Different terminology has been used to describe this group, including ultra-high risk (UHR), clinical high risk (CHR), and at-risk mental state (ARMS).

Two main methods for defining this putative prodromal state (which will be called UHR in this chapter) have been developed. The first entails using either the Comprehensive Assessment of the At-Risk Mental State (CAARMS) (Yung et al., 2005) or the Structured Interview for Prodromal Syndromes (SIPS) and Scale of Prodromal Symptoms (SOPS) (Miller et al., 2003). Both the CAARMS and the SIPS/SOPS identify three UHR groups, among individuals 14–25 years of age (previously 30 years) who were referred or were help-seeking for mental health problems: (1) the attenuated psychotic symptoms (APS) syndrome, which indicates the presence of subthreshold psychotic symptoms in terms of either intensity, frequency, or duration; (2) the brief limited intermittent psychotic symptoms (BLIPS) syndrome, which involves threshold psychotic symptoms for a subthreshold period of time; and (3) trait vulnerability plus a decline in psychosocial functioning, which identifies young people with a decline in functioning, or chronic low functioning, who also have a first-degree relative with a history of a psychotic disorder or

The Complex Connection between Cannabis and Schizophrenia. http://dx.doi.org/10.1016/B978-0-12-804791-0.00007-0

a schizotypal personality disorder. The second approach to identifying those at UHR is with "basic symptoms," which identifies subjectively experienced disturbances in different domains, including perception, thought processing, language, and attention; such symptoms can be measured using the Bonn Scale for the Assessment of Basic Symptoms (Klosterkötter, Hellmich, Steinmeyer, & Schultze-Lutter, 2001) and the Schizophrenia Proneness Instrument, Adult Version (Schultze-Lutter et al., 2007).

The ultimate purpose in identifying those at UHR for psychosis is to develop and then offer interventions to prevent the transition to a first episode of a psychotic disorder. A meta-analysis by van der Gaag et al. (2013) demonstrated that there has been provisional success on this front, with the overall effect of diverse interventions—cognitive-behavioral therapy, omega-3 fatty acids, and low-dose antipsychotic medications—having a risk reduction of 54% at 12 months, with a number needed to treat of 9 (van der Gaag et al., 2013). Another strategy has been attempting to identify subgroups of UHR individuals who are at even greater risk of developing a psychotic disorder. For example, certain baseline characteristics have been found to be associated with an increased risk of transition to psychosis, such as low functioning, a longer duration of symptoms (Nelson et al., 2013), unusual thought content such as suspiciousness (Cannon et al., 2008), and evidence of formal thought disorder (Thompson et al., 2013). UHR individuals with neurocognitive deficits are also more likely to go on to develop a psychotic disorder (Bora & Murray, 2014), as are UHR individuals with poor premorbid functioning or neurophysiological disturbances (Nieman et al., 2014). The role of cannabis use in the development of a psychotic disorder in the UHR population has received considerable attention due to cannabis use being linked to psychosis risk in a dose-dependent manner across numerous prospective, population-based studies (Moore et al., 2007), and because cannabis use is both widespread and a potentially modifiable factor.

DOES CANNABIS USE INCREASE RISK OF TRANSITION TO PSYCHOSIS IN THOSE IDENTIFIED AS ULTRA-HIGH RISK FOR PSYCHOSIS?

A meta-analysis was conducted by Kraan et al. (2016) on the association between cannabis use and transition to psychosis in individuals at UHR for psychosis; specifically, they examined whether either lifetime cannabis use or current cannabis abuse or dependence was associated with an increased risk of transition to psychosis. The study identified seven prospective studies, which included a total of 1171 UHR individuals. The meta-analysis demonstrated that lifetime cannabis use was not associated with an increased risk of transition to psychosis (OR = 1.14, 95% CI = 0.86–1.52). However, UHR individuals with current cannabis abuse or dependence had a greater risk of transitioning to a psychotic disorder (OR = 1.75, 95% CI = 1.14–2.71, $p = 0.01$). This was an important finding, as it demonstrated the importance of distinguishing between lifetime use and current-use disorder. As noted by the authors, it may also suggest a dose-dependent relationship between current cannabis use and transition to psychosis, mirroring findings seen in the general population (Moore et al., 2007).

More recently, McHugh and colleagues examined the role of cannabis use in the risk of transition in a cohort of 190 UHR individuals attending the Personal Assessment and Crisis Evaluation (PACE) UHR clinic in Melbourne, Australia (McHugh et al., 2017). The findings of this study replicated those of the meta-analysis, in that transition to a psychotic disorder was not predicted simply by a lifetime history of cannabis use, but by a greater severity of cannabis abuse, defined according to frequency of use, inability to cease use, and negative social consequences of the use of cannabis. Additionally, young people with a history of experiencing attenuated psychotic symptoms while using cannabis were nearly five times more likely to transition to a psychotic disorder.

The meta-analysis by Kraan et al. (2016) focused on whether lifetime cannabis use and/or current cannabis abuse or dependence were associated with an increased risk of transition to a psychotic disorder. However, there are other important factors relevant to this relationship, such as the proportion of UHR individuals who use cannabis, the age at first use, duration of treatment, and potential confounders of the relationship. Therefore, for the purposes of this chapter, findings relating to these factors were reviewed from the seven individual studies included in the meta-analysis, in addition to another study published by Buchy et al. (2015) and the aforementioned study by McHugh et al. (2017). The findings from all nine studies are presented in Table 1.

Age at First Cannabis Use and Risk of Transition

The age at first cannabis use has significance in the UHR population because it has been established that early cannabis use, before the age of 14, is a risk factor for the earlier onset of a psychotic disorder (Schimmelmann et al., 2011). In the general population, cannabis use has been linked to psychosis risk among individuals who display a specific genetic vulnerability, but only when the onset of cannabis use occurs before the age of 14 (Caspi et al., 2005). Four of the studies presented in Table 1 provided information on the mean age at first cannabis use, which ranged from 15 to 17 years; this appears to be outside of the window of highest vulnerability for developing psychosis following early cannabis use (i.e., before age 14). A study by Valmaggia et al. (2014) found that an earlier age at first cannabis use was associated with an increased risk of transitioning to a psychotic disorder (Valmaggia et al., 2014). In contrast, McHugh et al. found no relationship between age of first cannabis use and transition risk (McHugh et al., 2017). However, they did find that individuals with a history of cannabis-induced attenuated psychotic symptoms reported a younger age at first cannabis use than individuals with a history of cannabis use and no associated attenuated psychotic symptoms (McHugh et al., 2017). As noted earlier, those with a history

TABLE 1 Summary of Studies on the Association Between Cannabis Use and Transition to Psychosis in UHR Samples

Cohort or Service Name, Location, Reference	N	Lifetime Use of Cannabis [b] (%)	Frequent Cannabis Use (%)	Cannabis Abuse (%)	Cannabis Dependence (%)	Mean Age at First Use, Years (SD)	Mean Duration of Cannabis Use, Years (SD)	Increased Risk of Transition, Lifetime Cannabis Use	Increased Risk of Transition, Cannabis Abuse/Dependence	Potential Confounders/Interactions, and Notes
Personal Assessment and Crisis Evaluation (PACE), Melbourne, Australia, Phillips et al. (2002)	100	37.0	16.0	–	18.0	–	–	No	No	–
Center of Prevention and Evaluation (COPE), New York, USA, Corcoran et al. (2008)[a]	32	40.6	–	6.3	18.8 (past)	–	–	–	–	–

European Prediction of Psychosis Study (EPOS), Germany, Finland, the Netherlands, and England, Dragt et al. (2012)	245	42.0	10.6	18.4	–	17.3 (3.0)	–	No	No	
Recognition and Prevention Program (RAP), New York, USA, Auther et al. (2012)	101	36.5	–	10.4	–	–	–	No	No	Age, positive and negative symptoms—not associated
Outreach and Support in South London (OASIS), South London, UK, Valmaggia et al. (2014)	182	73.6	52.2	–	–	15.5 (3.1)	6.0 (4.6)	No	Yes	Heavy and early use of cannabis was associated with increased risk of transition

Continued

TABLE 1 Summary of Studies on the Association Between Cannabis Use and Transition to Psychosis in UHR Samples—cont'd

Cohort or Service Name, Location, Reference	N	Lifetime Use of Cannabis [b] (%)	Frequent Cannabis Use (%)	Cannabis Abuse (%)	Cannabis Dependence (%)	Mean Age at First Use, Years (SD)	Mean Duration of Cannabis Use, Years (SD)	Increased Risk of Transition, Lifetime Cannabis Use	Increased Risk of Transition, Cannabis Abuse/Dependence	Potential Confounders/Interactions, and Notes
Enhancing the Prospective Prediction of Psychosis (PREDICT), Toronto, Canada and North Carolina and Connecticut, USA, Buchy, Perkins, Woods, Liu, and Addington (2014)	170	30.6[c]	–	6.5	1.1			No	–	Lesser use of alcohol was associated with transition; tobacco use was not a confounder
North American Prodrome Longitudinal Study (NAPLS-1), North America, Auther et al. (2015)	370	38.1		13.2	6.5	–	–	No	Yes	Alcohol use was a confounder

North American Prodrome Longitudinal Study (NAPLS-2), North America, Buchy et al. (2015)	735	55.2	13.5	3.5	0.3	15.7 (2.8)		No		Tobacco use was not a confounder
Personal Assessment and Crisis Evaluation (PACE), Melbourne, Australia, McHugh et al. (2017)	190	57.9	27.0	8.5	4.3	16.3 (3.2)	–	No	–	History of cannabis-induced attenuated psychotic symptoms predicted transition

The definition of frequent cannabis use was taken from each individual study.
[a]*The study by Corcoran et al. (2008) was included in the meta-analysis by Kraan et al. (2016); however, it does not provide a definition of "transition" and only examines whether cannabis use is associated with an exacerbation of psychotic symptoms.*
[b]*Lifetime use defined as more than five times in their life.*
[c]*Calculated as the sum of use without impairment, abuse, and dependence.*

of cannabis-induced attenuated psychotic symptoms were also at significantly greater risk of transitioning to a psychotic disorder. These findings are consistent with earlier studies showing that adolescence may be a period of particular vulnerability to the effects of cannabis on psychosis risk (Caspi et al., 2005). Given that the relationship between age at first cannabis use and psychosis risk in UHR individuals has only received attention in these few studies, and with somewhat discrepant findings, this issue deserves further attention in future research.

Lifetime Cannabis Use and Risk of Transition: Caveats in Interpretation of Findings to Date

As displayed in Table 1, the proportion of UHR individuals who endorsed lifetime cannabis use varied from 30.6% to 73.6%, and the proportion reporting frequent cannabis use varied from 10.6% to 52.2%. These differences in the baseline rate of cannabis use in the UHR populations are important. In a debate article on whether the social environment can cause schizophrenia, van Os and McGuffin (2003) argued that a challenge in epidemiological research occurs when the risk factor of interest is highly prevalent. The example provided was that if everyone in the population smoked tobacco, then it would be impossible to detect an association between smoking and lung cancer because there would be no nonsmokers with which to compare the cancer rates. While this is an extreme example, the exposure rates of UHR individuals to cannabis were as high as nearly three-quarters in one study (Valmaggia et al., 2014), with current use present in over half of the sample. Therefore, if the baseline prevalence of cannabis use is very high in the UHR cohort, it may be statistically difficult to demonstrate a difference between users and nonusers.

Another factor that may account for the fact that most studies in Table 1 do not demonstrate a relationship between lifetime history of cannabis use and transition to psychosis is the variability in the effects of cannabis use on psychosis risk. For example, McHugh and colleagues found that cannabis use inflated transition risk only in a vulnerable subgroup of individuals: specifically, those reporting a history of cannabis-induced attenuated psychotic symptoms (McHugh et al., 2017). In fact, when individuals used cannabis without experiencing cannabis-induced attenuated psychotic symptoms, their risk of transition was similar to individuals with no history of cannabis use. The findings of McHugh and colleagues (2017) mirror earlier genetic studies showing that cannabis use was associated with enhanced risk of developing a psychotic disorder only among individuals who carried specific risk genes, in particular, genes that enhance striatal dopaminergic function (Caspi et al., 2005; Di Forti et al., 2012; Colizzi et al., 2015). Together, these findings provide compelling evidence that cannabis use may enhance risk for psychosis only in individuals with an underlying vulnerability (possibly genetic) to these effects. Consequently, it is possible that the studies to date reported in Table 1 have predominantly failed to observe a relationship between lifetime history of cannabis use and transition risk because the samples examined include both vulnerable and resilient subgroups.

DOES CANNABIS USE INCREASE THE LIKELIHOOD OF BEING AT ULTRA-HIGH RISK FOR PSYCHOSIS?

The focus of the meta-analysis by Kraan et al. (2016) was whether use of cannabis increases the risk of transition to psychosis in the population of young people identified as being UHR. As previously stated, this meta-analysis found that lifetime use of cannabis was not associated with an increased risk of transition. However, it is possible that the effect of early cannabis use could have an influence earlier in the development of a psychotic disorder. A hypothesis that has received less attention—possibly because a matched healthy control group is required to test it—is whether the early use of cannabis increases the risk of being UHR for psychosis. Auther et al. (2012) found that the rates of lifetime cannabis use were higher in a UHR cohort when compared with a healthy control group. This finding was replicated by Buchy et al., who found that UHR individuals were younger at first use of cannabis and had a higher lifetime prevalence and frequency of cannabis use when compared with healthy controls (Buchy et al., 2015). These studies suggest that cannabis use could have an influence on risk early in the development of a psychotic disorder. Fewer studies have addressed this question, as it requires a more complicated study design, requiring a representative cohort from the general population. Testing this hypothesis is further complicated by the lack of accurate data on the prevalence of cannabis use in young people in the general population in different global regions and countries.

POSSIBLE CONFOUNDERS IN THE RELATIONSHIP BETWEEN CANNABIS USE AND RISK OF TRANSITION

Multiple factors typically contribute to the onset of psychotic disorders, and therefore it is possible that there are factors either confounding the relationship between cannabis use and transition to psychosis in the UHR population, or interacting with cannabis use to modify the level of risk. For example, alcohol use has been examined as one potential confounding factor. A study by Auther and colleagues (2015) found that alcohol use was a confounder in the relationship between cannabis misuse and the development of a psychotic disorder. Buchy and colleagues (2014) found that lesser use of alcohol was associated with a greater risk of developing psychosis, while McHugh and colleagues found no relationship between heavy alcohol use and transition to psychosis (McHugh et al., 2017). This inconsistency across studies highlights the need for further research to address the ways in which use of alcohol may either confound or interact with the relationship between cannabis use and onset of psychosis.

FUTURE DIRECTIONS FOR RESEARCH

Improving the Identification of Ultra-High Risk Individuals Who Are at Greatest Risk of Transition

One of the main goals in UHR research has been attempting to identify demographic and clinical characteristics associated with an increased risk of transition to psychosis. For example, a validated risk prediction tool has been developed that can estimate the risk of transition for UHR individuals based on age, symptoms, and neuropsychological factors (Cannon et al., 2016). This has become increasingly important, as over the past decade a decline in the transition rates in UHR samples has been observed (Yung et al., 2007). There are a number of possible explanations for this decline, including a treatment effect or a dilution effect due to a change in the clinical characteristics of referred patients (Wiltink, Velthorst, Nelson, McGorry, & Yung, 2015; Hartmann et al., 2016). While a reduction in transition rates is a positive thing from a clinical point of view, it has meant that there are more false positives and that much higher numbers are required in trials for an intervention to demonstrate a benefit. Therefore, to "enrich" the UHR population, a method to identify a subgroup of those at greatest risk of transition may be to include additional factors in the eligibility criteria, such as current cannabis use disorder, and/or attenuated psychotic symptoms experienced during cannabis intoxication.

Another potential method for identifying factors associated with transition to psychosis in UHR individuals is the inclusion of genetic factors. Currently, 108 loci have been found to be associated with an increased risk of schizophrenia (Consortium, 2014), and one of the next steps involves understanding how environmental factors interact with these genes in the development of psychotic disorders (van Os, Rutten, & Poulton, 2008). As noted earlier, studies in the general population have demonstrated a cannabis-by-gene interactions that appear to partially drive the association between cannabis use and psychosis risk. Specifically, one identified interaction is that carriers of the catechol-*O*-methyltransferase (COMT) valine allele at the valine(158)methionine polymorphism are more likely to develop schizophrenia if they have smoked cannabis (Caspi et al., 2005). A study by Nieman et al. explored the gene-by-environment interaction between the COMT valine(158)methionine genotype and cannabis use in individuals at UHR for psychosis (Nieman et al., 2016). The study found that weekly cannabis use was associated with more severe positive symptoms and that this effect was increased in UHR individuals who were carriers of the COMT valine allele, and even more so in those homozygous for the allele. If replicated, these results could have significant clinical implications, in that it could be possible to identify UHR individuals who would be at greatest risk of developing a psychotic disorder if they were to commence or continue using cannabis. Additionally, whether and how this forewarning could lead to a change in behavior could be a focus of further research.

Including a Broader Array of Relevant Risk Factors

There are other established risk factors for psychotic disorders that could be potential confounders in the association between cannabis use and the risk of developing a psychotic disorder in the UHR population (or cannabis could be the confounder for the other factors associated with psychosis), such as social deprivation (O'Donoghue, Roche, & Lane, 2016), migrant status (Cantor-Graae & Selten, 2005), childhood adversity (Wicks, Hjern, Gunnell, Lewis, & Dalman, 2005), and medical disorders such as epilepsy (Clancy, Clarke, Connor, Cannon, & Cotter, 2014) or a history of head injury (Clarke, Kelleher, Clancy, & Cannon, 2012). A limitation of the literature to date is that it has tended to examine factors in isolation, and the studies presented in Table 1 examined a very limited range of potential confounders. Another possible pathway in the etiology of psychotic disorders that has received less attention in UHR research is how a number of these factors may interact with each other, as in the "two-hit" hypothesis (Maynard, Sikich, Lieberman, & LaMantia, 2001), which proposes that genetic or environmental factors disrupt early central nervous system development, which in turn leads to long-term vulnerabilities to a "second hit" that may then lead to a psychotic disorder. This has particular relevance to the cannabis-psychosis link, as cannabis use is associated with an earlier age at onset of psychosis by approximately 2–3 years (Large, Sharma, Compton, Slade, & Nielssen, 2011), and cannabis use during early adolescence is more likely to result in transition to psychosis (Caspi et al., 2005). Indeed, cannabis use in the presence of other environmental risk factors such as obstetric complications, urbanicity, or migrant status is associated with an even earlier age at psychosis onset, by nearly 10 years in two separate studies (Stepniak et al., 2014; O'Donoghue et al., 2015). Thus, combining multiple risk factors including cannabis use into "risk factor panels" poses another potential method of identifying UHR individuals who may be particularly vulnerable to the effects of cannabis use and at greatest risk of developing a psychotic disorder.

Developing Interventions for Cannabis Misuse in the Ultra-High Risk Group

The finding that current cannabis abuse or dependence was associated with an increased risk of transition to psychosis in the UHR population (Kraan et al., 2016) makes this a potential intervention target, since cannabis use is modifiable. A limited number of studies have been conducted to evaluate interventions aimed at reducing cannabis use among individuals with psychotic disorders, and the evidence is strongest for longer term interventions (Baker, Hides, & Lubman, 2010). A systematic review found that interventions including motivational interviewing reduced the quantity of cannabis consumed, but no intervention appeared to reduce the frequency of use (Baker et al., 2010). There have been no published studies on interventions aimed at reducing cannabis use in UHR samples. However, if findings related to motivational interviewing and reductions in cannabis quantities used hold for the UHR population, this might represent a method of reducing psychosis transition rates.

Cannabinoids as a Potential Intervention in Those at Ultra-High Risk

The endocannabinoid system has been identified as a potential therapeutic target for the treatment of psychotic disorders. Δ^9-Tetrahydrocannabinol (THC) is the component of cannabis believed to be responsible for inducing psychotic symptoms, while another component, cannabidiol (CBD), may have antipsychotic properties and has been a focus of research as a potential therapeutic agent for psychotic disorders (Leweke, Mueller, Lange, & Rohleder, 2016); see also Chapter 14. Randomized, controlled trials are currently in progress to determine whether CBD is effective as either an adjunctive therapy or as monotherapy in psychotic disorders. In a trial comparing CBD to amisulpride, it was found that both treatments led to significant clinical improvements in psychotic symptoms, but CBD had a more favorable side-effect profile (Leweke et al., 2012). The primary pharmacological mechanism by which CBD may exert an antipsychotic effect is unclear at present, though one theory is that it may enhance anandamide signaling, as participants in a CBD treatment group had an increase in serum anandamide levels, a fatty acid endocannabinoid neurotransmitter (Leweke et al., 2012). This has particular relevance to the UHR population, as UHR individuals with lower levels of anandamide in their cerebrospinal fluid are at higher risk of transition to a full-threshold psychotic disorder (Koethe et al., 2009). No studies to date have been conducted on the use of CBD in the UHR population; yet, there is a biologically plausible mechanism of action and the risk-benefit ratio appears acceptable, considering that the side-effect profile may be better than that of antipsychotic medications.

CONCLUSIONS

Further research to identify the mechanisms by which cannabis use increases the risk of transition to psychosis would lead to further insights into the etiology of psychotic disorders, as well as inform potential interventions to reduce the risk of transition. The UHR state offers an optimal time to intervene in risk factors such as cannabis use, to possibly mitigate the risk of developing a full-threshold psychotic disorder. Additionally, the effectiveness of CBD as a potential intervention needs to be evaluated in a randomized, controlled trial in a UHR sample.

Key Chapter Points

A lifetime history of cannabis use does not appear to increase the risk of transition to psychosis in the UHR population; however, current cannabis abuse or dependence does increase this risk.

In one study, young people with a history of experiencing attenuated psychotic symptoms while using cannabis were nearly five times more likely to transition to a psychotic disorder.

Young people identified as UHR have higher rates of cannabis use and an earlier age at first use when compared with the general population, suggesting that cannabis use increase the risk of being UHR.

Future directions of research include identifying which UHR individuals are at greater risk for developing a psychotic disorder by combining cannabis use with data on exposure to other environmental risk factors and genetics.

Cannabidiol, a component of cannabis, is currently being evaluated as an intervention for psychotic disorders, but could also be considered as a potential agent to reduce the risk of transition to psychosis in the UHR population.

REFERENCES

Auther, A. M., Cadenhead, K. S., Carrión, R. E., Addington, J., Bearden, C. E., Cannon, T. D., et al. (2015). Alcohol confounds relationship between cannabis misuse and psychosis conversion in a high-risk sample. *Acta Psychiatrica Scandinavica*, *132*, 60–68.

Auther, A. M., McLaughlin, D., Carrion, R. E., Nagachandran, P., Correll, C. U., & Cornblatt, B. A. (2012). Prospective study of cannabis use in adolescents at clinical high risk for psychosis: Impact on conversion to psychosis and functional outcome. *Psychological Medicine*, *42*, 2485–2497.

Baker, A. L., Hides, L., & Lubman, D. I. (2010). Treatment of cannabis use among people with psychotic or depressive disorders: A systematic review. *Journal of Clinical Psychiatry*, *71*, 247–254.

Bora, E., & Murray, R. M. (2014). Meta-analysis of cognitive deficits in ultra-high risk to psychosis and first-episode psychosis: Do the cognitive deficits progress over, or after, the onset of psychosis? *Schizophrenia Bulletin*, *40*, 744–755.

Buchy, L., Cadenhead, K. S., Cannon, T. D., Cornblatt, B. A., McGlashan, T. H., Perkins, D. O., et al. (2015). Substance use in individuals at clinical high risk of psychosis. *Psychological Medicine*, *45*, 2275–2284.

Buchy, L., Perkins, D., Woods, S. W., Liu, L., & Addington, J. (2014). Impact of substance use on conversion to psychosis in youth at clinical high risk of psychosis. *Schizophrenia Research*, *156*, 277–280.

Cannon, T. D., Cadenhead, K., Cornblatt, B., Woods, S. W., Addington, J., Walker, E., et al. (2008). Prediction of psychosis in youth at high clinical risk: a multisite longitudinal study in North America. *Archives of General Psychiatry*, *65*, 28–37.

Cannon, T. D., Yu, C., Addington, J., Bearden, C. E., Cadenhead, K. S., Cornblatt, B. A., et al. (2016). An individualized risk calculator for research in prodromal psychosis. *American Journal of Psychiatry*, *173*, 980–988, appi.ajp.2016.15070890.

Cantor-Graae, E., & Selten, J. P. (2005). Schizophrenia and migration: A meta-analysis and review. *American Journal of Psychiatry*, *162*, 12–24.

Caspi, A., Moffitt, T. E., Cannon, M., McClay, J., Murray, R., Harrington, H., et al. (2005). Moderation of the effect of adolescent-onset cannabis use on adult psychosis by a functional polymorphism in the catechol-O-methyltransferase gene: Longitudinal evidence of a gene X environment interaction. *Biological Psychiatry*, *57*, 1117–1127.

Clancy, M. J., Clarke, M. C., Connor, D. J., Cannon, M., & Cotter, D. R. (2014). The prevalence of psychosis in epilepsy; a systematic review and meta-analysis. *BMC Psychiatry*, *14*, 75.

Clarke, M. C., Kelleher, I., Clancy, M., & Cannon, M. (2012). Predicting risk and the emergence of schizophrenia. *The Psychiatric Clinics of North America*, *35*, 585–612.

Colizzi, M., Iyegbe, C., Powell, J., Blasi, G., Bertolino, A., Murray, R. M., et al. (2015). Interaction between DRD2 and AKT1 genetic variations on risk of psychosis in cannabis users: A case-control study. *NPJ Schizophrenia*, *1*, 15025.

Consortium SWGotPG. (2014). Biological insights from 108 schizophrenia-associated genetic loci. *Nature*, *511*, 421–427.

Corcoran, C. M., Kimhy, D., Stanford, A., Khan, S., Walsh, J., Thompson, J., et al. (2008). Temporal association of cannabis use with symptoms in individuals at clinical high risk for psychosis. *Schizophrenia Research*, *106*, 286–293.

Di Forti, M., Iyegbe, C., Sallis, H., Kolliakou, A., Falcone, M. A., Paparelli, A., et al. (2012). Confirmation that the AKT1 (rs2494732) genotype influences the risk of psychosis in cannabis users. *Biological Psychiatry*, *72*, 811–816.

Dragt, S., Nieman, D. H., Schultze-Lutter, F., van der Meer, F., Becker, H., de Haan, L., et al. (2012). Cannabis use and age at onset of symptoms in subjects at clinical high risk for psychosis. *Acta Psychiatrica Scandinavica*, *125*, 45–53.

Hartmann, J. A., Yuen, H. P., McGorry, P. D., Yung, A. R., Lin, A., Wood, S. J., et al. (2016). Declining transition rates to psychotic disorder in "ultra-high risk" clients: Investigation of a dilution effect. *Schizophrenia Research*, *170*, 130–136.

Klosterkötter, J., Hellmich, M., Steinmeyer, E. M., & Schultze-Lutter, F. (2001). Diagnosing schizophrenia in the initial prodromal phase. *Archives of General Psychiatry*, *58*, 158–164.

Koethe, D., Giuffrida, A., Schreiber, D., Hellmich, M., Schultze-Lutter, F., Ruhrmann, S., et al. (2009). Anandamide elevation in cerebrospinal fluid in initial prodromal states of psychosis. *The British Journal of Psychiatry*, *194*, 371–372.

Kraan, T., Velthorst, E., Koenders, L., Zwaart, K., Ising, H. K., van den Berg, D., et al. (2016). Cannabis use and transition to psychosis in individuals at ultra-high risk: Review and meta-analysis. *Psychological Medicine*, *46*, 673–681.

Large, M., Sharma, S., Compton, M. T., Slade, T., & Nielssen, O. (2011). Cannabis use and earlier onset of psychosis: a systematic meta-analysis. *Archives of General Psychiatry*, *68*, 555–561.

Leweke, F. M., Mueller, J. K., Lange, B., & Rohleder, C. (2016). Therapeutic potential of cannabinoids in psychosis. *Biological Psychiatry*, *79*, 604–612.

Leweke, F. M., Piomelli, D., Pahlisch, F., Muhl, D., Gerth, C. W., Hoyer, C., et al. (2012). Cannabidiol enhances anandamide signaling and alleviates psychotic symptoms of schizophrenia. *Translational Psychiatry*, *2*, e94.

Maynard, T. M., Sikich, L., Lieberman, J. A., & LaMantia, A. S. (2001). Neural development, cell-cell signaling, and the "two-hit" hypothesis of schizophrenia. *Schizophrenia Bulletin*, *27*, 457–476.

McHugh, M. J., McGorry, P. D., Yung, A. R., Lin, A., Wood, S. J., Hartmann, J. A., et al. (2017). Cannabis-induced attenuated psychotic symptoms: implications for prognosis in young people at ultra-high risk for psychosis. *Psychological Medicine*, *47*, 616–626.

Miller, T. J., McGlashan, T. H., Rosen, J. L., Cadenhead, K., Ventura, J., McFarlane, W., et al. (2003). Prodromal assessment with the structured interview for prodromal syndromes and the scale of prodromal symptoms: Predictive validity, interrater reliability, and training to reliability. *Schizophrenia Bulletin*, *29*, 703–715.

Moore, T. H., Zammit, S., Lingford-Hughes, A., Barnes, T. R., Jones, P. B., Burke, M., et al. (2007). Cannabis use and risk of psychotic or affective mental health outcomes: A systematic review. *Lancet*, *370*, 319–328.

Nelson, B., Yuen, H. P., Wood, S. J., Lin, A., Spiliotacopoulos, D., Bruxner, A., et al. (2013). Long-term follow-up of a group at ultra high risk ("prodromal") for psychosis: The PACE 400 study. *JAMA Psychiatry*, *70*, 793–802.

Nieman, D. H., Dragt, S., van Duin, E. D., Denneman, N., Overbeek, J. M., de Haan, L., et al. (2016). COMT Val(158)Met genotype and cannabis use in people with an At Risk Mental State for psychosis: Exploring gene x environment interactions. *Schizophrenia Research, 174*, 24–28.

Nieman, D. H., Ruhrmann, S., Dragt, S., Soen, F., van Tricht, M. J., Koelman, J. H., et al. (2014). Psychosis prediction: Stratification of risk estimation with information-processing and premorbid functioning variables. *Schizophrenia Bulletin, 40*, 1482–1490.

O'Donoghue, B., Lyne, J., Madigan, K., Lane, A., Turner, N., O'Callaghan, E., et al. (2015). Environmental factors and the age at onset in first episode psychosis. *Schizophrenia Research, 168*, 106–112.

O'Donoghue, B., Roche, E., & Lane, A. (2016). Neighbourhood level social deprivation and the risk of psychotic disorders: A systematic review. *Social Psychiatry and Psychiatric Epidemiology, 51*, 941–950.

Phillips, L. J., Curry, C., Yung, A. R., Yuen, H. P., Adlard, S., & McGorry, P. D. (2002). Cannabis use is not associated with the development of psychosis in an 'ultra' high-risk group. *The Australian and New Zealand Journal of Psychiatry, 36*, 800–806.

Schimmelmann, B. G., Conus, P., Cotton, S. M., Kupferschmid, S., Karow, A., Schultze-Lutter, F., et al. (2011). Cannabis use disorder and age at onset of psychosis—A study in first-episode patients. *Schizophrenia Research, 129*, 52–56.

Schultze-Lutter, F., Ruhrmann, S., Picker, H., von Reventlow, H., Brockhaus-Dumke, A., & Klosterkotter, J. (2007). Basic symptoms in early psychotic and depressive disorders. *The British Journal of Psychiatry, 191*, s31–s37.

Stepniak, B., Papiol, S., Hammer, C., Ramin, A., Everts, S., Hennig, L., et al. (2014). Accumulated environmental risk determining age at schizophrenia onset: A deep phenotyping-based study. *The Lancet Psychiatry, 1*, 444–453.

Thompson, A., Nelson, B., Bruxner, A., O'Connor, K., Mossaheb, N., Simmons, M. B., et al. (2013). Does specific psychopathology predict development of psychosis in ultra high-risk (UHR) patients? *The Australian and New Zealand Journal of Psychiatry, 47*, 380–390.

Valmaggia, L. R., Day, F. L., Jones, C., Bissoli, S., Pugh, C., Hall, D., et al. (2014). Cannabis use and transition to psychosis in people at ultra-high risk. *Psychological Medicine, 44*, 2503–2512.

van der Gaag, M., Smit, F., Bechdolf, A., French, P., Linszen, D. H., Yung, A. R., et al. (2013). Preventing a first episode of psychosis: Meta-analysis of randomized controlled prevention trials of 12 month and longer-term follow-ups. *Schizophrenia Research, 149*, 56–62.

van Os, J., & McGuffin, P. (2003). Can the social environment cause schizophrenia? *The British Journal of Psychiatry, 182*, 291–292.

van Os, J., Rutten, B. P., & Poulton, R. (2008). Gene-environment interactions in schizophrenia: Review of epidemiological findings and future directions. *Schizophrenia Bulletin, 34*, 1066–1082.

Wicks, S., Hjern, A., Gunnell, D., Lewis, G., & Dalman, C. (2005). Social adversity in childhood and the risk of developing psychosis: A national cohort study. *American Journal of Psychiatry, 162*, 1652–1657.

Wiltink, S., Velthorst, E., Nelson, B., McGorry, P. M., & Yung, A. R. (2015). Declining transition rates to psychosis: The contribution of potential changes in referral pathways to an ultra-high-risk service. *Early Intervention in Psychiatry, 9*, 200–206.

Yung, A. R., & Nelson, B. (2013). The ultra-high risk concept—A review. *Canadian Journal of Psychiatry Revue Canadienne de Psychiatrie, 58*, 5–12.

Yung, A. R., Yuen, H. P., Berger, G., Francey, S., Hung, T. C., Nelson, B., et al. (2007). Declining transition rate in ultra high risk (prodromal) services: Dilution or reduction of risk? *Schizophrenia Bulletin, 33*, 673–681.

Yung, A. R., Yuen, H. P., McGorry, P. D., Phillips, L. J., Kelly, D., Dell'Olio, M., et al. (2005). Mapping the onset of psychosis: The comprehensive assessment of at-risk mental states. *The Australian and New Zealand Journal of Psychiatry, 39*, 964–971.

Chapter 8

Cannabis-Induced Psychotic Disorders

Luca Pauselli
University of Perugia, Perugia, Italy
Columbia University Medical Center, New York, NY, United States
New York State Psychiatric Institute, New York, NY, United States

NOSOLOGY AND CLASSIFICATION

Cannabis, as all psychoactive drugs, changes one's behavior, cognition, and perceptions. In particular situations, cannabis use can lead to psychiatric states that can be considered pathological. These phenomena have been described in the literature since 1843 when O'Shaughnessy (1843) described acute cataleptic reactions among his patients who were prescribed cannabis, and 1845 when the French psychiatrist Jacques Moreau (1973) in his book "Hashish and Mental Illness" drew attention to hallucinatory phenomena. He reported that "cannabis resin" could precipitate "*acute psychotic reactions, generally lasting but a few hours, but occasionally as long as a week; the reaction seemed dose-related and its main features included paranoid ideation, illusions, hallucinations, delusions, depersonalization, confusion, restlessness, and excitement. There can be delirium, disorientation, and marked clouding of consciousness.*" At the beginning of the 20th century, the term *cannabis psychosis* was introduced to describe different situations in which cannabis intoxication and psychotic symptoms co-occur (Ames, 1958; Talbott & Teague, 1969; Warnock, 1903). According to the World Health Organization (WHO) (1971), the main features were described as including "*paranoid ideas, illusions, hallucinations, depersonalization, delusions, confusion, restlessness and excitement. In occasional instances there may be additional features of delirium, disorientation, and marked clouding of consciousness.*"

As highlighted by Leweke, Gerth, and Klosterkoetter (2004) in their review, there are many naturalistic data available on cannabis psychosis, but caution is needed when interpreting these data because of a lack of coherence in the nosological classification. One important aspect is the time-course of these psychotic states. The authors recommended a differentiation between

The Complex Connection between Cannabis and Schizophrenia. http://dx.doi.org/10.1016/B978-0-12-804791-0.00008-2

longer-lasting or persistent reactions to higher doses or longer exposure to natural cannabis preparations and intermittent or short-term psychotic reactions after short-term ingestion of cannabis or even synthetic Δ^9-tetrahydrocannabinol (THC). The available studies they examined differ significantly with respect to the characterization of patients and their clinical symptomatology, leading the authors to conclude that there is no convincing evidence from controlled and open studies that "cannabis psychosis" can be identified as a nosological entity of its own. For this reason, they suggested the more neutral term "cannabis-associated psychosis," implying that with regard to psychopathology, there are several states that may all be induced, or at least intensified, by cannabis use.

Ramström (2004), in his review of the literature on cannabis psychosis, proposed four main forms of cannabis-related psychosis:

a. Cannabis-caused Delirium—characterized by an acute, short-lasting, profound state of confusion occurring for a few hours to a number of days, usually arising where a large amount of cannabis had been consumed. There is cessation of symptoms if the individual stops cannabis use.
b. Cannabis-caused Toxic Psychosis (cannabis psychosis)—This is short-lived psychosis (lasting up to six weeks) with psychotic symptoms arising in the context of cannabis intoxication. Again, there is cessation of symptoms if the person stops cannabis use.
c. Cannabis-caused Functional, Non-schizophrenic Psychosis—This psychosis arises in clear consciousness and in the context of cannabis use and intoxication, but persisting beyond the elimination of cannabis intake. Although this entity has been described by some researchers, others suggest that it is probably schizophrenia.
d. Schizophrenia and other psychotic disorders following cannabis use, abuse, or dependence, in which cannabis is a precipitating factor for the onset or exacerbation of schizophrenia or a related psychotic disorder.

For the first time in the official diagnostic system, the diagnosis of cannabis delusional disorder appeared in the DSM-III-R (APA, 1987). It did not provide specific psychopathological criteria; instead, this diagnosis is made on the basis of its temporal relationship with the use of the substance. In the DSM-IV (APA, 1994) and DSM-IV-TR (APA, 2000), cannabis psychosis was included in the broader chapter on Substance-Induced Psychotic Disorders. The main criteria to make this diagnosis are: (a) prominent hallucinations or delusions, and (b) evidence from the history, physical examination, or laboratory findings that the psychotic symptoms develop during or within a month of substance intoxication or withdrawal. The DSM-5 (APA, 2015) did not provide major modifications to these two main criteria. In the text, recommendations are made to help in distinguishing substance-induced psychotic disorders from primary psychotic disorders, namely that the onset should arise during or soon after the intoxication or substance withdrawal, but can persist for weeks (whereas primary psychotic disorders may precede the onset of substance use or may occur during times of sustained abstinence). However, no clear-cut time criteria are specified for any substances.

Since the current diagnostic criteria are so nonspecific, the tools available for the assessment of clinical diagnoses make it difficult to identify substance-induced psychotic disorders, including cannabis-induced psychotic disorder, in a reliable and valid fashion. This translates into scarce possibilities for a study to show significant differences between groups when diagnoses are an important grouping or outcome variable. These considerations brought Hasin and her team to develop the Psychiatric Interview for Substance and Mental Disorders (PRISM), a tool to overcome the lack of diagnostic interviews that are suitable for comorbidity research (Hasin et al., 1996). The first version of PRISM shared some important general features with the Structured Clinical Interview for DSM-III (SCID). The authors relocated the alcohol and drug use disorders from the middle of the interview to the beginning, increased the level of accuracy in collecting alcohol and drug use history, and added separate organic exclusion items for alcohol-, drug-, and medication-induced disorders. New features included: (1) specific rating guidelines for duration requirements, exclusion specifications, and decision rules for frequently encountered sources of uncertainty; (2) separation of dissimilar components of individual criteria into different items, to ensure assessment of all components and to allow for simplified guidelines; (3) addition of probes to standardize the interview and to indicate questions to use in exploring an unclear response; (4) an introductory section with indirect indicators of alcohol, drug, or psychiatric problems; (5) computer programs developed to produce diagnoses from the interview data; and (6) training materials to improve inter-rater reliability. All of these improvements made the DSM-III PRISM and then the DSM-IV PRISM reliable tools to assess substance-induced psychotic disorders with proven psychometric properties (Torrens, Serrano, Astals, Pérez-Domínguez, & Martín-Santos, 2004). DSM-5 PRISM was developed and is currently undergoing validation through an NIH-funded clinical trial (https://clinicaltrials.gov/ct2/show/NCT02660619, last accessed 12/04/16).

Since the literature on this topic is so conflicting with regard to nosology, this chapter will focus on Cannabis-Induced Psychotic Disorder, as described in the DSM-5 (APA, 2015), which can be described as an acute, transient, self-limited psychotic reaction during cannabis intoxication or withdrawal in the absence of a primary psychotic disorder.

Laboratory evidence for this disorder has been collected through the past decade. D'Souza et al. (2004) designed an experiment to characterize the dose-related psychotomimetic effects of THC, in healthy individuals under double-blind, placebo-controlled laboratory conditions, using standardized behavioral and cognitive assessments. The group found that intravenous injection of THC produced transient effects including positive symptoms (suspiciousness, paranoid and grandiose delusions, conceptual disorganization, and illusions); negative symptoms (blunted affect, reduced rapport, lack of spontaneity, psychomotor retardation, and emotional withdrawal); perceptual alterations; euphoria; anxiety; and deficits in working memory, recall, and attention, without altering general orientation. Similar effects have been replicated in several studies from different research groups (Bhattacharyya et al., 2012; D'Souza et al., 2008; Kaufmann et al., 2010; Morrison et al., 2009).

PATHOGENESIS THEORIES

Most of the literature on cannabis-induced psychotic disorder comprises small case series or case reports, focused on symptomatology more than pathogenesis. Some studies have explored experimental evidence for a link between cannabis and psychosis, such as those focused on healthy controls developing psychotic-like symptoms after controlled administration of THC.

Sewell, Ranganathan, and D'Souza (2009), in their review, identified possible mechanisms by which cannabinoids cause psychotic symptoms. THC interacts as a partial agonist with cannabinoid (CB_1) receptors, which are distributed with high density in brain regions that have been implicated in the putative neural circuitry of psychosis, modulating neurotransmitters in dopaminergic, GABA-ergic, and glutamatergic circuits (Breivogel & Sim-Selley, 2009; Sewell et al., 2009). Dopamine, while unlikely to play a major role in mediating the psychotomimetic effects of THC, may be involved in working memory deficits. GABA, through a disruption in neural synchrony, could interfere with memory consolidation, associative functions, and normal sensorimotor gating mechanisms, eventually leading to psychotic symptoms. Cannabinoids reduce glutamatergic synaptic transmission in several brain regions involved in the regulation of gating functions.

Cortes-Briones et al. (2015) used the neural noise theory to test an alternative explanatory hypothesis for the psychosis-like effects of THC. According to this theory, perception, language, emotion, and memory rely on integrity of long-range functional networks formed by ensembles of brain areas, processing information in a coordinated manner (Bullmore & Sporns, 2009). Studies have shown abnormal neural connectivity in schizophrenia, which has been related to the presence and intensity of psychotic symptoms (Higashima et al., 2007; Lawrie et al., 2002). Cortes-Briones et al. (2015) found that at doses that produced psychosis-like effects, THC increased neural noise measured in the EEG of humans in a dose-dependent manner. Furthermore, increases in neural noise were positively related with THC-induced psychosis-like, but not negative-like, effects. These findings suggest that neural noise may contribute to the psychotomimetic effects of cannabis.

Dragogna et al. (2014) investigated neurometabolic differences between individuals with schizophrenia with comorbid cannabis abuse and those with substance-induced psychotic disorder. They enrolled patients experiencing an acute psychotic episode, affected either by schizophrenia with or without cannabis abuse with recent onset (<5 years of illness) or by substance-induced psychotic disorder (all were cannabis abusers). All patients were assessed with the Positive and Negative Syndrome Scale (PANSS), urinary toxicological tests, and brain 18-FDG-PET scanning in a resting condition. They found bilateral hypermetabolism in the posterior cingulum and the precuneus in cannabis-induced psychotic disorder patients compared to patients with schizophrenia, with or without cannabis abuse. The posterior cingulum and precuneus are two core regions of the default mode network in humans, the role of which, even though not completely understood, seem to be relevant for attending to external and internal stimuli (Gusnard, Akbudak, Shulman, & Raichle, 2001; Raichle et al., 2001) and for self-referential and reflective activity such as inner speech, recall of personal experiences, mental imagery, and planning of future events (Greicius, Krasnow, Reiss, & Menon, 2003). Dysfunction of the default mode network in schizophrenia has been studied, but there is not yet sufficient clear-cut evidence to build a coherent theory (Pomarol-Clotet et al., 2008; Salgado-Pineda et al., 2011; Wolf et al., 2011).

The authors of this study (Dragogna et al., 2014) critically interpreted their results. What one would have expected was the presence of a dysfunction primarily affecting the patients with schizophrenia (with or without cannabis abuse) in the implicated brain areas, and not necessarily in those with cannabis-induced psychotic disorder (i.e., the opposite of their findings). This would have been in line with theories regarding schizophrenia and substance-induced psychosis, where the former appears to be the more correlated with abnormalities in neuronal connectivity (Bullmore, Frangou, & Murray, 1997; Fletcher, McKenna, Friston, Frith, & Dolan, 1999; Friston, 1999; Stephan, Friston, & Frith, 2009).

PSYCHOPATHOLOGICAL FEATURES OF CANNABIS-INDUCED PSYCHOTIC DISORDER AND POSSIBLE DIFFERENCES WITH SCHIZOPHRENIA AND OTHER PRIMARY PSYCHOTIC DISORDERS

Identifying specific psychopathological features of cannabis-induced psychotic disorder is of fundamental importance. Being able to differentiate this disorder from schizophrenia-spectrum disorders would define cannabis-induced psychotic disorder as an independent nosological entity. From the clinical point of view, differentiating cannabis-induced psychotic disorder from schizophrenia-spectrum disorders when they co-occur with a cannabis use disorder could have a major impact on treatment approaches and outcomes. Despite the importance of the topic, few researchers have focused their attention on it.

Caton et al. (2005) were among the first to systematically focus on determining whether there are differences in demographic, family, and clinical characteristics among those with a diagnosis of substance-induced psychosis compared with those with a diagnosis of a primary psychotic disorder and concurrent substance use; and on identifying key predictors that could help clinicians to correctly classify early-phase psychotic disorders that co-occur with substance use. The authors did not focus on specific substances, but cannabis was the most abused substance in both substance-induced psychotic disorder and primary psychotic disorders. Compared to those with a primary psychotic disorder, subjects with substance-induced psychosis had a significantly later age at onset of psychosis, greater likelihood of being in a relationship, greater antisocial personality disorder comorbidity, more frequent homelessness, poorer family support, and greater likelihood of having a parent with substance abuse. Subjects with primary psychosis had less insight and more severe psychiatric symptoms, a finding that was not limited to positive symptoms but also included negative symptoms and general psychopathology. In contrast, subjects with substance-induced psychosis had more severe forms of substance use disorders, characterized by long periods of substance use, use of multiple drugs, severe psychosocial problems, and greater substance dependence. Another characteristic distinguishing the substance-induced psychosis group was visual hallucinations.

Baldacchino et al. (2012) performed a systematic and narrative review of the literature with the purpose of looking for evidence of a specific constellation of symptoms that are consistently characteristic of cannabis psychosis within an inpatient psychiatric setting, and to determine whether these combine to create a psychopathology that is distinct from that of other types of psychosis. They found only 13 studies (Basu, Malhotra, Bhagat, & Varma, 1999; Bersani, Orlandi, Kotzalidis, & Pancheri, 2002; Caspari, 1999; Dubertret, Bidard, Adès, & Gorwood, 2006; Imade & Ebie, 1991; Mathers & Ghodse, 1992; McGuire, Jones, & Harvey, 1994; Modestin, Gladen, & Christen, 2001; Nunez & Gurpegui, 2002; Onyango, 1986; Rottanburg, Ben-Arie, Robins, Teggin, & Elk, 1982; Thacore & Shukla, 1976; Thornicroft, Meadows, & Politi, 1992) that specifically compared the psychopathology of cannabis users with psychosis to non-cannabis users and substance nonusers in an inpatient psychiatric setting, and only eight met the quality threshold of either "medium" or "strong" internal and external validity levels (Bersani et al., 2002; Caspari, 1999; Dubertret et al., 2006; Mathers & Ghodse, 1992; McGuire et al., 1994; Modestin et al., 2001; Rottanburg et al., 1982; Thornicroft et al., 1992). Of these studies, one by McGuire et al. (1994) found no significant differences in psychopathology between the cannabis-induced psychotic disorder group and the control group. Across the seven studies finding differences, the cannabis-induced psychotic disorder groups scored significantly ($P < .05$) lower on measures of: (1) affective flattening, alogia, avolition-apathy, anhedonia (from the Scale for the Assessment of Negative Symptoms, SANS); (2) negative symptoms (measured with the Positive and Negative Syndrome Scale, PANSS); (3) avolition-apathy (from the Diagnostic Interview for Genetic Studies, DIGS); (4) flattening of affect, incoherent speech, hysteria, and auditory hallucinations (based on the

Present State Examination, PSE); and (5) thought disorder (from the Scale for the Assessment of Positive symptoms, SAPS). The cannabis-induced psychotic disorder groups scored significantly ($P < .05$) higher than the control groups on measures of: (1) thought disturbance (rated with the Brief Psychiatric Rating Scale, BPRS); (2) hostility and hallucinations (measured with the Association for Methodology and Documentation in Psychiatry System, AMDP); (3) changed perception, thought insertion, nonverbal auditory hallucinations, delusions of control, grandiose delusions, hypomania, and agitation (PSE); and (4) incoherent speech, grandiose and religious delusions, and agitation (from the Symptom Checklist, SCL). The reviewers (Baldacchino et al., 2012) concluded that the available evidence is mixed but suggests that cannabis psychosis may include more positive psychotic features than psychosis in the absence of cannabis use.

Rubio et al. (2012) conducted a study to establish the psychopathological differences between primary psychotic disorders with cannabis consumption and cannabis-induced psychotic disorder during the first weeks of treatment in inpatient facilities. A sample of subjects with psychotic symptoms and cannabis use, referred to the psychiatry inpatient units of three university general hospitals in Madrid over a 36-month period, were assessed (psychopathologic symptoms were evaluated using the SCL-90-R). The final sample included 50 subjects with a diagnosis of cannabis-induced psychotic disorder and 104 subjects with primary psychotic disorders and concurrent cannabis use disorder. Patients with cannabis-induced psychotic disorder were older and more likely to be employed. In terms of cannabis consumption, subjects with cannabis-induced psychotic disorder smoked larger amounts of cannabis. About 80% smoked more than eight joints per day, and the diagnosis of cannabis dependence was more prevalent in this group. There were clinical and psychopathological differences between cannabis-induced psychosis and primary psychosis with concurrent cannabis use in the short-term presentation. Subjects with cannabis-induced psychotic disorder had more psychopathological symptoms belonging to a neurotic profile (somatization, obsessive-compulsive, interpersonal sensitivity, depression, anxiety, and phobic anxiety). According to the authors, depressive symptoms could be used to distinguish cannabis-induced psychotic disorder from primary psychotic disorders, while clinical variables related to "neurotic" symptoms could be involved in the susceptibility to cannabis-induced psychosis. The same group (Morales-Muñoz et al., 2014) tried to look at pre-pulse inhibition of the startle reflex as a feature distinguishing between cannabis-induced psychotic disorder and schizophrenia. Both showed deficits at the most preattentional levels, whereas cannabis-induced psychotic disorder patients performed better than patients with schizophrenia at higher attentional levels. These results suggest that cannabis-induced psychotic disorder involves a different group of patients than schizophrenia, and that deficits in pre-pulse inhibition functioning at 30 ms could be a useful psychophysiological measure to detect cannabis-induced psychotic disorder.

TREATMENT

As reported by Crippa et al. (2012) in their review of the literature on the pharmacological management of psychotic, anxious, and affective disorders related to cannabis intoxication, the scientific literature in this area is limited. There is a paucity of substance-specific studies, since most are "substance use" research studies without specifying the individual substance, such as cannabis.

Two double-blind, randomized, controlled trials (Berk, Brook, & Trandafir, 1999, Berk, Brook, & Nur, 2000) are available in the literature. The authors compared two different atypical antipsychotic agents—olanzapine 10 mg/day and risperidone 6 mg/day—with haloperidol 10 mg/day. Both medications appeared to be as effective as haloperidol in the treatment of cannabis-induced psychotic disorder, though olanzapine was associated with a lower rate of extrapyramidal side effects compared to haloperidol.

Rimonabant is a selective, potent, and orally active cannabinoid CB_1 receptor antagonist that exhibits high affinity for the CB_1 receptor. Huestis et al. (2001) administered rimonabant to 63 healthy men with a history of marijuana use, finding that a single dose of 90 mg was able to block acute psychological and physiological effects of smoked marijuana without altering THC pharmacokinetics. The same group then replicated the experiment (Huestis et al., 2007), finding that repeated, lower daily rimonabant doses (40 mg) attenuated the acute psychological and physiological effects of smoked cannabis to a similar degree as a single 90 mg dose. The lack of clinical trials of rimonabant, and its important side effect of inducing depressive episodes (Doggrell, 2008), limits its use in this context. More research is needed on this agent.

Cannabidiol (CBD) is the other main psychoactive compound of cannabis, in addition to THC. As reported by Iseger and Bossong (2015) in their literature review, CBD treatment of patients with psychotic symptoms confirms the potential of CBD as an effective, safe, and well-tolerated antipsychotic compound. It is as of yet unknown if the agent might treat cannabis-induced psychotic disorder.

Morgan, Schafer, Freeman, and Curran (2010) conducted a naturalistic study to assess if varying levels of CBD and THC have an impact on the memory and psychotomimetic effects of cannabis. A sample of cannabis was collected from each user and analyzed for levels of cannabinoids. On the basis of highest and lowest CBD content of cannabis, two groups of individuals were directly compared, finding no differences in the psychotomimetic effects, while a reduction in memory impairment was present in the high CBD content group. Englund et al. (2013) in a laboratory-based study, tested the hypothesis that pretreatment with CBD inhibited THC-elicited psychosis and cognitive impairment in healthy participants. The authors found that pretreatment with CBD inhibited THC-induced paranoia, as well as the detrimental effects of THC on episodic memory. In addition, CBD decreased the proportion of participants who experienced clinically significant acute THC psychosis. Even though CBD may reduce THC-related psychotic-like symptoms, no clinical studies have been conducted to prove efficacy in subjects affected by cannabis-induced psychotic disorder.

At present, patients in need of treatment should receive a symptom-guided care plan based on the use of acute psychopharmacological interventions, mainly benzodiazepines and second-generation antipsychotics. Such medications can be gently tapered (and discontinued, with close follow-up) upon complete resolution of psychotic symptoms.

EPIDEMIOLOGY AND CONVERSION OF CANNABIS-INDUCED PSYCHOTIC DISORDER TO PRIMARY PSYCHOTIC DISORDERS

It is difficult to find reliable data on the specific epidemiology of cannabis-induced psychotic disorder due to inconsistency in the classification system. Most of the available data pertain to substance-induced psychotic disorders more generally as a group. The most up-to-date study was published by Chen, Hsieh, Chang, Hung, and Chan (2015), who used the National Health Insurance Database to investigate the incidence and prevalence of alcohol-induced psychotic disorder and substance-induced psychotic disorders in Taiwan, and to determine the timeframes in which these two disorders develop into persistent psychotic conditions. No specific data were reported on the incidence of cannabis-induced psychotic disorder, but the incidence for substance-induced psychotic disorders in general was 3.093 per 100,000 person-years, similar to that derived by Kirkbride et al. (2009), which was 5 per 100,000 person-years, and by Weibell et al. (2013), which was 6 per 100,000 person-years.

Two important studies (Arendt, Rosenberg, Foldager, Perto, & Munk-Jørgensen, 2005; Niemi-Pynttäri et al., 2013) not only addressed the epidemiology of cannabis-induced psychotic disorder, but also focused on the rate of conversion to a schizophrenia-spectrum disorder. Arendt et al. (2005) studied inpatients and outpatients treated for cannabis-induced psychotic disorder (code F12.5 in ICD-10; World Health Organization, 1992) between 1994 and 1999; those with previous psychotic symptoms were excluded. Data were extracted from the Danish Psychiatric Central Register. They found an average incidence ratio of 2.7 per 100,000 person-years. The mean age at time of first treatment was 27.0 years. The mean length of follow-up was 5.9 years, during which time 44.5% of those diagnosed with cannabis-induced psychotic disorder were later diagnosed with a schizophrenia-spectrum disorder. Paranoid schizophrenia was the most common condition, followed by acute and transient psychotic disorders, personality disorders, and unspecified schizophrenia. The first-episode of schizophrenia-spectrum disorder occurred for 47.1% of the sample greater than one year after the diagnosis of cannabis-induced psychotic disorder. Male gender and younger age were related to increased risk of conversion. In line with these findings are also results from another study (Crebbin, Mitford, Paxton, & Turkington, 2009) investigating the incidence, course, diagnostic stability, and outcomes of first-episode, drug-induced psychosis. Data were collected in a county in Northern England between 1998 and 2005, at presentation and annual follow-up, for patients aged 16 years and older with a first episode of either drug-induced psychosis or schizophrenia. In this study, the authors found that one-third of patients who presented with a first-episode, drug-induced psychosis developed schizophrenia or a schizophrenia-like psychosis within two years. All were reported as using cannabis around the time of presentation.

Niemi-Pynttäri et al. (2013) investigated substance-induced psychosis and conversion to a schizophrenia-spectrum disorder. They used the nationwide Finnish Hospital Discharge Register to collect data on 18,478 inpatients discharged with a diagnosis of substance-induced psychotic disorder between 1987 and 2003 in Finland. Patients were followed until first occurrence of a schizophrenia-spectrum disorder, death, or December 2003. Within the sample, 125 persons (0.7%) had a diagnosis of cannabis-induced psychotic disorder, and this diagnosis had the highest conversion rate (12.5 per 100 person-years) and a 46% chance of being diagnosed with a schizophrenia-spectrum disorder in the 8 years following admission, much higher than amphetamine-, hallucinogen-, opioid-, and alcohol-induced psychoses.

CONCLUSIONS AND FUTURE DIRECTIONS

Cannabis, and more specifically THC, has been proven, through clinical observation and laboratory confirmation, to cause psychiatric symptoms as a result of intoxication and withdrawal. These transient effects include positive and negative psychotic symptoms. In clinical settings, the official nosological system identifies this entity as cannabis-induced psychotic disorder. Epidemiologic studies found an incidence for this disorder of about 2.7 per 100,000 person-years. Rate of conversion to schizophrenia-spectrum disorders ranges between 44.5% and 46% in 5–8 years of follow-up.

Diagnostic tools to identify cannabis-induced psychotic disorder have limited reliability, especially if used outside of research settings with highly trained interviewers. This is due to a lack of clear-cut criteria to define this disorder. Pathogenesis theories still allow for at least several pathways to a coherent model that explains the role of THC in causing psychotic symptoms. Psychopathological features of cannabis-induced psychotic disorder broadly overlap with symptoms of schizophrenia and other psychotic disorders. Differential diagnosis becomes particularly difficult with regard to the first episode of a psychotic disorder with concurrent cannabis use/abuse, which is very common. Currently, the treatment for cannabis-induced psychotic disorder is the same as for schizophrenia-spectrum disorders, though only two clinical trials have confirmed the efficacy of this approach.

This overview of cannabis-induced psychotic disorder highlights possible research questions that still need to be addressed, including:

- Does the type, strain, and THC concentration of cannabis affect the risk of cannabis-induced psychotic disorder?
- Does cannabis-induced psychotic disorder share common pathogenic pathways with primary psychotic disorders?
- What are the predictors of who will develop a cannabis-induced psychotic disorder after use of cannabis?
- What are the predictors of who will go on to "convert" to a primary psychotic disorder?
- How long should antipsychotics be continued after resolution of psychotic symptoms?
- Are alternative drugs—other than antipsychotics—available, effective, and safe in the treatment of cannabis-induced psychotic disorder?

Key Chapter Points

- The association between cannabis use and the occurrence of psychotic symptoms has been observed and described since the mid-19th century, however only in the last decade has causality been scientifically proven in laboratory settings.
- Current nosology does not provide clear-cut criteria to identify and accurately investigate the phenomenon of cannabis-induced psychosis in real clinical settings. This translates to a lack of studies that explore several aspects of cannabis-associated psychosis.
- Differential diagnosis between cannabis-induced psychotic disorder and primary psychosis is clinically challenging due to the overlap of clinical characteristics and the co-occurrence of cannabis use in subjects diagnosed with primary psychosis.
- No evidence-based guidelines exist for the treatment of cannabis-induced psychotic disorder. Psychopharmacological interventions should be symptom-guided and preference should be given to benzodiazepines and second-generation antipsychotics.
- Among all substance-induced psychotic disorders, cannabis-induced psychotic disorder has the highest conversion rate to a schizophrenia-spectrum disorders.

REFERENCES

American Psychiatric Association. (1987). *Diagnostic and statistical manual of mental disorders* (3rd ed., revised (DSM-III-R)). Washington, DC: American Psychiatry Press.

American Psychiatric Association. (1994). *Diagnostic and statistical manual of mental disorders* (4th ed. (DSM-IV-TR)). Washington, DC: American Psychiatry Press.

American Psychiatric Association. (2000). *Diagnostic and statistical manual of mental disorders* (4th ed., Text Revision (DSM-IV-TR)). Washington, DC: American Psychiatric Press.

American Psychiatric Association. (2015). *Diagnostic and statistical manual of mental disorders* (5th ed., Text Revision (DSM 5)). Washington, DC: American Psychiatric Press.

Ames, F. (1958). A clinical and metabolic Study of cute intoxication with cannabis sative and its role in the model psychoses. *Journal of Mental Science*, *104*, 972–999.

Arendt, M., Rosenberg, R., Foldager, L., Perto, G., & Munk-Jørgensen, P. (2005). Cannabis-induced psychosis and subsequent schizophrenia-spectrum disorders: Follow-up study of 535 incident cases. *The British Journal of Psychiatry*, *187*(6), 510–515.

Baldacchino, A., Hughes, Z., Kehoe, M., Blair, A., The, Y., Windeatt, S., et al. (2012). Cannabis psychosis: Examining the evidence for a distinctive psychopathology in a systematic narrative review. *The American Journal of Addictions*, *21*, S88–S98.

Basu, D., Malhotra, A., Bhagat, A., & Varma, V. K. (1999). Cannabis psychosis and acute schizophrenia. *European Addiction Research*, *5*(2), 71–73.

Berk, M., Brook, S., & Nur, F. (2000). Risperidone compared to haloperidol in cannabis-induced psychotic disorder: A double blind randomized controlled trial. *International Journal of Psychiatry in Clinical Practice*, *4*(2), 139–142.

Berk, M., Brook, S., & Trandafir, A. I. (1999). A comparison of olanzapine with haloperidol in cannabis-induced psychotic disorder: A double-blind randomized controlled trial. *International Clinical Psychopharmacology*, *14*(3), 177–180.

Bersani, G., Orlandi, V., Kotzalidis, G. D., & Pancheri, P. (2002). Cannabis and schizophrenia: Impact on onset, course, psychopathology and outcomes. *European Archives of Psychiatry and Clinical Neuroscience*, *252*(2), 86–92.

Bhattacharyya, S., Crippa, J. A., Allen, P., Martin-Santos, R., Borgwardt, S., Fusar-Poli, P., et al. (2012). Induction of psychosis by Δ^9-tetrahydrocannabinol reflects modulation of prefrontal and striatal function during attentional salience processing. *Archives of General Psychiatry*, *69*(1), 27–36.

Breivogel, C. S., & Sim-Selley, L. J. (2009). Basic neuroanatomy and neuropharmacology of cannabinoids. *International Review of Psychiatry*, *21*(2), 113–121.

Bullmore, E. T., Frangou, S., & Murray, R. M. (1997). The dysplastic net hypothesis: An integration of developmental and dysconnectivity theories of schizophrenia. *Schizophrenia Research*, *28*, 143–156.

Bullmore, E., & Sporns, O. (2009). Complex brain networks: Graph theoretical analysis of structural and functional systems. *Nature Reviews Neuroscience*, *10*, 186–198.

Caspari, D. (1999). Cannabis and schizophrenia: Results of a follow-up study. *European Archives of Psychiatry and Clinical Neuroscience*, *249*, 45–49.

Caton, C. L., Drake, R. E., Hasin, D. S., Dominguez, B., Shrout, P. E., Samet, S., et al. (2005). Differences between early-phase primary psychotic disorders with concurrent substance use and substance-induced psychoses. *Archives of General Psychiatry*, *62*(2), 137–145.

Chen, W. L., Hsieh, C. H., Chang, H. T., Hung, C. C., & Chan, C. H. (2015). The epidemiology and progression time from transient to permanent psychiatric disorders of substance-induced psychosis in Taiwan. *Addictive Behaviors*, *47*, 1–4.

Cortes-Briones, J. A., Cahill, J. D., Skosnik, P. D., Mathalon, D. H., Williams, A., Sewell, R. A., et al. (2015). The psychosis-like effects of Δ^9-tetrahydrocannabinol are associated with increased cortical noise in healthy humans. *Biological Psychiatry*, *78*(11), 805–813.

Crebbin, K., Mitford, E., Paxton, R., & Turkington, D. (2009). First-episode drug-induced psychosis: A medium term follow up study reveals a high-risk group. *Social Psychiatry and Psychiatric Epidemiology*, *44*(9), 710–715.

Crippa, J. A., Derenusson, G. N., Chagas, M. H., Atakan, Z., Martín-Santos, R., Zuardi, A. W., et al. (2012). Pharmacological interventions in the treatment of the acute effects of cannabis: A systematic review of literature. *Harm Reduction Journal*, *9*(1), 1.

Doggrell, S. A. (2008). Is rimonabant efficacious and safe in the treatment of obesity? *Expert Opinion on Pharmacotherapy*, *9*, 2727–2731.

Dragogna, F., Mauri, M. C., Marotta, G., Armao, F. T., Brambilla, P., & Altamura, A. C. (2014). Brain metabolism in substance-induced psychosis and schizophrenia: A preliminary PET study. *Neuropsychobiology*, *70*(4), 195–202.

D'Souza, D. C., Braley, G., Blaise, R., Vendetti, M., Oliver, S., Pittman, B., et al. (2008). Effects of haloperidol on the behavioral, subjective, cognitive, motor, and neuroendocrine effects of delta-9-tetrahydrocannabinol in humans. *Psychopharmacology*, *198*(4), 587–603.

D'Souza, D. C., Perry, E., MacDougall, L., Ammerman, Y., Cooper, T., Wu, Y. T., et al. (2004). The psychotomimetic effects of intravenous delta-9-tetrahydrocannabinol in healthy individuals: Implications for psychosis. *Neuropsychopharmacology*, *29*, 1558–1572.

Dubertret, C., Bidard, I., Adès, J., & Gorwood, P. (2006). Lifetime positive symptoms in patients with schizophrenia and cannabis abuse are partially explained by co-morbid addiction. *Schizophrenia Research*, *86*(1), 284–290.

Englund, A., Morrison, P. D., Nottage, J., Hague, D., Kane, F., Bonaccorso, S., et al. (2013). Cannabidiol inhibits THC-elicited paranoid symptoms and hippocampal-dependent memory impairment. *Journal of Psychopharmacology*, *27*(1), 19–27.

Fletcher, P., McKenna, P. J., Friston, K. J., Frith, C. D., & Dolan, R. J. (1999). Abnormal cingulate modulation of fronto-temporal connectivity in schizophrenia. *NeuroImage*, *9*, 337–342.

Friston, K. J. (1999). Schizophrenia and the disconnection hypothesis. *Acta Psychiatrica Scandinavica Supplement*, *395*, 68–79.

Greicius, M. D., Krasnow, B., Reiss, A. L., & Menon, V. (2003). Functional connectivity in the resting brain: A network analysis of the default mode hypothesis. *Proceedings of the National Academy of Sciences*, *100*, 253–258.

Gusnard, D. A., Akbudak, E., Shulman, G. L., & Raichle, M. E. (2001). Medial pre-frontal cortex and self-referential mental activity: Relation to a default mode of brain function. *Proceedings of the National Academy of Sciences*, *98*, 4259–4264.

Hasin, D. S., Trautman, K. D., Miele, G. M., & Samet, S. (1996). Psychiatric Research Interview for Substance and Mental Disorders (PRISM): Reliability for substance abusers. *The American Journal of Psychiatry*, *153*(9), 1195–1201.

Higashima, M., Takeda, T., Kikuchi, M., Nagasawa, T., Hirao, N., Oka, T., et al. (2007). State-dependent changes in intrahemispheric EEG coherence for patients with acute exacerbation of schizophrenia. *Psychiatry Research*, *149*, 41–47.

Huestis, M. A., Boyd, S. J., Heishman, S. J., Preston, K. L., Bonnet, D., Le Fur, G., et al. (2007). Single and multiple doses of rimonabant antagonize acute effects of smoked cannabis in male cannabis users. *Psychopharmacology*, *194*(4), 505–515.

Huestis, M. A., Gorelick, D. A., Heishman, S. J., Preston, K. L., Nelson, R. A., Moolchan, E. T., et al. (2001). Blockade of effects of smoked marijuana by the CB1-selective cannabinoid receptor antagonist SR141716. *Archives of General Psychiatry*, *58*(4), 322–328.

Imade, A., & Ebie, J. (1991). A retrospective study of symptom patterns of cannabis-induced psychosis. *Acta Psychiatrica Scandinavica*, *83*, 134–136.

Iseger, T. A., & Bossong, M. G. (2015). A systematic review of the antipsychotic properties of cannabidiol in humans. *Schizophrenia Research*, *162*(1), 153–161.

Kaufmann, R. M., Kraft, B., Frey, R., Winkler, D., Weiszenbichler, S., Bäcker, C., et al. (2010). Acute psychotropic effects of oral cannabis extract with a defined content of Delta9-tetrahydrocannabinol (THC) in healthy volunteers. *Pharmacopsychiatry*, *43*(1), 24–32.

Kirkbride, J. B., Croudace, T., Brewin, J., Donoghue, K., Mason, P., Glazebrook, C., et al. (2009). Is the incidence of psychotic disorder in decline? Epidemiological evidence from two decades of research. *International Journal of Epidemiology*, *38*(5), 1255–1264.

Lawrie, S. M., Buechel, C., Whalley, H. C., Frith, C. D., Friston, K. J., & Johnstone, E. C. (2002). Reduced frontotemporal functional connectivity in schizophrenia associated with auditory hallucinations. *Biological Psychiatry*, *51*, 1008–1011.

Leweke, F. M., Gerth, C. W., & Klosterkoetter, J. (2004). Cannabis-associated psychosis. Current status of research. *CNS Drugs*, *18*(13), 895–910.

Mathers, D., & Ghodse, A. (1992). Cannabis and psychotic illness. *British Journal of Psychiatry*, *161*, 648–653.

Moreau, J. (1973). *Hashish and mental illness*. New York, NY: Raven.

McGuire, P., Jones, P., & Harvey, I. (1994). Cannabis and acute psychosis. *Schizophrenia Research*, *13*, 161–167.

Modestin, J., Gladen, C. J., & Christen, S. (2001). A comparative study on schizophrenic patients with dual diagnosis. *Journal of Addictive Diseases*, *20*, 41–51.

Morales-Muñoz, I., Jurado-Barba, R., Ponce, G., Martínez-Gras, I., Jiménez-Arriero, M.Á., Moratti, S., et al. (2014). Characterizing cannabis-induced psychosis: A study with prepulse inhibition of the startle reflex. *Psychiatry Research*, *220*(1), 535–540.

Morgan, C. J., Schafer, G., Freeman, T. P., & Curran, H. V. (2010). Impact of cannabidiol on the acute memory and psychotomimetic effects of smoked cannabis: Naturalistic study. *The British Journal of Psychiatry*, *197*(4), 285–290.

Morrison, P. D., Zois, V., McKeown, D. A., Lee, T. D., Holt, D. W., Powell, J. F., et al. (2009). The acute effects of synthetic intravenous Δ9-tetrahydrocannabinol on psychosis, mood and cognitive functioning. *Psychological Medicine*, *39*, 1607–1616.

Niemi-Pynttäri, J. A., Sund, R., Putkonen, H., Vorma, H., Wahlbeck, K., & Pirkola, S. P. (2013). Substance-induced psychoses converting into schizophrenia: A register-based study of 18,478 Finnish inpatient cases. *The Journal of Clinical Psychiatry*, *74*(1), 1–478.

Nunez, L. A., & Gurpegui, M. (2002). Cannabis induced psychosis: A cross-sectional comparison with acute schizophrenia. *Acta Psychiatrica Scandinavica*, *105*, 173–178.

O'Shaughnessy, W. B. (1843). On the preparations of the Indian Hemp, or Gunjah: Cannabis Indica their effects on the animal system in health, and their utility in the treatment of tetanus and other convulsive diseases. *Provincial Medical Journal and Retrospect of the Medical Sciences*, *5*(123), 363.

Onyango, R. (1986). Cannabis psychosis in young psychiatric inpatients. *British Journal of Addiction*, *81*, 419–423.

Pomarol-Clotet, E., Salvador, R., Sarro, S., Gomar, J., Vila, F., Martinez, A., et al. (2008). Failure to deactivate in the prefrontal cortex in schizophrenia: Dysfunction of the default mode network? *Psychological Medicine*, *38*, 1185–1193.

Raichle, M. E., MacLeod, A. M., Snyder, A. Z., Powers, W. J., Gusnard, D. A., & Shulman, G. L. (2001). A default mode of brain function. *Proceedings of the National Academy of Sciences*, *98*, 676–682.

Ramström, J. (2004). *Adverse health consequences of cannabis use*. Sweden: National Institute of Public Health.

Rottanburg, D., Ben-Arie, O., Robins, A., Teggin, A., & Elk, R. (1982). Cannabis-associated psychosis with hypomanic features. *The Lancet*, *320*(8312), 1364–1366.

Rubio, G., Marín-Lozano, J., Ferre, F., Martínez-Gras, I., Rodriguez-Jimenez, R., Sanz, J., et al. (2012). Psychopathologic differences between cannabis-induced psychoses and recent-onset primary psychoses with abuse of cannabis. *Comprehensive Psychiatry*, *53*(8), 1063–1070.

Salgado-Pineda, P., Fakra, E., Delaveau, P., McKenna, P. J., Pomarol-Clotet, E., & Blin, O. (2011). Correlated structural and functional brain abnormalities in the default mode network in schizophrenia patients. *Schizophrenia Research*, *125*, 101–109.

Sewell, R. A., Ranganathan, M., & D'Souza, D. C. (2009). Cannabinoids and psychosis. *International Review of Psychiatry*, *21*(2), 152–162.

Stephan, K. E., Friston, K. J., & Frith, C. D. (2009). Dysconnection in schizophrenia: From abnormal synaptic plasticity to failures of self-monitoring. *Schizophrenia Bulletin*, *35*, 509–527.

Talbott, J. A., & Teague, J. W. (1969). Marijuana psychosis: Acute toxic psychosis associated with the use of cannabis derivatives. *JAMA*, *210*, 299–302.

Thacore, V., & Shukla, S. (1976). Cannabis psychosis and paranoid schizophrenia. *Archives of General Psychiatry*, *33*, 383–386.

Thornicroft, G., Meadows, G., & Politi, P. (1992). Is cannabis psychosis a distinct category? *European Psychiatry*, *7*, 277–282.

Torrens, M., Serrano, D., Astals, M., Pérez-Domínguez, G., & Martín-Santos, R. (2004). Diagnosing comorbid psychiatric disorders in substance abusers: Validity of the Spanish versions of the Psychiatric Research Interview for Substance and Mental Disorders and the Structured Clinical Interview for DSM-IV. *American Journal of Psychiatry*, *161*(7), 1231–1237.

Warnock, J. (1903). Insanity from Hasheesh. *Journal of Mental Science*, *49*, 96–110.

Weibell, M. A., Joa, I., Bramness, J., Johannessen, J. O., McGorry, P. D., ten Velden Hegelstad, W., et al. (2013). Treated incidence and baseline characteristics of substance induced psychosis in a Norwegian catchment area. *BMC Psychiatry*, *13*(1), 1.

Wolf, N. D., Sambataro, F., Vasic, N., Frasch, K., Schmid, M., Schönfeldt-Lecuona, C., et al. (2011). Dysconnectivity of multiple resting-state networks in patients with schizophrenia who have persistent auditory verbal hallucinations. *Journal of Psychiatry & Neuroscience*, *36*, 366–374.

World Health Organization. (1971). *Technical report series no. 478: The use of cannabis*. Geneva: World Health Organization.

World Health Organization. (1992). *The ICD-10 classification of mental and behavioral disorders*. Geneva: World Health Organization.

Chapter 9

Synthetic Cannabinoids and Synthetic Cannabinoid-Induced Psychotic Disorders

Papanti Duccio[*,†], **Schifano Fabrizio**[†], **Orsolini Laura**[†,‡]
[*]*Drug Addiction Centre, Latisana, Italy* [†]*University of Hertfordshire, Hatfield, United Kingdom*
[‡]*Polyedra Research Group, Teramo, Italy*

INTRODUCTION

Both phytocannabinoids, the psychoactive compounds naturally present in the cannabis plant, and synthetic cannabinoids (SCs; also known as synthetic cannabimimetics) exert their effects through alteration of the endocannabinoid system (Szabo, 2014). There are at least 104 different phytocannabinoids in cannabis, and their combination contributes to the psychoactive effects of cannabis (ElSohly & Gul, 2014). In 1967, the United States (US) National Institute of Mental Health warned law enforcement authorities that, in the future, organized crime might begin producing and marketing "synthetic marijuana" products (Burnham, 1967). Although interest in the discovery and consumption of cannabis-like compounds started during the 1970s (Brown & Malone, 1978; Siegel, 1976), the production of SCs was largely ignored by underground, clandestine chemists over the next two decades, essentially due to the relative availability of natural Δ^9-tetrahydrocannabinol (THC) (the main psychoactive cannabinoid present in cannabis) (Karel & Arrizabalaga, 1998). Most SCs were initially synthesized during the 1990s for biomedical research purposes, in parallel with an increasing understanding of the endocannabinoid system. At present, SCs (see Fig. 1) constitute the largest group of novel psychoactive substances (NPSs) monitored by the European Monitoring Centre for Drugs and Drug Addiction (EMCDDA, 2015). NPSs are typically defined as new narcotic/psychotropic drugs that are not controlled by the United Nations' 1961 Narcotic Drugs and 1971 Psychotropic Substances Conventions, but which may pose a public health threat. It has been suggested that over 700 SC compounds are in fact available, and it is likely that even this is an underestimate (ACMD, 2014). Current high levels of SC availability probably reflect both the overall demand

The Complex Connection between Cannabis and Schizophrenia. http://dx.doi.org/10.1016/B978-0-12-804791-0.00009-4

FIG. 1 Synthetic cannabinoids in their powder/pure form, after being synthesized. *(Image taken from the Deep Web.)*

for cannabis-like products and the rapidity with which manufacturers produce and supply new SCs to avoid ever-changing drug controls (EMCDDA, 2015).

SYNTHETIC CANNABINOIDS

SC-based preparations (see Fig. 2) are presumably used to achieve a cannabis-like intoxication. SCs have been available on the recreational drug market since 2004, and their psychoactive effects were first described in 2008 (Auwärter et al., 2009; Steup, 2008). Typically, SC compounds are marketed to potential customers as substances that are not controlled (e.g., "legal highs"). They are

FIG. 2 Synthetic cannabinoid products as they appear after the vegetable substrate has been mixed with synthetic cannabinoid compounds. *(Image taken from the Deep Web.)*

often "reformulated" with subtle structural variations into new compounds as soon as they are scheduled as controlled to attempt to avoid penalties to producers and dealers. Furthermore, SCs are usually contained in packaging labeled "not for human consumption" and lacking information about the compounds in attempt to circumvent food, drug, and medicine safety laws (see Fig. 3). SCs can be purchased, either wholesale or retail, from a range of sources, including head shops, gas stations, street dealers, and the internet (both at the surface and deep/darknet levels) (Daly, 2013; Papanti, Orsolini, Francesconi, & Schifano, 2014; Power, 2013).

SC preparations are often composed of a dried plant base, to mimic the look and texture of marijuana, and sprayed with a mixture of SC molecules and/or other entirely different compounds involved in endocannabinoids' metabolism/transportation (Schifano, Papanti, Orsolini, & Corkery, 2016). SCs are usually identified by their brands, often reminiscent of street names for cannabis strains (e.g., Spice, K2, Northern Lights, Skunk, Amnesia, Kronic, Kush), and sold in 0.5–3-gram metal-foil sachets. Other available SC preparations look like hashish, or are found in capsules, e-liquids (i.e., vaporizable liquid solutions for electronic cigarettes), or tablets, or are sprayed on herbal cannabis (EMCDDA, 2015). SC intake typically occurs by inhalation from a joint, bong, or pipe (World Health Organization, 2014), but insufflation, oral ingestion, rectal

FIG. 3 Appearance of synthetic cannabinoid product packages. *(Adapted from Bonavigo, T., Papanti, D., & Schifano, F. (2013). (Fake) pot art. In Second international conference on novel psychoactive substances (NPS), Swansea, UK; September 12–13th, 2013 (p. 281). Swansea, UK. Available at: http://www.mededlive.co.uk/nps_social/documents/posters/Duccio.pdf.)*

administration, and injection have been described as well (Schifano, Orsolini, Duccio Papanti, & Corkery, 2015).

Synthetic Cannabinoid Use

Over the past several years, and especially in male adolescents, SCs have gained popularity (EMCDDA, 2015; Palamar & Acosta, 2015; Papanti et al., 2014; Substance Abuse and Mental Health Services Administration, Center for Behavioral Health Statistics and Quality, October 16, 2014; Winstock & Barratt, 2013a, 2013b), possibly due to a number of factors, including their wide availability in a quasi-legal market, difficulty of detection, and affordability (Papanti et al., 2013). While cannabis use can be detected with a simple screening test, there are no readily available point-of-care tests for SCs (Naviglio, Papanti, Moressa, & Ventura, 2015). Indeed, due to the lack of appropriate reference samples, it is difficult to identify SC compounds even with more sophisticated gas chromatography detection tests (Assi, Fergus, Stair, Corazza, & Schifano, 2011). The challenges in detecting SCs make them attractive for people in monitored settings where regular drug tests are being carried out, including forensic psychiatric wards, residential drug treatment facilities, prisons, probation services, the military, and certain workplaces. SCs might also be attractive for both athletes undergoing anti-doping testing and driving license candidates (Abdulrahim & Bowden-Jones, 2015; Schifano et al., 2015; see Table 1).

As SCs were synthesized for preclinical and nonclinical research purposes, most compounds had never been tested in animals or humans prior to their emergence on the recreational drug scene (Thomas, Wiley, Pollard, & Grabenauer, 2014). Therefore, thus far, most available SC-related information is anecdotal, e.g., from internet fora and surveys, calls to Poison Control Centers, and case reports in the literature (Papanti et al., 2013; Spaderna, Addy, & D'Souza, 2013). Because of this lack of information, it is not surprising that many SC users do not seem aware of the potential serious adverse effects; these compounds may be perceived to be similar to marijuana and hence "safe" and "natural" (Schifano et al., 2009).

TABLE 1 Reasons for Synthetic Cannabinoid Consumption
Relative availability compared with cannabis
Misperception of safety
Misperception of legality, and circumvention of drug-related prosecution
Undetectability in drug screening tests, and circumvention of detection
Curiosity
Desirable psychoactive effects
Cannabis withdrawal relief/cannabis use disorder self-medication

Composition and Pharmacology

Within any given SC-containing package, at least a few different SC compounds are typically identified (Kikura-Hanajiri, Uchiyama, Kawamura, & Goda, 2012), with batches of the same brand possibly exhibiting high variability in terms of specific SC molecules and concentrations (Choi et al., 2013). SCs include compounds with completely different structures, chemical properties, potencies, pharmacokinetics, and pharmacodynamics. The natural phytocannabinoid that SCs are meant to mimic, THC, has a dibenzofuran structure, which is not identified in most SC molecules. Conversely, a range of SC molecules contain nitrogen atoms in their scaffolds, which are not found in phytocannabinoids. While THC is a partial agonist at cannabinoid receptors, CB_1 and CB_2, SCs can be full or even super-agonists (De Luca, Bimpisidis, et al., 2015; De Luca, Castelli, et al., 2015; Pertwee, 2008), hence exhibiting high levels of affinity while eliciting maximal activity at cannabinoid receptors (Schifano et al., 2016). Furthermore, THC's effects are modulated or dampened by the presence of other natural compounds identified in cannabis; e.g., terpenoids, cannabidiol (CBD), and tetrahydocannabivarin (Papanti et al., 2014), while no such “modulating” compounds are detected in SC-containing products (Brents & Prather, 2014). Apart from the CB_1/CB_2 effects, several SCs have been found to possess both monoamine oxidase (MAO) and serotonin reuptake inhibiting properties (Fišar, 2012), while possibly interacting with a range of other receptors, including nicotinic acetylcholine, glycine, and/or ionotropic glutamate/NMDA receptors (Adis R&D Profile, 2003; Pertwee, 2010; Pertwee et al., 2010). As components of the structure or as substituents, a number of SC molecules incorporate indole-derived moieties (Lewin, Seltzman, Carroll, Mascarella, & Reddy, 2014), typically identified in indoleamine hallucinogens such as dimethyltryptamine (Halberstadt & Geyer, 2011). From this point of view, it could be argued that the ingestion of indole SC compounds may be associated with significant levels of serotonin (5-HT) receptor activation (Wells & Ott, 2011; Yip & Dart, 2014). Overall, SC activity at noncannabinoid receptors may well contribute to the complex clinical effects observed (Schifano et al., 2016).

As SC preparations almost always contain multiple SCs in a single batch (Abdulrahim & Bowden-Jones, 2015; Schifano et al., 2015), there is a potential for single-product drug–drug interactions. Furthermore, some SC metabolites retain both affinity and activity at CB_1 receptors, hence delaying and/or intensifying receptor activation and contributing to the toxicity of these products (Fantegrossi, Moran, Radominska-Pandya, & Prather, 2014). The recent trend of SC fluorination, commonly applied in medicinal chemistry, may increase the compounds' lipophilicity, hence promoting their absorption through biological membranes and the blood–brain barrier (Ismail, 2002; Wilkinson, Banister, & Kassiou, 2015), and therefore possibly enhancing their overall toxicity (Fantegrossi et al., 2014). A recent trend has also been reported whereby other substances with psychoactive properties have been identified in SC-containing preparations, including: *Mitragyna speciosa* (Ogata, Uchiyama,

Kikura-Hanajiri, & Goda, 2013); the beta-2 adrenergic agonist clenbuterol (Ashton, 2012); benzodiazepines such as phenazepam (Park et al., 2013); tryptamines (Park et al., 2013); phenethylamine derivatives (Uchiyama, Shimokawa, et al., 2013); cathinones (Uchiyama, Matsuda, Kawamura, Kikura-Hanajiri, & Goda, 2013); MAO inhibitors such as harmine and harmaline (Dresen et al., 2010); opioid receptor agonists such as AH-7921 and *O*-desmethyltramadol (Dresen et al., 2010; Uchiyama, Matsuda, et al., 2013); anesthetics such as benzocaine (Lonati et al., 2014); and NMDA receptor antagonists (Wurita, Hasegawa, Minakata, Watanabe, & Suzuki, 2014). Other factors potentially contributing to SC product toxicity include: the pharmacological activity of SC pyrolysis by-products (Bell & Nida, 2015; Fattore, 2016); the presence of contaminants, side-products, and solvents in the packaging (Abdulrahim & Bowden-Jones, 2015); the total lack of product quality control, leading to significant differences in concentration ("hot-spots") of SCs within herbal and e-liquid formulations (Baggaley, 2015; Papanti et al., 2014); and an increased vulnerability to SC-related adverse effects in some users, due to either preexisting conditions (e.g., psychiatric disorders) and/or concurrent intake of other drugs (Fantegrossi et al., 2014; Papanti et al., 2013).

ACUTE EFFECTS, LETHALITY, TOLERANCE, AND WITHDRAWAL

Although there are similarities in terms of psychoactive effects between low-dosage SC and cannabis/THC intake, both visual/auditory hallucinations and intense paranoia typically occur only with higher levels of SC intake. SC-related perceptual disturbances have been described as "fractals/geometric patterns," "trails," and "flashes of color," often combined with relaxation and increased creativity (Spaderna et al., 2013). In comparison to cannabis, SC intake may also be associated with a shorter duration of action and more intense "hangover" effects (Winstock & Barratt, 2013b). High SC dosages are associated with significant anxiety (Wessinger, Moran, & Seely, 2015), together with a range of other unpleasant experiences (e.g., "bad trips," "death trips," Çoban, 2014). "Bad trips" can be characterized by a range of symptoms including suspiciousness/paranoia, altered experience of self, and dissociative effects, such as sensations of living in different/parallel realities (Bilgrei, 2016; Kjellgren, Henningsson, & Soussan, 2013; Soussan & Kjellgren, 2014).

The psychoactive effects of SCs may be more intense in individuals with minimal previous exposure to cannabis (Hermanns-Clausen, Kneisel, Szabo, & Auwärter, 2013). Adverse SC-related effects may be severe; indeed, SC intake is associated with a 30-fold higher risk of seeking emergency room care as compared to natural cannabis (Winstock, Lynskey, Borschmann, & Waldron, 2015). Acute SC intoxication seems to be at times similar to the clinical picture associated with the use of stimulant/sympathomimetic recreational drugs (Baumann et al., 2014; Naviglio et al., 2015; Wood & Dargan, 2012). On other occasions, a short-acting but potentially life-threatening serotonin syndrome-like clinical presentation may be observed, with reported signs and symptoms of elevated heart rate and blood pressure, mydriasis, agitation/anxiety, hyperglycemia, dyspnea or tachypnea, nausea or vomiting, and seizures (Hermanns-Clausen et al., 2013; Naviglio et al., 2015; Schifano et al., 2016). Other SC-related acute adverse effects include self-injurious or aggressive behavior, hyperemesis, nystagmus, stroke, myocardial infarction, and acute kidney injury (Mir, Obafemi, Young, & Kane, 2011; Centers for Disease Control and Prevention (CDC) (CDC), 2013; Freeman et al., 2013; Hopkins & Gilchrist, 2013; Louh & Freeman, 2014; Rose et al., 2015; see Table 2). Presentations of agitated/excited delirium have recently been described in association with SC intoxication (Berry-Caban, Ee, Ingram, Berry, & Kim, 2013; Kasper et al., 2015; Trecki, Gerona, & Schwartz, 2015). Individuals experiencing this may be very aggressive, hallucinating, combative, and tachycardic for up to several days (Schwartz et al., 2015; Spaderna et al., 2013; Vilke et al., 2012).

TABLE 2 Signs, Symptoms, and Adverse Effects of Synthetic Cannabinoid Intoxication

Agitation
Auditory/visual hallucinations
Delusions
Tachycardia
Tachypnea
Hypertension
Mydriasis
Seizures
Hypokalaemia
Nausea
Vomiting/hyperemesis
Toxic hepatitis/liver failure
Acute kidney injuries
Dysrhythmia
Acute coronary syndrome
Stroke
Rhabdomyolysis
Serotonin syndrome

TABLE 3 Signs and Symptoms of Synthetic Cannabinoid Withdrawal

Drug craving
Agitation
Anxiety/irritability
Feelings of emptiness/depressive symptoms
Nightmares/insomnia
Tachycardia
Tremor
Profuse sweating
Headache
Vomiting
Malaise

Overall, most clinicians do not appear to be well trained in terms of treatment/management issues related to SC use and intoxication (Lank, Pines, & Mycyk, 2013). A number of deaths have been related to SC ingestion, either on their own or in combination with other substances (Corkery et al., 2013; Kronstrand, Roman, Andersson, & Eklund, 2013; Saito et al., 2013; Schaefer et al., 2013; Shanks, Dahn, & Terrell, 2012; Wikström, Thelander, Dahlgren, & Kronstrand, 2013). In contrast, there are no reports of fatal cannabis overdoses in the epidemiological literature (Calabria, Degenhardt, Hall, & Lynskey, 2010).

Both tolerance and symptoms of withdrawal have been described in association with SC intake, suggesting that SCs may have a relatively high misuse and dependence liability (Gunderson, Haughey, Ait-Daoud, Joshi, & Hart, 2012; Spaderna et al., 2013; Vandrey, Dunn, Fry, & Girling, 2012). The SC withdrawal syndrome may be characterized by drug craving, tachycardia, tremor, profuse sweating, nightmares or insomnia, headache, anxiety/irritability, feelings of emptiness or depressive symptoms, and somatic complaints (Nacca et al., 2013; Zimmermann et al., 2009; Rominger et al., 2013; Macfarlane & Christie, 2015; see Table 3).

Δ^9-TETRAHYDROCANNABINOL AND PSYCHOSIS

THC intake has been shown to produce dose-related onset of and/or increases in positive and negative psychotic symptoms in both individuals with and without a psychotic disorder (D'Souza et al., 2004; Henquet et al., 2005; Mason et al., 2009; Miettunen et al., 2008; Morrison et al., 2009; Murray, Paparelli, Morrison, Marconi, & Di Forti, 2013; van Os et al., 2002; Verdoux, Sorbara, Gindre, Swendsen, & van Os, 2003; Kelley et al., 2016; WHO, 2016). Indeed, intravenous THC administration constitutes one of the current pharmacological models of psychosis (Morrison et al., 2009; Steeds, Carhart-Harris, & Stone, 2015). Furthermore, the available evidence points to a causal role for THC in schizophrenia-spectrum disorders (WHO, 2016), with findings pointing to a dose–response relationship between levels of THC ingested and risk for psychosis (Marconi, Di Forti, Lewis, Murray, & Vassos, 2016). In contrast, the consumption of cannabis with relatively high CBD content (putatively a weak CB_1 antagonist with anxiolytic, antipsychotic, and anticraving properties; Pertwee, 2008; Leweke et al., 2012; Morgan, Das, Joye, Curran, & Kamboj, 2013) seems to be associated with fewer psychotic experiences (Schubart et al., 2011).

"Skunk" (cannabis with THC concentrations of 12%–18% and virtually no CBD; Morgan & Curran, 2008), on the other hand, is associated with an increased risk and severity of cannabis dependence (Freeman & Winstock, 2015), higher risk for psychotic disorders (Di Forti et al., 2014, 2015), and a lower age at onset of psychosis (WHO, 2016). A recent trend has been observed with the use of butane hash oil, a cannabis-derived product approaching 80% THC concentration (Stogner & Miller, 2015). This preparation has been associated with new-onset psychotic episodes in otherwise healthy individuals (Pierre, Gandal, & Son, 2016), rapid development of tolerance, and withdrawal symptoms (Loflin & Earleywine, 2014).

SYNTHETIC CANNABINOIDS AND PSYCHOSIS

SC preparations do not contain any CBD (or other modulating compounds), and many SCs are high potency, full agonists at CB_1 (De Luca, Castelli, et al., 2015; Fattore, 2016). Some SCs have been tested as medications, but their use has been associated with adverse psychotogenic effects. Levonantradol, an anti-nausea and analgesic SC medication, was withdrawn from the market because of unacceptable psychiatric side effects (e.g., hallucinations, perceptual and thought disturbances, paranoia, dissociation, depersonalization; D'Souza, Sewell, & Ranganathan, 2009). Another SC compound, Org 26828, has been tested in humans for its sedative, amnestic, analgesic, and anti-emetic properties. However, the trial was suspended due to the occurrence of a range of severe psychotic-like adverse effects (Zuurman et al., 2010).

So far, no longitudinal studies have evaluated the long-term effects of SC ingestion in humans. Psychotic disorders associated with SC intake can be conceptualized as transient, acute psychotic episodes (Bebarta, Ramirez, & Varney, 2012; Hermanns-Clausen et al., 2013; Hurst, Loeffler, & McLay, 2011; Peglow, Buchner, & Briscoe, 2012; Rodgman, Kinzie, & Leimbach, 2011; Vearrier & Osterhoudt, 2010); "ex novo," long-standing/persistent psychotic disorders ("Spiceophrenia"; Benford & Caplan, 2011; Hurst et al., 2011; Van Der Veer & Friday, 2011; Papanti et al., 2013); and relapse or worsening of a preexisting psychosis (Every-Palmer, 2011; Hermanns-Clausen et al., 2013; Müller et al., 2010; Tung, Chiang, & Lam, 2012).

Acute psychotic reactions can occur following a single exposure to or with repeated use of SCs, and include a wide range of clinically relevant psychopathological symptoms, including perceptual alterations, depersonalization, dissociation, illusions, auditory and visual hallucinations, paranoid delusions, bizarre/disorganized behavior and speech, catatonia, agitation/aggression, and suicidal ideation or behavior (Papanti et al., 2013). Following SC consumption, new psychopathological reactions superimposed on a preexisting mental disorder (e.g., the emergence of manic-like symptoms in individuals with schizophrenia and no previous affective symptoms) have been described as well (Celofiga, Koprivsek, & Klavz, 2014; Müller et al., 2010).

The occurrence of hallucinations and/or delusions is less likely with cannabis than with SC consumption, having been observed in 2% and 11.2% of users, respectively (Forrester, Kleinschmidt, Schwarz, & Young, 2012). Furthermore, in comparison to cannabis, SC-related psychotic episodes are associated with more frequent agitation and behavioral dyscontrol (Brakoulias, 2012). Overall, comparative studies of SC versus cannabis users admitted to psychiatric units show that SC users are generally younger, and present with higher rates of involuntary admission, greater severity of disease, more frequent aggression, and longer lengths of stay (Glue et al., 2013; Shoval, 2015; see Table 4). Interestingly, a definitive "challenge—re-challenge" relationship between SC intake and psychotic reactions has been described, such that the consumption of SCs temporally precedes the psychotic reaction, which attenuates or largely regresses after the discontinuation, and reappears on repeated exposure, suggesting a cause–effect relationship in the elicitation of psychosis (Begaud, 1984; Karch & Lasagna, 1975; Papanti et al., 2013; Peglow et al., 2012). Persistent hallucinations after the repeated use of SCs has also been described (Lerner, Goodman, Bor, & Lev-Ran, 2014). Finally, a number of deaths by suicide following SC intake have been documented in the literature (Lászik et al., 2015; Patton et al., 2013; Rosenbaum, Carreiro, & Babu, 2012; Shanks et al., 2012).

TABLE 4 Features Associated With Synthetic Cannabinoid- Versus Cannabis-Related, Psychoses

Higher frequency of hallucinations/delusions
Higher occurrence of agitation/behavioral dyscontrol
Higher rates of involuntary admission
Greater severity of disease
Higher levels of aggression
Longer lengths of inpatient stay

TREATMENT OF SYNTHETIC CANNABINOID-INDUCED PSYCHOSIS

In SC-induced psychosis associated with agitation, the use of oral or parenteral benzodiazepines is a reasonable option, with the additional advantage of preventing the possible occurrence of seizures, which have been associated with SC intoxication (Manseau, 2016). According to existing systematic reviews and guidelines, there is no specific recommended antipsychotic for the treatment of SC-induced psychosis (Papanti et al., 2013; Tait, Caldicott, Mountain, Hill, & Lenton, 2015). In the case of antipsychotic administration, an electrocardiogram may need to be performed to monitor parameters such as the QTc interval, because SC users may present with cardiac abnormalities as well as vomiting and electrolyte alterations such as hypokalaemia, further complicating the clinical picture. In addition, it is important to note that antipsychotics may also lower the threshold for seizures (Monte et al., 2014; Naviglio et al., 2015).

More generally, in treating substance-induced psychotic disorders, second-generation antipsychotics may confer some advantages over first-generation antipsychotics (San, Arranz, & Martinez-Raga, 2007). Many second-generation antipsychotics rapidly dissociate from dopamine receptors, which may dampen responses in the reward system less, hence not heightening cravings (because of a differential effect on dopaminergic neurotransmission) (Juckel et al., 2006; Kapur & Seeman, 2000). In fact, some second-generation agents may even increase dopamine levels in the nucleus accumbens shell (Tanda et al., 2015). Studies have shown that both clozapine and olanzapine may present distinct advantages in reducing substance-induced psychotic symptoms without increasing craving in patients with concomitant substance use disorders (Green, Noordsy, Brunette, & O'Keefe, 2008; Machielsen et al., 2012). Additionally, second-generation antipsychotics act as antagonists at serotonin 5-HT_{2A} receptors, the main target of most hallucinogenic drugs (Valeriani et al., 2015).

CONCLUSIONS

SCs represent a large class of NPSs mainly used by young adults. Over the past decade, their intake has been increasingly associated with morbidity and mortality (Fantegrossi et al., 2014). It appears that SC use may be a significant and relevant factor in precipitating and/or perpetuating psychosis in both healthy and vulnerable individuals (Papanti et al., 2013). On the other hand, due to the relatively recent appearance of SC products and sparse scientific literature, the exact risk of developing psychosis following SC use cannot be calculated. Furthermore, the likelihood of comorbid substance use typically described in SC users and the lack of adequate toxicological screening tools present significant barriers to understanding the role of SCs in directly causing psychotic disturbances. Finally, there is large chemical/pharmacological heterogeneity among SC compounds, which may further limit the interpretation of the available, mostly anecdotal, evidence (Papanti et al., 2013). Adolescents may be particularly vulnerable to drugs that interact with cannabinoid receptors, and so parents and educators should be cautioned to monitor for SC use (Chadwick, Miller, & Hurd, 2013; Keshavan, Giedd, Lau, Lewis, & Paus, 2014; Malone, Hill, & Rubino, 2010). Physicians need to be familiar with the presenting signs and symptoms of both THC and SC intoxication, and the public needs to be educated about the risks of using these psychoactive substances, including the impact on mental health (Papanti et al., 2014). Appropriate, nonjudgmental prevention campaigns with a special focus on the differences between SCs and natural cannabis should be organized on a large scale. At the same time, clinicians need to be regularly updated about NPSs, including SCs, to recognize promptly the signs and symptoms of intoxication, the medical and psychiatric risks, and potential longer-term harms related to their use.

Key Chapter Points

- Over the last decade, illicit drug markets have changed substantially, with the emergence of a range of NPSs, the largest class of these being SCs. The internet/web may play a major role in shaping this unregulated market.
- In contrast to the phytocannabinoid THC, found in natural cannabis and a partial agonist at cannabinoid receptors, available SCs are often full- or even super-agonists at the same receptors. This may at least partially explain their different and often more intense effects when compared to THC.
- Adverse effects of SCs may be severe, and their consumption has been associated with: addiction, psychotic episodes, onset and relapse of psychotic disorders, a multitude of harmful consequences for physical health, and death.
- SC intake usually cannot be identified with readily available drug screening tests. Advanced tests can be performed, but they are not generally available in acute care settings and are used mainly for research and forensic investigations.

ACKNOWLEDGMENTS

We thank Tommaso Bonavigo, MD, for giving us permission to reproduce and adapt his work, as it appears in Fig. 3. This chapter was supported in part by grants from the European Commission (Drug Prevention and Information Programme 2014–16; contract no. JUST/2013/DPIP/AG/4823; *EU-MADNESS project*). Further financial support was provided by the EU Commission-targeted call on cross border law enforcement cooperation in the field of drug trafficking—DG Justice/DG Migrations and Home Affairs (JUST/2013/ISEC/DRUGS/AG/6429) *Project EPS/NPS* (Enhancing Police Skills concerning Novel Psychoactive Substances; NPS).

REFERENCES

Abdulrahim, D., & Bowden-Jones, O. (2015). *Guidance on the clinical management of acute and chronic harms of club drugs and novel psychoactive substances*. Available at: http://www.drugsandalcohol.ie/24292/.

ACMD. (2014). *"Third generation" synthetic cannabinoids*. Available at: https://www.gov.uk/government/uploads/system/uploads/attachment_data/file/380161/CannabinoidsReport.pdf.

Adis R&D Profile. (2003). Dexanabinol. *Drugs in R&D*, *4*, 185–187.

Ashton, J. C. (2012). Synthetic cannabinoids as drugs of abuse. *Current Drug Abuse Reviews*, *5*, 158–168. Available at http://www.ncbi.nlm.nih.gov/pubmed/22530798. Accessed 21.05.16.

Assi, S., Fergus, S., Stair, J., Corazza, O., & Schifano, F. (2011). Emergence and identification of new products of designer drug products from the internet. *European Pharmaceutical Review*, *6*, 68–72. Available at: http://uhra.herts.ac.uk/handle/2299/9600. Accessed 21.05.16.

Auwärter, V., Dresen, S., Weinmann, W., Muller, M., Putz, M., & Ferreiros, N. (2009). "Spice" and other herbal blends: Harmless incense or cannabinoid designer drugs? *Journal of Mass Spectrometry*, *44*, 832–837. Available at http://www.ncbi.nlm.nih.gov/pubmed/19189348.

Baggaley, K. (2015). Corrupt chemists tweak compounds faster than law enforcement can call them illegal. *Science News*, *187*, 22–25.

Baumann, M. H., Solis, E., Watterson, L. R., Marusich, J. A., Fantegrossi, W. E., & Wiley, J. L. (2014). Baths salts, spice, and related designer drugs: The science behind the headlines. *Journal of Neuroscience*, *34*, 15150–15158. Available at: http://www.pubmedcentral.nih.gov/articlerender.fcgi?artid=4228124&tool=pmcentrez&rendertype=abstract. Accessed 22.05.16.

Bebarta, V. S., Ramirez, S., & Varney, S. M. (2012). Spice: A new "legal" herbal mixture abused by young active duty military personnel. *Substance Abuse*, *33*, 191–194.

Begaud, B. (1984). Standardized assessment of adverse drug reactions: The method used in France Special Workshop—Clinical. *Drug Information Journal*, *18*, 275–281.

Bell, S., & Nida, C. (2015). Pyrolysis of drugs of abuse: A comprehensive review. *Drug Testing and Analysis*, *7*, 445–456.

Benford, D. M., & Caplan, J. P. (2011). Psychiatric sequelae of Spice, K2, and synthetic cannabinoid receptor agonists. *Psychosomatics*, *52*, 295. Available at: http://www.ncbi.nlm.nih.gov/pubmed/21565605. Accessed 11.04.16.

Berry-Caban, C. S., Ee, J., Ingram, V., Berry, C. E., & Kim, E. H. (2013). Synthetic cannabinoid overdose in a 20-year-old male US soldier. *Substance Abuse*, *34*, 70–72. Available at: http://www.ncbi.nlm.nih.gov/pubmed/23327506.

Bilgrei, O. R. (2016). From "herbal highs" to the "heroin of cannabis": Exploring the evolving discourse on synthetic cannabinoid use in a Norwegian Internet drug forum. *The International Journal on Drug Policy*. Available at: http://linkinghub.elsevier.com/retrieve/pii/S0955395916000347.

Brakoulias, V. (2012). Products containing synthetic cannabinoids and psychosis. *The Australian and New Zealand Journal of Psychiatry*, *46*, 281–282. Available at: http://www.ncbi.nlm.nih.gov/pubmed/22391292. Accessed 23.05.16.

Brents, L. K., & Prather, P. L. (2014). The K2/Spice phenomenon: Emergence, identification, legislation and metabolic characterization of synthetic cannabinoids in herbal incense products. *Drug Metabolism Reviews*, *46*, 72–85. Available at: http://www.pubmedcentral.nih.gov/articlerender.fcgi?artid=4100246&tool=pmcentrez&rendertype=abstract. Accessed 22.05.16.

Brown, J. K., & Malone, M. H. (1978). "Legal highs"—Constituents activity, toxicology, and herbal folklore. *Clinical Toxicology*, *12*, 1–31.

Burnham, D. (1967). synthetic-marijuana-report-.pdf. Psychol Fears Underworld Will Sell Synth Marijuana.

Calabria, B., Degenhardt, L., Hall, W., & Lynskey, M. (2010). Does cannabis use increase the risk of death? Systematic review of epidemiological evidence on adverse effects of cannabis use. *Drug and Alcohol Review*, *29*, 318–330. Available at: http://www.ncbi.nlm.nih.gov/pubmed/20565525. Accessed 22.05.16.

Celofiga, A., Koprivsek, J., & Klavz, J. (2014). Use of synthetic cannabinoids in patients with psychotic disorders: Case series. *Journal of Dual Diagnosis*, *10*, 168–173. Available at: http://www.tandfonline.com/doi/abs/10.1080/15504263.2014.929364.

Centers for Disease Control and Prevention (CDC). (2013). Acute kidney injury associated with synthetic cannabinoid use—Multiple states, 2012. *MMWR Morbidity and Mortality Weekly Report*, *62*, 93–98. Available at: http://www.ncbi.nlm.nih.gov/pubmed/23407124. Accessed 28.04.16.

Chadwick, B., Miller, M. L., & Hurd, Y. L. (2013). Cannabis use during adolescent development: Susceptibility to psychiatric illness. *Frontiers in Psychology*, *4*, 129. Available at: http://www.pubmedcentral.nih.gov/articlerender.fcgi?artid=3796318&tool=pmcentrez&rendertype=abstract. Accessed 13.04.16.

Choi, H., Heo, S., Choe, S., Yang, W., Park, Y., Kim, E., et al. (2013). Simultaneous analysis of synthetic cannabinoids in the materials seized during drug trafficking using GC-MS. *Analytical and Bioanalytical Chemistry*, *405*, 3937–3944. Available at: http://www.ncbi.nlm.nih.gov/pubmed/23208283. Accessed 21.05.16.

Çoban, M. (2014). The rise of synthetic marijuana in Turkey: The Bonzai phenomenon of the 2010s. *Addicta: The Turkish Journal on Addictions*, *1*, 1–22.

Corkery, J., Claridge, H., Loi, B., Goodair, C., Schifano, F., & Deaths, S. A. (2013). *Drug-related deaths in the UK : 2012 Annual Report 2013 National Programme on Substance Abuse Deaths Annual Report 2013 on deaths between*. Available at: http://www.drugsandalcohol.ie/21379/1/National_Programme_on_Substance_Abuse_Deaths_-_Annual_Report_2013_on_Drug-related_Deaths_in_the_UK_January-December_2012.pdf.

Daly, M. (2013). Streets legal. *Druglink*, *28*, 7.

De Luca, M. A., Bimpisidis, Z., Melis, M., Marti, M., Caboni, P., Valentini, V., et al. (2015). Stimulation of in vivo dopamine transmission and intravenous self-administration in rats and mice by JWH-018, a Spice cannabinoid. *Neuropharmacology*, *99*, 705–714. Available at: http://dx.doi.org/10.1016/j.neuropharm.2015.08.041.

De Luca, M. A., Castelli, M. P., Loi, B., Porcu, A., Martorelli, M., Miliano, C., et al. (2015). Native CB1 receptor affinity, intrinsic activity and accumbens shell dopamine stimulant properties of third generation SPICE/K2 cannabinoids: BB-22, 5F-PB-22, 5F-AKB-48 and STS-135. *Neuropharmacology*, *105*, 630–638. Available at: http://www.ncbi.nlm.nih.gov/pubmed/26686391. Accessed 12.04.16.

Di Forti, M., et al. (2014). Daily use, especially of high-potency cannabis, drives the earlier onset of psychosis in cannabis users. *Schizophrenia Bulletin*, *40*, 1509–1517. Available at: http://schizophreniabulletin.oxfordjournals.org/content/early/2013/12/14/schbul.sbt181.abstract. Accessed 06.05.16.

Di Forti, M., et al. (2015). Proportion of patients in south London with first-episode psychosis attributable to use of high potency cannabis: A case-control study. *The Lancet Psychiatry*, *2*, 233–238. Available at: http://www.thelancet.com/article/S2215036614001175/fulltext. Accessed 10.02.16.

Dresen, S., Ferreirós, N., Pütz, M., Westphal, F., Zimmermann, R., & Auwärter, V. (2010). Monitoring of herbal mixtures potentially containing synthetic cannabinoids as psychoactive compounds. *Journal of Mass Spectrometry*, *45*, 1186–1194. Available at: http://www.ncbi.nlm.nih.gov/pubmed/20857386. Accessed 22.05.16.

D'Souza, D. C., Perry, E., MacDougall, L., Ammerman, Y., Cooper, T., Wu, Y.-T., et al. (2004). The psychotomimetic effects of intravenous delta-9-tetrahydrocannabinol in healthy individuals: Implications for psychosis. *Neuropsychopharmacology*, *29*, 1558–1572. Available at: http://www.ncbi.nlm.nih.gov/pubmed/15173844. Accessed 17.04.16.

D'Souza, D. C., Sewell, R. A., & Ranganathan, M. (2009). Cannabis and psychosis/schizophrenia: Human studies. *European Archives of Psychiatry and Clinical Neuroscience*, *259*, 413–431. Available at: http://www.pubmedcentral.nih.gov/articlerender.fcgi?artid=2864503&tool=pmcentrez&rendertype=abstract. Accessed 07.05.16.

ElSohly, M., & Gul, W. (2014). Constituents of *Cannabis sativa*. In R. G. Pertwee (Ed.), *Handbook of Cannabis* (pp. 3–22). Oxford: Oxford University Press.

EMCDDA. (2015). *Synthetic cannabinoids in Europe*. Available at: www.emcdda.europa.eu/topics/pods/synthetic-cannabinoids.

Every-Palmer, S. (2011). Synthetic cannabinoid JWH-018 and psychosis: An explorative study. *Drug and Alcohol Dependence*, *117*, 152–157. http://dx.doi.org/10.1016/j.drugalcdep.2011.01.012.

Fantegrossi, W. E., Moran, J. H., Radominska-Pandya, A., & Prather, P. L. (2014). Distinct pharmacology and metabolism of K2 synthetic cannabinoids compared to Δ(9)-THC: Mechanism underlying greater toxicity? *Life Sciences*, *97*, 45–54. Available at: http://www.sciencedirect.com/science/article/pii/S0024320513005523. Accessed 03.03.16.

Fattore, L. (2016). Synthetic cannabinoids—Further evidence supporting the relationship between cannabinoids and psychosis. *Biological Psychiatry*. Available at: http://linkinghub.elsevier.com/retrieve/pii/S0006322316000834.

Fišar, Z. (2012). Cannabinoids and monoamine neurotransmission with focus on monoamine oxidase. *Progress in Neuropsychopharmacology and Biological Psychiatry*, *38*, 68–77. Available at: http://www.sciencedirect.com/science/article/pii/S0278584611003757. Accessed 22.05.16.

Forrester, M. B., Kleinschmidt, K., Schwarz, E., & Young, A. (2012). Synthetic cannabinoid and marijuana exposures reported to poison centers. *Human and Experimental Toxicology*, *31*, 1006–1011. Available at: http://www.ncbi.nlm.nih.gov/pubmed/22859662. Accessed 17.04.16.

Freeman, M. J., Rose, D. Z., Myers, M. A., Gooch, C. L., Bozeman, A. C., & Burgin, W. S. (2013). Ischemic stroke after use of the synthetic marijuana "spice". *Neurology*, *81*, 2090–2093. Available at: http://www.pubmedcentral.nih.gov/articlerender.fcgi?artid=3863350&tool=pmcentrez&rendertype=abstract. Accessed 22.05.16.

Freeman, T. P., & Winstock, A. R. (2015). Examining the profile of high-potency cannabis and its association with severity of cannabis dependence. *Psychological Medicine*, *45*, 3181–3189. Available at: http://www.ncbi.nlm.nih.gov/pubmed/26213314.

Glue, P., Al-Shaqsi, S., Hancock, D., Gale, C., Strong, B., & Schep, L. (2013). Hospitalisation associated with use of the synthetic cannabinoid K2. *New Zealand Medical Journal*, *126*, 18–23. Available at: http://www.ncbi.nlm.nih.gov/pubmed/23831873. Accessed 23.05.16.

Green, A. I., Noordsy, D. L., Brunette, M. F., & O'Keefe, C. (2008). Substance abuse and schizophrenia: Pharmacotherapeutic intervention. *Journal of Substance Abuse Treatment*, *34*, 61–71.

Gunderson, E. W., Haughey, H. M., Ait-Daoud, N., Joshi, A. S., & Hart, C. L. (2012). "Spice" and "k2" herbal highs: A case series and systematic review of the clinical effects and biopsychosocial implications of synthetic cannabinoid use in humans. *American Journal on Addictions*, *21*, 320–326.

Halberstadt, A. L., & Geyer, M. A. (2011). Multiple receptors contribute to the behavioral effects of indoleamine hallucinogens. *Neuropharmacology*, *61*, 364–381. Available at: http://www.sciencedirect.com/science/article/pii/S0028390811000207. Accessed 12.05.16.

Henquet, C., Krabbendam, L., Spauwen, J., Kaplan, C., Lieb, R., Wittchen, H.-U., et al. (2005). Prospective cohort study of cannabis use, predisposition for psychosis, and psychotic symptoms in young people. *BMJ*, *330*, 11. Available at: http://www.pubmedcentral.nih.gov/articlerender.fcgi?artid=539839&tool=pmcentrez&rendertype=abstract. Accessed 09.05.16.

Hermanns-Clausen, M., Kneisel, S., Szabo, B., & Auwärter, V. (2013). Acute toxicity due to the confirmed consumption of synthetic cannabinoids: Clinical and laboratory findings. *Addiction*, *108*, 534–544. Available at: http://www.ncbi.nlm.nih.gov/pubmed/22971158. Accessed 22.05.16.

Hopkins, C. Y., & Gilchrist, B. L. (2013). A case of cannabinoid hyperemesis syndrome caused by synthetic cannabinoids. *Journal of Emergency Medicine*, *45*, 544–546. Available at: http://www.ncbi.nlm.nih.gov/pubmed/23890687. Accessed 08.05.16.

Hurst, D., Loeffler, G., & McLay, R. (2011). Psychosis associated with synthetic cannabinoid agonists: A case series. *American Journal of Psychiatry*, *168*, 1119.

Ismail, F. M. D. (2002). Important fluorinated drugs in experimental and clinical use. *Journal of Fluorine Chemistry*, *118*, 27–33. Available at: http://www.sciencedirect.com/science/article/pii/S0022113902002014. Accessed 11.05.16.

Juckel, G., Schlagenhauf, F., Koslowski, M., Filonov, D., Wüstenberg, T., Villringer, A., et al. (2006). Dysfunction of ventral striatal reward prediction in schizophrenic patients treated with typical, not atypical, neuroleptics. *Psychopharmacology*, *187*, 222–228. Available at: http://www.ncbi.nlm.nih.gov/pubmed/16721614. Accessed 21.10.16.

Kapur, S., & Seeman, P. (2000). Antipsychotic agents differ in how fast they come off the dopamine D2 receptors. Implications for atypical antipsychotic action. *Journal of Psychiatry and Neuroscience*, *25*, 161–166. Available at: http://www.ncbi.nlm.nih.gov/pubmed/10740989. Accessed 21.10.16.

Karch, F. E., & Lasagna, L. (1975). Adverse drug reactions: A critical review. *JAMA*, *234*, 1236–1241.

Karel, V., & Arrizabalaga, P. (1998). *Designer drugs directory*. Lausanne: Elsevier.

Kasper, A. M., Ridpath, A. D., Arnold, J. K., Chatham-Stephens, K., Morrison, M., Olayinka, O., et al. (2015). Severe illness associated with reported use of synthetic cannabinoids—Mississippi, April 2015. *MMWR Morbidity and Mortality Weekly Report*, *64*, 1121–1122. Available at: http://www.cdc.gov/mmwr/preview/mmwrhtml/mm6439a7.htm/npapers3://publication/doi/10.15585/mmwr.mm6439a7.

Kelley, M. E., Ramsay Wan, C., Broussard, B., Crisafio, A., Cristofaro, S., Johnson, S., et al. (2016). Marijuana use in the immediate 5-year premorbid period is associated with increased risk of onset of schizophrenia and related psychotic disorders. *Schizophrenia Research*, *171*, 62–67. Available at: http://ovidsp.ovid.com/ovidweb.cgi?T=JS&PAGE=reference&D=psyc11&NEWS=N&AN=2016-03128-001.

Keshavan, M. S., Giedd, J., Lau, J. Y. F., Lewis, D. A., & Paus, T. (2014). Changes in the adolescent brain and the pathophysiology of psychotic disorders. *The Lancet Psychiatry*, *1*, 549–558. http://dx.doi.org/10.1016/S2215-0366(14)00081-9.

Kikura-Hanajiri, R., Uchiyama, N., Kawamura, M., & Goda, Y. (2012). Changes in the prevalence of synthetic cannabinoids and cathinone derivatives in Japan until early 2012. *Forensic Toxicology*, *31*, 44–53. Available at: http://link.springer.com/10.1007/s11419-012-0165-2. Accessed 14.04.16.

Kjellgren, A., Henningsson, H., & Soussan, C. (2013). Fascination and social togetherness—Discussions about spice smoking on a swedish internet forum. *Substance Abuse: Research and Treatment*, *7*, 191–198.

Kronstrand, R., Roman, M., Andersson, M., & Eklund, A. (2013). Toxicological findings of synthetic cannabinoids in recreational users. *Journal of Analytical Toxicology*, *37*, 534–541. Available at: http://www.ncbi.nlm.nih.gov/pubmed/23970540. Accessed 17.04.16.

Lank, P. M., Pines, E., & Mycyk, M. B. (2013). Emergency physicians' knowledge of cannabinoid designer drugs. *The Western Journal of Emergency Medicine*, *14*, 467–470.

Lászik, A., Törő, K., Vannai, M., Sára-Klausz, G., Kócs, T., Farkas, R., et al. (2015). *Self inflicted fatal injuries in association with synthetic cannabinoid abuse*. In: *24th international meeting on forensic medicine Alpe-Adria-Pannonia* pp 27.

Lerner, A., Goodman, C., Bor, O., & Lev-Ran, S. (2014). Synthetic cannabis substances (SPS) use and hallucinogen persisting perception disorder (HPPD): Two case reports. *Israel Journal of Psychiatry and Related Sciences*, *51*, 277–280. Available at: http://proxy2cobimet.net:2048/login?user=gdrtyhfsekiog&pass=Dr5KlW@zx1!7(6$Nt&url=http://search.ebscohost.com/login.aspx?direct=true&db=mnh&AN=25841224&lang=es&site=ehost-live.

Leweke, F. M., Piomelli, D., Pahlisch, F., Muhl, D., Gerth, C. W., Hoyer, C., et al. (2012). Cannabidiol enhances anandamide signaling and alleviates psychotic symptoms of schizophrenia. *Translational Psychiatry*, *2*, e94. Available at: http://www.pubmedcentral.nih.gov/articlerender.fcgi?artid=3316151&tool=pmcentrez&rendertype=abstract. Accessed 05.05.16.

Lewin, A. H., Seltzman, H. H., Carroll, F. I., Mascarella, S. W., & Reddy, P. A. (2014). Emergence and properties of spice and bath salts: A medicinal chemistry perspective. *Life Sciences*, *97*, 9–19. Available at: http://www.sciencedirect.com/science/article/pii/S0024320513005791. Accessed 14.04.16.

Loflin, M., & Earleywine, M. (2014). A new method of cannabis ingestion: The dangers of dabs? *Addictive Behaviors*, *39*, 1430–1433.

Lonati, D., Buscaglia, E., Papa, P., Valli, A., Coccini, T., Giampreti, A., et al. (2014). MAM-2201 (analytically confirmed) intoxication after "Synthacaine" consumption. *Annals of Emergency Medicine*, *64*, 629–632. Available at: http://www.scopus.com/inward/record.url?eid=2-s2.0-84914112407&partnerID=tZOtx3y1. Accessed 22.05.16.

Louh, I. K., & Freeman, W. D. (2014). A "spicy" encephalopathy: Synthetic cannabinoids as cause of encephalopathy and seizure. *Critical Care*, *18*, 553. Available at: http://www.pubmedcentral.nih.gov/articlerender.fcgi?artid=4201992&tool=pmcentrez&rendertype=abstract. Accessed 22.05.16.

Macfarlane, V., & Christie, G. (2015). Synthetic cannabinoid withdrawal: A new demand on detoxification services. *Drug and Alcohol Review*, *34*, 147–153. Available at: http://www.ncbi.nlm.nih.gov/pubmed/25588420. Accessed 13.05.16.

Machielsen, M., Beduin, A. S., Dekker, N., Kahn, R. S., Linszen, D. H., van Os, J., et al. (2012). Differences in craving for cannabis between schizophrenia patients using risperidone, olanzapine or clozapine. *Journal of Psychopharmacology*, *26*, 189–195. Available at: http://www.ncbi.nlm.nih.gov/pubmed/21768161.

Malone, D. T., Hill, M. N., & Rubino, T. (2010). Adolescent cannabis use and psychosis: Epidemiology and neurodevelopmental models. *British Journal of Pharmacology*, *160*, 511–522. Available at: http://www.pubmedcentral.nih.gov/articlerender.fcgi?artid=2931552&tool=pmcentrez&rendertype=abstract. Accessed 22.05.16.

Manseau, M. W. (2016). Synthetic cannabinoids emergence, epidemiology, clinical effects, and management. In M. T. Compton (Ed.), *Marijuana and mental health* (pp. 149–169). Arlington, VA: American Psychiatric Association Publishing.

Marconi, A., Di Forti, M., Lewis, C., Murray, R. M., & Vassos, E. (2016). Meta-analysis of the association between the level of cannabis use and risk of psychosis. *Schizophrenia Bulletin*, *42*, 1262–1269.

Mason, O., Morgan, C. J. A., Dhiman, S. K., Patel, A., Parti, N., & Curran, H. V. (2009). Acute cannabis use causes increased psychotomimetic experiences in individuals prone to psychosis. *Psychological Medicine*, *39*, 951–956. Available at: http://www.ncbi.nlm.nih.gov/pubmed/19017430. Accessed 22.05.16.

Miettunen, J., Törmänen, S., Murray, G. K., Jones, P. B., Mäki, P., Ebeling, H., et al. (2008). Association of cannabis use with prodromal symptoms of psychosis in adolescence. *British Journal of Psychiatry*, *192*, 470–471. Available at: http://www.ncbi.nlm.nih.gov/pubmed/18515902. Accessed 20.05.16.

Mir, A., Obafemi, A., Young, A., & Kane, C. (2011). Myocardial infarction associated with use of the synthetic cannabinoid K2. *Pediatrics*, *128*, e1622–e1627. Available at: http://www.ncbi.nlm.nih.gov/pubmed/22065271. Accessed 22.05.16.

Monte, A. A., Bronstein, A. C., Cao, D. J., Heard, K. J., Hoppe, J. A., Hoyte, C. O., et al. (2014). An outbreak of exposure to a novel synthetic cannabinoid. *New England Journal of Medicine*, *370*, 389–390. Available at: http://www.nejm.org/doi/abs/10.1056/NEJMc1313655. Accessed 21.10.16.

Morgan, C. J. A., & Curran, H. V. (2008). Effects of cannabidiol on schizophrenia-like symptoms in people who use cannabis. *British Journal of Psychiatry*, *192*, 306–307. Available at: http://www.ncbi.nlm.nih.gov/pubmed/18378995. Accessed 22.05.16.

Morgan, C. J. A., Das, R. K., Joye, A., Curran, H. V., & Kamboj, S. K. (2013). Cannabidiol reduces cigarette consumption in tobacco smokers: Preliminary findings. *Addictive Behaviors*, *38*, 2433–2436. Available at: http://www.ncbi.nlm.nih.gov/pubmed/23685330. Accessed 22.05.16.

Morrison, P. D., Zois, V., McKeown, D. A., Lee, T. D., Holt, D. W., Powell, J. F., et al. (2009). The acute effects of synthetic intravenous Delta9-tetrahydrocannabinol on psychosis, mood and cognitive functioning. *Psychological Medicine*, *39*, 1607–1616. Available at: http://www.ncbi.nlm.nih.gov/pubmed/19335936. Accessed 16.04.16.

Müller, H., Sperling, W., Köhrmann, M., Huttner, H. B., Kornhuber, J., & Maler, J. M. (2010). The synthetic cannabinoid Spice as a trigger for an acute exacerbation of cannabis induced recurrent psychotic episodes. *Schizophrenia Research*, *118*, 309–310. http://dx.doi.org/10.1016/j.schres.2009.12.001.

Murray, R. M., Paparelli, A., Morrison, P. D., Marconi, A., & Di Forti, M. (2013). What can we learn about schizophrenia from studying the human model, drug-induced psychosis? *American Journal of Medical Genetics Part B, Neuropsychiatric Genetics*, *162B*, 661–670. Available at: http://www.ncbi.nlm.nih.gov/pubmed/24132898. Accessed 22.05.16.

Nacca, N., Vatti, D., Sullivan, R., Sud, P., Su, M., & Marraffa, J. (2013). The synthetic cannabinoid withdrawal syndrome. *Journal of Addiction Medicine*, *7*, 296–298. Available at: http://www.ncbi.nlm.nih.gov/pubmed/23609214. Accessed 22.05.16.

Naviglio, S., Papanti, D., Moressa, V., & Ventura, A. (2015). An adolescent with an altered state of mind. *BMJ*, *350*, h299. Available at: http://www.bmj.com/content/350/bmj.h299.

Ogata, J., Uchiyama, N., Kikura-Hanajiri, R., & Goda, Y. (2013). DNA sequence analyses of blended herbal products including synthetic cannabinoids as designer drugs. *Forensic Science International*, *227*, 33–41. Available at: http://www.ncbi.nlm.nih.gov/pubmed/23092848. Accessed 12.04.16.

Palamar, J. J., & Acosta, P. (2015). Synthetic cannabinoid use in a nationally representative sample of US high school seniors. *Drug and Alcohol Dependence*, *149*, 194–202. Available at: http://www.pubmedcentral.nih.gov/articlerender.fcgi?artid=4361370&tool=pmcentrez&rendertype=abstract. Accessed 28.04.16.

Papanti, D., Orsolini, L., Francesconi, G., & Schifano, F. (2014). "Noids" in a nutshell: Everything you (don't) want to know about synthetic cannabimimetics. *Advances in Dual Diagnosis*, *7*, 137–148. Available at: http://www.emeraldinsight.com/doi/abs/10.1108/ADD-02-2014-0006.

Papanti, D., Schifano, F., Botteon, G., Bertossi, F., Mannix, J., Vidoni, D., et al. (2013). "Spiceophrenia": A systematic overview of "spice"-related psychopathological issues and a case report. *Human Psychopharmacology*, *28*, 379–389. Available at: http://www.ncbi.nlm.nih.gov/pubmed/23881886. Accessed 19.05.16.

Park, Y., Lee, C., Lee, H., Pyo, J., Jo, J., Lee, J., et al. (2013). Identification of a new synthetic cannabinoid in a herbal mixture: 1-butyl-3-(2-methoxybenzoyl)indole. *Forensic Toxicology*, *31*, 187–196. Available at: http://link.springer.com/10.1007/s11419-012-0173-2. Accessed 21.05.16.

Patton, A. L., Chimalakonda, K. C., Moran, C. L., McCain, K. R., Radominska-Pandya, A., James, L. P., et al. (2013). K2 toxicity: Fatal case of psychiatric complications following AM2201 exposure. *Journal of Forensic Sciences*, *58*, 1676–1680. Available at: http://www.pubmedcentral.nih.gov/articlerender.fcgi?artid=4319529&tool=pmcentrez&rendertype=abstract. Accessed 23.05.16.

Peglow, S., Buchner, J., & Briscoe, G. (2012). Synthetic cannabinoid induced psychosis in a previously nonpsychotic patient. *American Journal on Addictions*, *21*, 287–288.

Pertwee, R. G. (2008). The diverse CB1 and CB2 receptor pharmacology of three plant cannabinoids: Delta9-tetrahydrocannabinol, cannabidiol and delta9-tetrahydrocannabivarin. *British Journal of Pharmacology*, *153*, 199–215. Available at: http://www.pubmedcentral.nih.gov/articlerender.fcgi?artid=2219532&tool=pmcentrez&rendertype=abstract. Accessed 15.01.15.

Pertwee, R. (2010). Receptors and channels targeted by synthetic cannabinoid receptor agonists and antagonists. *Current Medicinal Chemistry*, *17*, 1360–1381. Available at: http://www.ncbi.nlm.nih.gov/pubmed/20166927/nhttp://www.ncbi.nlm.nih.gov/pmc/articles/PMC3013229/pdf/nihms259799.pdf/nhttp://www.eurekaselect.com/openurl/content.php?genre=article&issn=0929-8673&volume=17&issue=14&spage=1360.

Pertwee, R. G., Howlett, A. C., Abood, M. E., Alexander, S. P. H., Di, Marzo V., Elphick, M. R., et al. (2010). International union of basic and clinical pharmacology. LXXIX. Cannabinoid receptors and their ligands: Beyond CB 1 and CB 2. *Pharmacological Reviews*, *62*, 588–631.

Pierre, J. M., Gandal, M., & Son, M. (2016). Cannabis-induced psychosis associated with high potency "wax dabs". *Schizophrenia Research*, *172*, 211–212. Available at: http://linkinghub.elsevier.com/retrieve/pii/S0920996416300561.

Power, M. (2013). *Drugs 2.0: The web revolution that's changing how the world gets high*. London: Portobello Books.

Rodgman, C., Kinzie, E., & Leimbach, E. (2011). Bad Mojo: Use of the new marijuana substitute leads to more and more ED visits for acute psychosis. *American Journal of Emergency Medicine*, *29*, 232. http://dx.doi.org/10.1016/j.ajem.2010.07.020.

Rominger, A., Cumming, P., Xiong, G., Koller, G., Förster, S., Zwergal, A., et al. (2013). Effects of acute detoxification of the herbal blend "Spice Gold" on dopamine D2/3 receptor availability: A [18F]fallypride PET study. *European Neuropsychopharmacology*, *23*, 1606–1610. Available at: http://www.europeanneuropsychopharmacology.com/article/S0924977X13000485/fulltext. Accessed 22.05.16.

Rose, D. Z., Guerrero, W. R., Mokin, M. V., Gooch, C. L., Bozeman, A. C., Pearson, J. M., et al. (2015). Hemorrhagic stroke following use of the synthetic marijuana "spice". *Neurology*, *85*, 1177–1179. Available at: http://www.ncbi.nlm.nih.gov/pubmed/26320200. Accessed 22.05.16.

Rosenbaum, C. D., Carreiro, S. P., & Babu, K. M. (2012). Here today, gone tomorrow…and back again? A review of herbal marijuana alternatives (K2, Spice), synthetic cathinones (bath salts), kratom, Salvia divinorum, methoxetamine, and piperazines. *Journal of Medical Toxicology*, *8*, 15–32. Available at: http://www.pubmedcentral.nih.gov/articlerender.fcgi?artid=3550220&tool=pmcentrez&rendertype=abstract. Accessed 23.05.16.

Saito, T., Namera, A., Miura, N., Ohta, S., Miyazaki, S., Osawa, M., et al. (2013). A fatal case of MAM-2201 poisoning. *Forensic Toxicology*, *31*, 333–337. Available at: http://link.springer.com/10.1007/s11419-013-0190-9. Accessed 17.04.16.

San, L., Arranz, B., & Martinez-Raga, J. (2007). Antipsychotic drug treatment of schizophrenic patients with substance abuse disorders. *European Addiction Research*, *13*, 230–243.

Schaefer, N., Peters, B., Bregel, D., Kneisel, S., Schmidt, P. H., & Ewald, A. H. (2013). A fatal case involving several synthetic cannabinoids. *Toxichem Krimtech*, *80*, 248–251.

Schifano, F., Corazza, O., Deluca, P., Davey, Z., Di Furia, L., Farre', M., et al. (2009). Psychoactive drug or mystical incense? Overview of the online available information on Spice products. *International Journal of Culture and Mental Health*, *2*, 137–144. http://dx.doi.org/10.1080/17542860903350888.

Schifano, F., Orsolini, L., Duccio Papanti, G., & Corkery, J. M. (2015). Novel psychoactive substances of interest for psychiatry. *World Psychiatry*, *14*, 15–26.

Schifano, F., Papanti, G. D., Orsolini, L., & Corkery, J. M. (2016). Novel psychoactive substances: The pharmacology of stimulants and hallucinogens. *Expert Review of Clinical Pharmacology*, *2433*, 1–12. Available at: http://www.tandfonline.com/doi/full/10.1586/17512433.2016.1167597.

Schubart, C. D., Sommer, I. E. C., van Gastel, W. A., Goetgebuer, R. L., Kahn, R. S., & Boks, M. P. M. (2011). Cannabis with high cannabidiol content is associated with fewer psychotic experiences. *Schizophrenia Research*, *130*, 216–221. Available at: http://www.ncbi.nlm.nih.gov/pubmed/21592732. Accessed 22.05.16.

Schwartz, M. D., Trecki, J., Edison, L. A., Steck, A. R., Arnold, J. K., & Gerona, R. R. (2015). A common source outbreak of severe delirium associated with exposure to the novel synthetic cannabinoid ADB-PINACA. *Journal of Emergency Medicine*, *48*, 573–580. Available at: http://www.ncbi.nlm.nih.gov/pubmed/25726258. Accessed 05.05.16.

Shanks, K. G., Dahn, T., & Terrell, A. R. (2012). Detection of JWH-018 and JWH-073 by UPLC-MS-MS in postmortem whole blood casework. *Journal of Analytical Toxicology*, *36*, 145–152. Available at: http://jat.oxfordjournals.org/content/36/3/145. Accessed 22.05.16.

Shoval, G. (2015). *Clinical characteristics of hospitalized synthetic cannabinoid users*. In: *IV international congress of dual disorders* p. 38.

Siegel, R. K. (1976). Herbal intoxication. Psychoactive effects from herbal cigarettes, tea, and capsules. *JAMA*, *236*, 473–476. Available at: http://www.ncbi.nlm.nih.gov/entrez/query.fcgi?cmd=Retrieve&db=PubMed&dopt=Citation&list_uids=947067.

Soussan, C., & Kjellgren, A. (2014). The flip side of "Spice": The adverse effects of synthetic cannabinoids as discussed on a Swedish Internet forum. *Nordic Studies on Alcohol and Drugs*, *31*, 207–219. Available at: http://www.scopus.com/inward/record.url?eid=2-s2.0-84899967933&partnerID=40&md5=db27c7d8994b22706dd923592e1bff1e.

Spaderna, M., Addy, P. H., & D'Souza, D. C. (2013). Spicing things up: Synthetic cannabinoids. *Psychopharmacology*, *228*, 525–540.

Steeds, H., Carhart-Harris, R. L., & Stone, J. M. (2015). Drug models of schizophrenia. *Therapeutic Advances in Psychopharmacology*, *5*, 43–58. Available at: http://tpp.sagepub.com/content/5/1/43/nhttp://tpp.sagepub.com/content/5/1/43.full.pdf/nhttp://tpp.sagepub.com/content/5/1/43.short/nhttp://www.ncbi.nlm.nih.gov/pubmed/25653831.

Steup, C. (2008). Untersuchung des Handelsproduktes "Spice" (Investigation of the commercial product "Spice"). *Main*, 1–6. Available at: http://usualredant.de/downloads/analyse-thc-pharm-spice-jwh-018.pdf.

Stogner, J. M., & Miller, B. L. (2015). Assessing the dangers of "dabbing": Mere marijuana or harmful new trend? *Pediatrics*, *136*, 1–3.

Substance Abuse and Mental Health Services Administration, Center for Behavioral Health Statistics and Quality. (October 16, 2014). *Update: Drug-related emergency department visits involving synthetic cannabinoids*. Rockville, MD: SAMHSA. Available at: https://www.samhsa.gov/data/sites/default/files/SR-1378/SR-1378.pdf.

Szabo, B. (2014). *Effects of phytocannabinoids on neurotransmission in the central and peripheral nervous systems*. In *Handbook of cannabis*. Oxford: Oxford University Press. (pp. 157–162).

Tait, R. J., Caldicott, D., Mountain, D., Hill, S. L., & Lenton, S. (2015). A systematic review of adverse events arising from the use of synthetic cannabinoids and their associated treatment. *Clinical Toxicology*, *3650*, 1–13. Available at: http://www.tandfonline.com/doi/full/10.3109/15563650.2015.1110590.

Tanda, G., Valentini, V., De Luca, M. A., Perra, V., Pietro, Serra G., & Di Chiara, G. (2015). A systematic microdialysis study of dopamine transmission in the accumbens shell/core and prefrontal cortex after acute antipsychotics. *Psychopharmacology*, *232*, 1427–1440. Available at: http://www.ncbi.nlm.nih.gov/pubmed/25345736. Accessed 21.10.16.

Thomas, B. F., Wiley, J. L., Pollard, G. T., & Grabenauer, M. (2014). Cannabinoid designer drugs: Effects and forensics. In R. G. Pertwee (Ed.), *Handbook of cannabis* (pp. 710–729). Oxford: Oxford University Press.

Trecki, J., Gerona, R. R., & Schwartz, M. D. (2015). Synthetic cannabinoid-related illnesses and deaths. *New England Journal of Medicine*, *373*, 103–107.

Tung, C. K., Chiang, T. P., & Lam, M. (2012). Acute mental disturbance caused by synthetic cannabinoid: A potential emerging substance of abuse in Hong Kong. *East Asian Archives of Psychiatry*, *22*, 31–33.

Uchiyama, N., Matsuda, S., Kawamura, M., Kikura-Hanajiri, R., & Goda, Y. (2013). Two new-type cannabimimetic quinolinyl carboxylates, QUPIC and QUCHIC, two new cannabimimetic carboxamide derivatives, ADB-FUBINACA and ADBICA, and five synthetic cannabinoids detected with a thiophene derivative α-PVT and an opioid receptor agonist AH-79. *Forensic Toxicology*, *31*, 223–240. Available at: http://www.scopus.com/inward/record.url?eid=2-s2.0-84879793220&partnerID=tZOtx3y1. Accessed 22.05.16.

Uchiyama, N., Shimokawa, Y., Matsuda, S., Kawamura, M., Kikura-Hanajiri, R., & Goda, Y. (2013). Two new synthetic cannabinoids, AM-2201 benzimidazole analog (FUBIMINA) and (4-methylpiperazin-1-yl)(1-pentyl-1H-indol-3-yl)methanone (MEPIRAPIM), and three phenethylamine derivatives, 25H-NBOMe 3,4,5-trimethoxybenzyl analog, 25B-NBOMe, and 2C-N-NBOMe, id. *Forensic Toxicology*, *32*, 105–115. Available at: http://link.springer.com/10.1007/s11419-013-0217-2. Accessed 21.05.16.

Valeriani, G., Corazza, O., Bersani, F. S., Melcore, C., Metastasio, A., Bersani, G., et al. (2015). Olanzapine as the ideal "trip terminator"? Analysis of online reports relating to antipsychotics' use and misuse following occurrence of novel psychoactive substance-related psychotic symptoms. *Human Psychopharmacology*, *30*, 249–254.

Van Der Veer, N., & Friday, J. (2011). Persistent psychosis following the use of Spice. *Schizophrenia Research*, *130*, 285–286. http://dx.doi.org/10.1016/j.schres.2011.04.022.

van Os, J., Bak, M., Hanssen, M., Bijl, R. V., de Graaf, R., & Verdoux, H. (2002). Cannabis use and psychosis: A longitudinal population-based study. *American Journal of Epidemiology*, *156*, 319–327. Available at: http://www.ncbi.nlm.nih.gov/pubmed/12181101. Accessed 12.04.16.

Vandrey, R., Dunn, K. E., Fry, J. A., & Girling, E. R. (2012). A survey study to characterize use of Spice products (synthetic cannabinoids). *Drug and Alcohol Dependence*, *120*, 238–241. Available at: http://www.drugandalcoholdependence.com/article/S0376871611003152/fulltext. Accessed 20.05.16.

Vearrier, D., & Osterhoudt, K. C. (2010). A teenager with agitation: Higher than she should have climbed. *Pediatric Emergency Care*, *26*, 462–465. Available at: http://www.ncbi.nlm.nih.gov/pubmed/20531137.

Verdoux, H., Sorbara, F., Gindre, C., Swendsen, J. D., & van Os, J. (2003). Cannabis use and dimensions of psychosis in a nonclinical population of female subjects. *Schizophrenia Research*, *59*, 77–84. Available at: http://www.ncbi.nlm.nih.gov/pubmed/12413646. Accessed 20.05.16.

Vilke, G. M., Debard, M. L., Chan, T. C., Ho, J. D., Dawes, D. M., Hall, C., et al. (2012). Excited delirium syndrome (ExDS): Defining based on a review of the Literature. *Journal of Emergency Medicine*, *43*, 897–905. Available at: http://dx.doi.org/10.1016/j.jemermed.2011.02.017.

Wells, D. L., & Ott, C. A. (2011). The "new" marijuana. *Annals of Pharmacotherapy*, *45*, 414–417. Available at: http://aop.sagepub.com/content/45/3/414.full. Accessed 22.05.16.

Wessinger, W. D., Moran, J. H., & Seely, K. A. (2015). Synthetic cannabinoid effects on behavior and motivation. In P. Campolongo & L. Fattore (Eds.), *Cannabinoid modulation of emotion, memory, and motivation* (pp. 205–224). New York: Springer.

WHO. (2016). *The health and social effects of nonmedical cannabis use*. Available at: http://www.who.int/substance_abuse/publications/cannabis_report/en/.

Wikström, M., Thelander, G., Dahlgren, M., & Kronstrand, R. (2013). An accidental fatal intoxication with methoxetamine. *Journal of Analytical Toxicology*, *37*, 43–46. Available at: http://jat.oxfordjournals.org/content/37/1/43.abstract. Accessed 09.05.16.

Wilkinson, S. M., Banister, S. D., & Kassiou, M. (2015). Bioisosteric fluorine in the clandestine design of synthetic cannabinoids. *Australian Journal of Chemistry*, *68*, 4–8.

Winstock, A. R., & Barratt, M. J. (2013a). The 12-month prevalence and nature of adverse experiences resulting in emergency medical presentations associated with the use of synthetic cannabinoid products. *Human Psychopharmacology*, *28*, 390–393. Available at: http://www.ncbi.nlm.nih.gov/pubmed/23881887. Accessed 20.05.16.

Winstock, A. R., & Barratt, M. J. (2013b). Synthetic cannabis: A comparison of patterns of use and effect profile with natural cannabis in a large global sample. *Drug and Alcohol Dependence*, *131*, 106–111. Available at: http://www.ncbi.nlm.nih.gov/pubmed/23291209. Accessed 29.04.16.

Winstock, A., Lynskey, M., Borschmann, R., & Waldron, J. (2015). Risk of emergency medical treatment following consumption of cannabis or synthetic cannabinoids in a large global sample. *Journal of Psychopharmacology*, *29*, 698–703. Available at: http://www.embase.com/search/results?subaction=viewrecord&from=export&id=L604701805/nhttp://dx.doi.org/10.1177/0269881115574493/nhttp://sfx.metabib.ch/sfx_locater?sid=EMBASE&issn=14617285&id=doi:10.1177/0269881115574493&atitle=Risk+of+emergency+medical+tr.

Wood, D. M., & Dargan, P. I. (2012). Novel psychoactive substances: How to understand the acute toxicity associated with the use of these substances. *Therapeutic Drug Monitoring*, *34*, 363–367. Available at: http://www.ncbi.nlm.nih.gov/pubmed/22673201.

World Health Organization. (2014). *JWH-018 critical review report agenda item 4.5*. In: *Expert committee on drug dependence, thirty-sixth meeting*.

Wurita, A., Hasegawa, K., Minakata, K., Watanabe, K., & Suzuki, O. (2014). A large amount of new designer drug diphenidine coexisting with a synthetic cannabinoid 5-fluoro-AB-PINACA found in a dubious herbal product. *Forensic Toxicology*, *32*, 331–337. Available at: http://link.springer.com/10.1007/s11419-014-0240-y. Accessed 22.05.16.

Yip, L., & Dart, R. C. (2014). Is there something more about synthetic cannabinoids? *Forensic Toxicology*, *32*, 340–341. Available at: http://link.springer.com/10.1007/s11419-013-0224-3. Accessed 22.05.16.

Zimmermann, U. S., Winkelmann, P. R., Pilhatsch, M., Nees, J. A., Spanagel, R., & Schulz, K. (2009). Withdrawal phenomena and dependence syndrome after the consumption of "spice gold". *Deutsches Ärzteblatt International*, *106*, 464–467. Available at: http://www.pubmedcentral.nih.gov/articlerender.fcgi?artid=2719097&tool=pmcentrez&rendertype=abstract. Accessed 13.05.16.

Zuurman, L., Passier, P.C.C.M., de Kam, M. L., Kleijn, H. J., Cohen, A. F., & van Gerven, J. M. (2010). Pharmacodynamic and pharmacokinetic effects of the intravenous CB1 receptor agonist Org 26828 in healthy male volunteers. *Journal of Psychopharmacology*, *24*, 1689–1696.

Chapter 10

Cannabis Use as an Independent Risk Factor for, or Component Cause of, Schizophrenia and Related Psychotic Disorders

Jodi M. Gilman, Sara M. Sobolewski, Anne Eden Evins
Massachusetts General Hospital, Boston, MA, United States
Harvard Medical School, Boston, MA, United States

EVIDENCE FOR DIRECT CAUSALITY

An etiological claim for causality can be made when the following criteria are met (Rothman & Greenland, 1998): (1) there is a probabilistic association between an x-variable (the assumed "cause," in this case, cannabis use) and a y-variable (the "effect," in this case, schizophrenia); (2) x precedes y in time; and (3) the influence of a confounder, z, that causes both x and y, can be ruled out. An etiological claim can be further strengthened when (4) a dose-response relationship is found. Therefore, in people who use more cannabis, a proportionally greater risk of developing schizophrenia must be observed. Finally, a claim for causality should include a biologically plausible, mechanistic explanation; that is, how could the use of cannabis set off a series of biological processes that could lead to schizophrenia? We will address each of these claims, highlighting their strengths and their limitations.

Association Between Cannabis Use and Schizophrenia

One of the first studies to identify an association between cannabis use and schizophrenia was published in *The Lancet* in 1987, based on an epidemiological study by Andreasson and colleagues examining the risk of hospitalization for schizophrenia in a cohort of Swedish cannabis users and nonusers over a 15-year period. Psychiatric admissions were 6.0 times more common in regular cannabis smokers and 2.3 times more frequent in occasional smokers than in subjects who had never smoked cannabis (Andreasson, Allebeck, Engstrom, & Rydberg, 1987). In 2002, this Swedish sample was reanalyzed, and after

The Complex Connection between Cannabis and Schizophrenia. http://dx.doi.org/10.1016/B978-0-12-804791-0.00010-0

adjusting the results by the dose of cannabis used, the authors have reported that the group of study subjects with the highest cannabis use (over 50 times) had an odds ratio of 3.1 for developing schizophrenia (Zammit, Allebeck, Andreasson, Lundberg, & Lewis, 2002). After 35 years of follow-up, the risk of developing schizophrenia was again 3.7 among frequent cannabis users compared with nonusers (Manrique-Garcia et al., 2012).

Other longitudinal studies have also shown an increased risk of developing psychosis after using cannabis (Arendt, Rosenberg, Foldager, Perto, & Munk-Jorgensen, 2005; van Os et al., 2002), even after adjustment for factors such as age, sex, social class, ethnicity, urbanicity, and use of other drugs (Arseneault et al., 2002; Henquet et al., 2006). Therefore, an association between cannabis use and schizophrenia is well established, and is rarely debated in the scientific community. In fact, the largest meta-analysis of all available published data through 2013 (involving 66,816 individuals) confirmed a positive association between the extent of cannabis use and the risk for psychosis. The pooled analysis reported an approximately fourfold increase in risk for the heaviest users and a twofold increase for the average cannabis user in comparison to nonusers. This observation remained stable irrespective of the study design (cohort or cross-sectional) or the outcome measure (broad definition of psychosis or narrow diagnosis of a psychotic disorder) (Marconi, Di Forti, Lewis, Murray, & Vassos, 2016).

Cannabis Use Precedes the Development of Schizophrenia

A second key piece of evidence for causation would be a temporal relationship between the initiation of cannabis use and the onset of schizophrenia. Many studies have shown that marijuana use predates onset of psychotic disorders, providing some evidence of a possible causal link (Allebeck, Adamsson, Engstrom, & Rydberg, 1993; Arseneault et al., 2002; Buhler, Hambrecht, Loffler, Ander Heiden, & Hafner, 2002; Mauri et al., 2006; Semple, McIntosh, & Lawrie, 2005; Zammit et al., 2002). However, even when cannabis use unequivocally predates a diagnosis, schizophrenia is a neurodevelopmental disorder, with signs or symptoms often emerging before the diagnosis is made. There is often a prodromal period during which evidence of an emerging disorder is present, though not yet clinically manifest.

More recent studies have focused specifically on the link between marijuana use and age at onset of psychosis (Barnes, 2006; Compton, Kelley, et al., 2009; Large, Sharma, Compton, Slade, & Nielssen, 2011; Sevy et al., 2010; Van Mastrigt, Addington, & Addington, 2004; Veen et al., 2004), rather than a diagnosis of a psychotic disorder. One study, designed to provide a thorough retrospective assessment of premorbid marijuana use from age 12 until the onset of psychosis in a sample of first-episode patients, found that escalation of premorbid use in the five years prior to the onset was highly predictive of an increased risk for onset (Kelley et al., 2016). This study found that daily use approximately doubled the rate of onset, even after controlling for simultaneous alcohol and tobacco use, providing evidence of a clear temporal relationship between escalations in use in the five years pre-onset and an increased rate of onset. Interestingly, this study found that the strength of the association was similar among those who used cannabis during adolescence (before the age 17) and those who used it in early adulthood (after the age 17), suggesting that young adult use may be just as important as adolescent use regarding these associations. Thus, there is strong evidence to support the claim that cannabis use often precedes the onset of psychotic symptoms in schizophrenia, and also hastens onset.

The Influence of Confounder Variables That Cause Both Cannabis Use and Schizophrenia Can Be Ruled Out

There are many confounding variables that may lead to the development of schizophrenia, and many studies have attempted to control for these variables. Studies have shown that controlling for familial risk of schizophrenia attenuates but does not eliminate the association between cannabis use and schizophrenia, supporting the hypothesis that cannabis exposure, particularly early and frequent exposure to high-Δ^9-tetrahydrocannabinol (THC)-potency cannabis, is a causal factor in the development of schizophrenia (Giordano, Ohlsson, Sundquist, Sundquist, & Kendler, 2015). In addition, there is evidence that cannabis exposure precedes onset of psychosis by up to seven years, even after controlling for other drug use (Andreasson et al., 1987; Di Forti et al., 2009; Fergusson, Horwood, & Swain-Campbell, 2003). An association remains even after adjustment for factors such as age, sex, social class, ethnicity, urbanicity, and use of other drugs (Arseneault et al., 2002; Henquet et al., 2006). However, it must be acknowledged that large epidemiological cohort studies do not lend themselves very well to proving that cannabis use directly causes the onset of schizophrenia, because an etiological claim can always be challenged by (unobserved) confounders. In other words, it is impossible to control for every possible confounder.

To better understand the potential causal role of cannabis use in the development of psychotic disorders, biomedical research on the effects of cannabinoids on neurotransmitter systems can elucidate mechanisms. There is evidence that individuals with or vulnerable to psychosis have a neurobiological response to the main psychoactive compound in cannabis, THC, which renders them more vulnerable to the psychotogenic effects of cannabis. Those with a psychotic disorder and their siblings are more sensitive than matched controls to the psychotogenic effects of acute THC administration (D'Souza et al., 2004; Schizophrenia Working Group of the Psychiatric Genomics Consortium, 2014). This may explain why even among those who are under medications for psychotic disorders, cannabis use is associated with an increased risk of relapse of psychotic symptoms (Alvarez-Jimenez et al., 2012). Inhalation of vaporized

THC (8 mg) by cannabis users with a psychotic disorder (not on medication) and those with first-degree relatives with a psychotic disorder has been associated with significantly greater striatal dopamine release than in control cannabis users without psychiatric illness, despite having a similar subjective response to THC (Kuepper et al., 2013). Such laboratory studies provide evidence—though not proof—that confounder variables do not fully account for the association between cannabis use and schizophrenia.

A Dose-Response Relationship Between Extent of Cannabis Use and Rate of Developing Schizophrenia Can Be Found

Finally, a possible causal association would be further supported if there were a dose-response relationship between the amount of cannabis smoked and the likelihood of developing schizophrenia. Studies have indeed shown that higher amounts of cannabis use at baseline are associated with: (1) greater risk of developing schizophrenia or psychotic symptoms (Fergusson et al., 2003; van Os et al., 2002; Zammit et al., 2002), (2) greater likelihood that psychotic symptoms at baseline persist (Henquet et al., 2006), and (3) earlier onset of psychosis (Gonzalez-Pinto et al., 2008). In addition, it has been shown that faster *progression* to high levels of use is associated with increased risk of psychosis (Boydell et al., 2006), and earlier onset of psychosis (Compton, Kelley, et al., 2009; Kelley et al., 2016). Age of initiation of marijuana use is also associated with age at onset of psychosis (Arseneault et al., 2002; Di Forti et al., 2014; Kelley et al., 2016; Leeson, Harrison, Ron, Barnes, & Joyce, 2012; Stefanis et al., 2013) indicating a possible cumulative dose effect. Thus, there is a well-replicated dose effect, such that daily use and use of high-THC-potency cannabis further increase the risk of developing a psychotic illness and of earlier onset of psychosis (Di Forti et al., 2009; Di Forti et al., 2014; Di Forti, Marconi, et al., 2015).

EVIDENCE FOR REVERSE CAUSATION: DO EARLY MANIFESTATIONS OF PSYCHOTIC DISORDERS CAUSE CANNABIS USE? (THE SELF-MEDICATION HYPOTHESIS)

Some researchers have argued that instead of cannabis preceding schizophrenia, schizophrenia and its related early signs and symptoms cause cannabis use. This argument purports that those already experiencing symptoms of schizophrenia use cannabis in an attempt to cope with the negative symptoms such as social withdrawal, anhedonia, or blunted affect, which tend to develop early and are poorly managed by antipsychotics.

The self-medication hypothesis hinges on the temporal sequence between the onset of cannabis use and the onset of detectable symptoms, and this hypothesis has largely been discarded as an explanation for the association between cannabis use and schizophrenia. Many studies have used strategies to test the self-medication hypothesis, such as measuring the temporal sequence between the two events, and conducting analyses on only those participants who reported no prior history of diagnosis or symptoms before the use of cannabis. For example, Zammit and colleagues excluded any participant who showed signs of a psychotic disorder at the time of recruitment; furthermore, an additional subgroup analysis was conducted in the group that developed schizophrenia within five years of recruitment (an attempt to exclude those who might have used cannabis because they experienced prodromal symptoms of schizophrenia) (Zammit et al., 2002). Other studies have specifically recruited participants who reported no history of psychotic symptoms (van Os et al., 2002; Weiser, Knobler, Noy, & Kaplan, 2002). In addition, some groups have used statistical techniques to control for any prior history of symptoms of any mental disorder (Arseneault et al., 2002; Fergusson et al., 2003). In each of these studies, results demonstrated that the onset of schizophrenia and its symptoms was preceded by prior use of cannabis, implying that cannabis use increases the risk of schizophrenia rather than schizophrenia increasing the risk of cannabis use.

An uncertainty regarding the ability of researchers to accurately measure negative or subtle mood symptoms that may emerge in childhood even prior to the prodromal phase of schizophrenia still exists. If patients are self-medicating very subtle mood, negative, or neurocognitive signs or symptoms, this correlation could be missed by retrospective reports or prospective monitoring of positive psychotic symptoms. However, most researchers are in agreement that this theory of reverse causality is not strongly supported by the existing literature.

EVIDENCE FOR SHARED ETIOLOGY: COMMON UNDERLYING FACTORS OR CONFOUNDING MECHANISMS

Several environmental factors have been shown to increase the risk of both psychosis and cannabis use, such as childhood trauma, urbanicity, prenatal environment, socioeconomic status, and the use of other substances such as tobacco. Given practical considerations, it is nearly impossible for studies to control for all of these confounding factors, and some researchers have argued that because of these potential confounders, causal associations between cannabis use and psychosis may be overestimated. With this in mind, we will discuss some of the underlying factors that may lead to both cannabis use and psychosis, and explore their roles in moderating the cannabis-psychosis link.

Childhood Trauma

Childhood adversity has been associated with an increased risk of psychosis in adulthood (Matheson, Shepherd, Pinchbeck, Laurens, & Carr, 2013). Stress, especially prolonged and pathological stress, adversely impacts psychological development, and is likely to be an important effect of childhood adversity and a component cause of a variety of adult psychiatric disorders, including but not limited to schizophrenia. Childhood adversity has been found to be associated with persistent sensitization and increased activation of the hypothalamic-pituitary-adrenal (HPA) axis, and hyperactivity of the HPA axis has been observed in individuals with first-episode psychosis (Mondelli et al., 2010), as well as in those with depression or anxiety (Heim et al., 2000). Furthermore, a history of childhood trauma is more prevalent in patients with psychosis compared with healthy controls (Bendall, Jackson, Hulbert, & McGorry, 2008), and several studies suggest that this relationship is causal (Larkin & Read, 2008; Read, van Os, Morrison, & Ross, 2005). For example, one study reported that childhood trauma in the form of bullying predicted the development of psychosis in a dose-dependent manner, and cessation of bullying reduced the frequency of psychotic symptoms in children compared to those who continued to experience the adversity (Kelleher et al., 2013). It is important to note, however, that the relationship between childhood trauma and psychosis has not been universally reported (Chen et al., 2010).

Childhood trauma is also known to be a strong risk factor for the development of problems with alcohol and drugs (Annerback, Sahlqvist, Svedin, Wingren, & Gustafsson, 2012; Goldstein et al., 2013; Rogosch, Oshri, & Cicchetti, 2010; Schwandt, Heilig, Hommer, George, & Ramchandani, 2013). Adults with addictions are more likely than the general population to have experienced childhood maltreatment (Schwandt et al., 2013), and among adolescents and young adults, a history of childhood maltreatment has been linked to heavy episodic drinking and alcohol use disorders in large, nationally representative studies (Goldstein et al., 2013; Shin, Edwards, & Heeren, 2009). Although few studies have specifically measured cannabis use, there is evidence that this association holds for cannabis as well as other drugs (Rogosch et al., 2010; Vilhena-Churchill & Goldstein, 2014).

Given the association between childhood trauma, cannabis use, and psychosis, recent studies have attempted to adjust for childhood trauma when conceptualizing the cannabis-psychosis link. One study found that cannabis use did not increase the risk of psychosis when controlling for early sexual trauma (Houston, Murphy, Adamson, Stringer, & Shevlin, 2008). The same study did, however, find a significant interaction between cannabis use and childhood sexual trauma such that exposure to both variables increased the risk of psychosis by a factor of 12 in those who used cannabis before 16 years of age. Another study found main effects for both factors independently, reporting that both childhood trauma and cannabis use interact synergistically, so that exposure to both childhood trauma and cannabis use increased the risk of psychosis beyond the risk associated with either factor alone (Harley et al., 2010). Consistent with these findings, Konings et al. (2012) analyzed prospective data from two independent, population-based studies—the Greek National Perinatal Study ($n=1636$) and The Netherlands Mental Health Survey and Incidence Study (NEMESIS) ($n=4842$)—and found that childhood maltreatment moderated the effects of cannabis use on psychosis in a dose-dependent, extra-linear manner, with earlier, more severe maltreatment being associated with the strongest effect of cannabis use on psychosis outcomes. It is important to mention that some studies have not found an interaction between cannabis use and childhood trauma (Kuepper, Henquet, Lieb, Wittchen, & van Os, 2011; Sideli et al., 2015); however, these differences may be due to sampling variation and differences in follow-up time.

As a whole, the evidence suggests an interaction between childhood trauma and cannabis use on the development of psychosis. Each on its own may contribute to psychosis, with a possible additive or even multiplicative effect, such that more severe trauma is associated with the strongest effects of cannabis use on the development of psychosis. Larger, longitudinal cohort studies with longer follow-up times are needed to fully understand this undoubtedly complex relationship.

Urbanicity

Being born and growing up in an urban environment has been consistently associated with an increased incidence of schizophrenia in a dose-response manner (Kelly et al., 2010; Krabbendam & van Os, 2005; March et al., 2008). This is consistent with the finding that living in urban areas is associated with increased risk of adolescents and young adults reporting at least one psychotic experience (Spauwen, Krabbendam, Lieb, Wittchen, & van Os, 2004). In addition, urbanicity is associated with an earlier age at onset of schizophrenia, with urban residents showing onset of psychotic symptoms one year earlier, on average, than rural residents (Stompe, Ortwein-Swoboda, Strobl, & Friedmann, 2000). Evidence suggests that this relationship may be mediated by social fragmentation (Allardyce et al., 2005; Zammit, Lewis, Dalman, & Allebeck, 2010), though this remains unclear since many other characteristics, such as ethnicity and immigrant status, are associated with both urban residence and schizophrenia (Kelly et al., 2010). Nonetheless, there is evidence that exposure to an urban environment between the age of 5 and 15 is associated with the greatest effect on later psychosis (Pedersen & Mortensen, 2001), suggesting a mediation by factors that have an impact specifically during development as opposed to at birth (such as obstetric complications or prenatal infections) or at onset of symptoms (Marcelis, Takei, & van Os, 1999).

The relationship between psychosis, urbanicity, and cannabis use is complicated by the fact that large studies (e.g., German Early Developmental Stages of Psychopathology [EDSP] cohort study) have found a significant association between urbanicity and cannabis use (Kuepper, van Os, Lieb, Wittchen, & Henquet, 2011), indicating that individuals living in an urban environment are more likely to use cannabis than individuals from a rural area. The interaction between cannabis use and urbanicity, therefore, may represent not only an underlying mechanism of moderation in which the effect of cannabis on psychosis is stronger in urban areas, but also of mediation.

One of the most thorough studies examined two independent general population samples in order to determine whether common, nonclinical developmental expressions of psychosis may become abnormally persistent when synergistically combined with developmental exposures such as cannabis use, trauma, and urbanicity (Cougnard et al., 2007). This study found that the level of environmental risk (cannabis use, childhood trauma, and growing up in an urban environment) acts additively to increase the risk of persistent psychosis in those with baseline psychotic experiences. In other words, they found that environmental load acted synergistically on persistent psychosis. The authors proposed that cumulative exposure to these developmental environmental risk factors in subjects with proneness to psychosis (as evidenced by psychotic experiences) may result in cumulative changes in the functioning of the dopaminergic system, affecting the persistence and the deterioration of developmental psychotic features. Furthermore, all of these factors (urbanicity, cannabis use, and stress from childhood trauma) may be explained by behavioral sensitization (Collip, Myin-Germeys, & Van Os, 2008), and the interaction of these factors may alter dopaminergic signaling in the mesolimbic system and prefrontal cortex, resulting in enduring sensitization to dopamine agonists (Kuepper et al., 2010).

Prenatal Environment

It is worth mentioning that along with cannabis use, childhood trauma, and the use of other drugs such as nicotine and alcohol, psychosis has also been linked to suboptimal prenatal environment. The neurodevelopmental model of schizophrenia proposes that disturbances to the development of the nervous system during gestation increase the risk of schizophrenia in later life; other factors, acting during upbringing, adolescence, or early adult life (e.g., cannabis use, childhood trauma) may also modify risk (McGrath, Mortensen, Visscher, & Wray, 2013; McGrath, Cornelis, et al., 2013; Murray, Jones, O'Callaghan, Takei, & Sham, 1992; Murray, O'Callaghan, Castle, & Lewis, 1992; Waddington, Lane, Larkin, & O'Callaghan, 1999; Waddington, Lane, Scully, et al., 1999; Weinberger, 1996).

Various mechanisms have been proposed to explain the association between a disturbed prenatal development and schizophrenia, including obstetric complications, infections, and low prenatal vitamin D. Each of these is more common in urban environments, prompting scientists to hypothesize that these factors may explain the relationship between urbanicity and the development of schizophrenia. This, however, does not appear to be the case. Harrison et al. (2003), in a large case-register study from Sweden, found that the effect of urbanicity remained significant even after controlling for age, birth weight, birth length, gestational age, season of the birth, age of the mother, Apgar score at 1 minute, maternal parity, caesarean section, and gender (Harrison et al., 2003). Torrey, Mortensen, Pedersen, Wohlfahrt, and Melbye (2001) studied 264 mothers of individuals with psychosis and 528 mothers of matched controls and found that urban/suburban residence at birth still made an independent contribution to risk of psychosis even after controlling for fever during pregnancy, complications during pregnancy, and cat ownership (which is associated with the *Toxoplasma gondii* parasite) between birth and age 13 years (Torrey et al., 2001).

Socioeconomic Status

Low social class or variables linked to low socioeconomic status (SES) have been associated with increased risk of schizophrenia (Eaton, 1974), prompting some researchers to propose that low SES explains the relationship between urbanicity and schizophrenia. However, in a large Danish study of over 7700 patients with schizophrenia, researchers found that while schizophrenia was associated with unemployment, low educational attainment, low income, and being single, the risk of schizophrenia for any given individual was also associated with birth in an urban area, even after taking these factors into account (Byrne, Agerbo, Eaton, & Mortensen, 2004). This suggests that individual-level socioeconomic factors are unlikely to fully account for the association between urbanicity and psychosis; this conclusion is supported by other studies that controlled for factors related to socioeconomic status at the individual level (Harrison, 1990).

Another study of 1923 individuals from the population-based German EDSP cohort found that adolescents who grew up in the city of Munich were much more likely to develop psychotic symptoms after cannabis use than individuals who grew up in the rural surroundings of Munich. This interaction effect was independent of confounding factors such as age, sex, SES, use of other drugs, and childhood trauma, suggesting an independent role for urbanicity. Importantly, these analyses revealed that the effect of cannabis use on follow-up incident psychotic symptoms was much stronger in individuals who grew up in an urban environment compared with individuals from rural surroundings; in other words, the majority of those who developed psychosis were exposed to both cannabis and urbanicity (Kuepper, van Os, et al., 2011).

Tobacco Use

It is important to consider tobacco use as a potential confounding factor in the cannabis-psychosis link because of its high rate of co-occurrence with cannabis use. As cannabis users are likely to smoke tobacco, and tobacco smokers are more likely to use cannabis than nonsmokers, separating the two factors is complicated (Gage et al., 2014). There have been few longitudinal studies specifically focusing on the relationship between tobacco use and psychosis, and these generally report an increased risk of psychosis in tobacco smokers (Sorensen, Mortensen, Reinisch, & Mednick, 2011; Weiser et al., 2004), though some studies actually found a lower risk of psychosis among tobacco smokers (Kelley et al., 2016; Zammit et al., 2003). A recent meta-analysis found that daily tobacco use was associated with an increased risk of psychotic disorder and an earlier age at onset of psychotic illness, but that the effect of smoking was modest, and when the analysis was restricted to studies in which daily cigarette smoking was specified, the association disappeared (Gurillo, Jauhar, Murray, & MacCabe, 2015).

Apart from having different biological actions from cannabis use, tobacco use may be associated with several environmental factors like SES and family adversity that would confound the relationship between cannabis use and schizophrenia if not considered (Hiscock, Bauld, Amos, Fidler, & Munafo, 2012). A recent study, however, determined that the daily use of cannabis approximately doubled the rate of onset of psychosis even after controlling for simultaneous alcohol and tobacco use (Kelley et al., 2016). This report is consistent with other longitudinal studies investigating the cannabis-psychosis relationship, which found an association between cannabis and psychosis even after adjusting for tobacco use (Henquet et al., 2005; Rossler, Hengartner, Angst, & Ajdacic-Gross, 2012; Wiles et al., 2006). However, other studies have found that adjusting for tobacco use resulted in a significant attenuation of the cannabis-psychosis association [decrease in the odds ratio (OR) from 3.2 to 1.2 (Gage et al., 2014)].

Perhaps the most thorough study was a meta-analysis of 25 articles from 29 samples that investigated the relationship between tobacco use and psychosis, conducted by Myles et al. (2012). In contrast to cannabis use (Large et al., 2011), they found no evidence that tobacco use was associated with a statistically significant or clinically relevant earlier age at onset of psychosis. The results suggest that the association between cannabis use and earlier onset of psychosis is robust and is not the result either of tobacco smoking by cannabis-using patients or other potential confounding factors (Myles et al., 2012). A note of caution in interpreting this lack of a relationship is that the association of psychosis with cannabis use versus cigarette use is rarely assessed in the same samples, and when they are, their effects are difficult to disentangle because of high rates of co-occurrence (Ksir & Hart, 2016). This co-occurrence is particularly problematic because, often, there is a reduced frequency of heavy cannabis users in light smokers or nonsmokers. This highlights the importance of using comprehensive measures of both cannabis and tobacco use. Finally, there is evidence that the use of other drugs such as alcohol (Auther et al., 2015) and stimulants (Sara, Burgess, Malhi, Whiteford, & Hall, 2014) may confound the relationship between cannabis use and psychosis.

There is little doubt that common etiological factors underlie both cannabis use and psychosis, and this highlights the need to control for other factors, so as to not overstate the cannabis-psychosis connection. However, the evidence suggests that these factors may be synergistic and additive with cannabis use, so that those with risk factors such as childhood trauma, urbanicity, and low SES may be more susceptible to cannabis-related effects on the development of psychosis. Future studies will need to consider, tease apart, and perhaps focus on specific subgroups of vulnerable individuals in order to better understand the contribution of all factors to this relationship.

EVIDENCE FOR GENE-BY-CANNABIS INTERACTIONS

The link between cannabis and psychosis may be explained by one or more gene-by-environment interactions, in which cannabis is an environmental trigger of genetic vulnerability to psychotic illness. Since the exposure to exogenous cannabinoids such as THC has been proposed to influence the dopaminergic pathways that are thought to underpin psychosis, genetic variants affecting those pathways have been suggested to combine synergistically with cannabis use in the development of psychotic symptoms.

Catechol-*O*-Methyltransferase

One of the most commonly implicated genes in potentially catalyzing the relationship between cannabis use and schizophrenia is the gene encoding catechol-*O*-methyltransferase (COMT). COMT is a gene that codes for an enzyme that degrades endogenous amines, including dopamine, which is well known to mediate the psychoactive effects of THC. This gene, particularly the Met (methionine) to Val (valine) substitution at codon 158, has been shown to interact with cannabis use to influence the development of psychosis by regulating dopaminergic transmission in the midbrain. The Val variant in this polymorphism has been associated with a higher likelihood of psychotic symptoms during adulthood, but only if individuals had used cannabis during adolescence (Caspi et al., 2005). Other studies have shown that Met(158) homozygotes show later onset of psychosis compared to Val(158) homozygotes, a relationship that is mediated by cannabis use. Val(158) homozygotes who are cannabis users also show reduced duration of untreated psychosis compared to nonusers (Pelayo-Teran, Suarez-Pinilla, Chadi, & Crespo-Facorro, 2012). A possible explanation for this may be the relationship between the Val(158) allele and more severe symptoms; the increase of psychotic symptomatology (particularly hallucinations) in Val(158) homozygotes could possibly encourage patients or their families to seek earlier medical attention.

In a psychiatric population ($n = 157$), the COMT Val(158)Met polymorphism showed an interaction with both age at first cannabis use and lifetime cannabis use, to predict age at onset of schizophrenia, suggesting that Val allele carriers may be more susceptible to the effect of cannabis use on regulation of the dopaminergic system (Estrada et al., 2011). In a randomized clinical trial, COMT Val allele carriers experienced greater cognitive impairment with THC (300 μg/kg), and more psychotic symptoms compared with placebo than Met allele carriers, but only among those who had reported a prior history of psychotic symptoms (Henquet et al., 2006).

While intuitively appealing, the finding that the COMT Val(158)Met polymorphism is a moderator of the association between cannabis use and psychosis has not always been replicated. For instance, Zammit, Owen, Evans, Heron, and Lewis (2011) examined whether cannabis use at age 14 was associated with psychotic symptoms at age 16 in the Avon Longitudinal Study of Parents and

Children ($n=2630$), and did not find an interaction between cannabis exposure and COMT genotype. A recent study ($n=533$) suggested that the predictive effect of the interaction between cannabis use and COMT genotype on psychosis is exerted only in those with a history of childhood abuse (Alemany et al., 2014). Finally, contradictory evidence from a recent study ($n=748$ patients) of COMT haplotypes in two Spanish samples indicated a higher degree of association between lifetime cannabis use and schizophrenia in Met (not Val) carriers (Costas et al., 2011).

Cannabinoid Receptor 1

Another genetic locus that has been implicated in the relationship between cannabis and psychosis involves polymorphisms in the cannabinoid receptor itself. Cannabinoid receptor 1 (CNR1, also known as CB_1 or CB_1R) regulates striatal dopamine and modulates the effects of exogenous cannabis. It is widely expressed in brain areas including the prefrontal cortex and medial temporal lobe (Pazos, Nunez, Benito, Tolon, & Romero, 2005). A recent study found that heavy cannabis use in the context of specific CB_1 genotypes contributed to greater white matter volume deficits and cognitive impairment, which could in turn increase schizophrenia risk (Ho, Wassink, Ziebell, & Andreasen, 2011). In addition, CB_1 has been shown to interact with another cannabinoid-related gene, mitogen-activated protein kinase 14 (MAPK14), to confer brain volume abnormalities in schizophrenia with heavy cannabis use (Onwuameze et al., 2013; Suarez-Pinilla et al., 2015). However, other schizophrenia-CB_1 genetic association studies, which have examined two CB_1 variants (the rs1049353 single nucleotide polymorphism (SNP) and an AAT trinucleotide repeat), have found mixed results (Chavarria-Siles et al., 2008; Seifert, Ossege, Emrich, Schneider, & Stuhrmann, 2007), highlighting the need for larger samples.

AKT1

AKT1, a protein kinase involved in molecules downstream of the dopamine D_2 receptor (Ozaita, Puighermanal, & Maldonado, 2007), has also been associated with schizophrenia and cannabis use. AKT1 codes for the protein kinase that forms part of the striatal dopamine receptor signaling cascade, which presents a plausible biological mechanism for interacting with cannabis use to confer increased risk for schizophrenia. Researchers have found that those having cytosine homozygote (C/C) rs2494732 genotypes who used cannabis had a two-fold higher chance of having a psychotic disorder (Van Winkel, van Beveren, Simons, & Genetic, 2011). In another study, those with the AKT1 C/C genotype with ever cannabis use and daily use showed two- and seven-fold increased likelihoods of a psychotic disorder, respectively, compared with ever users and daily users who were thymine homozygote (T/T) carriers (Di Forti et al., 2012). Thus, the AKT1 gene remains one of the most plausible to moderate the relationship between cannabis use and the development of psychosis, though this too is in need of replication in larger samples.

Brain-Derived Neurotrophic Factor

Another gene of interest in understanding psychosis and cannabis use is the brain-derived neurotrophic factor (BDNF) gene, a neurotrophin implicated in the modulation of various neurotransmitters including dopamine, serotonin, and γ-aminobutyric acid (GABA). Notably, a Val to Met substitution at codon 66 (rs6265) of BDNF results in less efficient intracellular trafficking and decreased activity-dependent BDNF secretion (Egan et al., 2003). Met carriers were found to have a significantly earlier age at onset of psychosis than Val carriers in a cohort of 159 Japanese patients with a diagnosis of schizophrenia (Numata et al., 2006) and in a cohort of 42 African American patients (Chao, Kao, & Porton, 2008). Yet, efforts by other groups have failed to replicate this association (Gourion et al., 2005; Naoe et al., 2007).

Animal and human studies suggest that cannabis use may directly influence BDNF-regulated physiological mechanisms (Derkinderen et al., 2003; D'Souza, Sewell, & Ranganathan, 2009; Jockers-Scherubl et al., 2004). For example, a study in mice reported a significant upregulation of BDNF mRNA in the hippocampus within 1 hour of THC injection (Derkinderen et al., 2003). A study in first-episode psychosis patients suggested that cannabis-using patients had significantly higher BDNF serum levels than both nonusing patients and matched controls (Jockers-Scherubl et al., 2004). Since BDNF upregulation in response to cannabis use may be less efficient in BDNF Met carriers, it has been proposed that cannabis use and BDNF Val66Met genotype interact to contribute to psychosis (D'Souza et al., 2009). In a recent study in which BDNF Val66Met and cannabis use before onset of illness were retrospectively assessed in a sample of 585 patients with schizophrenia, cannabis use was significantly associated with earlier age at onset of psychotic disorder, showing dose-response effects with higher frequency and earlier age at first use; however, no evidence was found for BDNF-by-cannabis use interaction (Decoster et al., 2011). However, a statistically significant BDNF-by-cannabis-by-sex interaction was found, suggesting the effects of cannabis use and BDNF Val66Met genotype on age at onset of psychotic disorders in female, but not male, patients with schizophrenia.

Dopamine Receptor D_2

Dopamine receptor D_2 (DRD2) genotype influenced the likelihood of a psychotic disorder in people who used cannabis, such that among cannabis users and daily cannabis users, carriers of the DRD2 rs1076560 thymine (T) allele had a three- and five-fold higher likelihood of a psychotic disorder, respectively (Colizzi et al., 2015). The authors concluded that these results indicate a model of interaction known as "qualitative gene-by-environment interaction" with a crossover pattern: carriers of the DRD2 rs1076560 T allele, compared to homozygote guanine (GG) subjects, have a lower probability of a psychotic disorder if they had never used cannabis, but a higher probability if they had a history of cannabis use, especially daily use (Colizzi et al., 2015).

Summary

Together, these genetic findings indicate that specific minor alleles (i.e., alleles other than the most common "major" alleles) may reduce or increase the risk for psychosis depending on the presence of a history of cannabis use. Such findings require validation in experimental designs and animal studies where both changes in the exposure and in the genotype can be modeled. It must be cautioned, however, that many of these findings are tentative, with small numbers of subjects, and require replication. An alternative explanation is that individuals at high genetic risk for schizophrenia may be more likely to use cannabis through a shared genetic risk for schizophrenia and cannabis use. Indeed, a recent report from a large genome-wide association study (GWAS) of an association between schizophrenia risk alleles and cannabis use suggests that part of the association between schizophrenia and cannabis use may be due to a shared genetic etiology (Power et al., 2014). It should be noted, however, that high-THC cannabis use was strongly associated with later development of schizophrenia in one study (Di Forti, Marconi, et al., 2015), and the polygenic risk score for schizophrenia (Schizophrenia Working Group of the Psychiatric Genomics Consortium, 2014) was unrelated to cannabis use or potency of cannabis used (Di Forti, Vassos, Lynskey, Craig, & Murray, 2015). Therefore, the proposition that some gene variants influence the likelihood of developing schizophrenia contingent on certain environmental exposure (e.g., cannabis use) is a strong, evidence-based, and scientifically feasible hypothesis.

PLAUSIBLE BIOLOGICAL MECHANISMS SUPPORTING CAUSATION

Epidemiological studies can only take us so far; in order to more directly establish causation, we need to focus on the neurobiological pathways affected by THC and how it mechanistically impacts the brain circuitry underlying schizophrenia. One explanation for how THC might be a component cause of schizophrenia regards the ability of THC to interfere with the strengthening and pruning of synaptic connections in the prefrontal cortex (Bossong & Niesink, 2010), which are linked to latent psychotic disorders (Sekar et al., 2016). According to this hypothesis, adolescent exposure to THC may transiently disturb physiological control of the endogenous cannabinoid system over glutamate and GABA release. As a result, THC may adversely affect adolescent experience-dependent maturation of neural circuitries within prefrontal cortical areas. Brain development is largely focused on the creation and elimination of neurons; in infancy and childhood, as many brain cells and connections as possible are created, and from childhood to adolescence, the brain undergoes a "pruning" process to make it as efficient as possible. This efficiency is thought to be achieved by synaptic refinement, the process by which some connections between brain cells are pruned and eliminated, and "useful" neurons, synapses, and dendrites are selected and preserved for the adult brain (Cohen-Cory, 2002; Katz & Shatz, 1996; Luna, 2009; Purves et al., 2008; Whitford et al., 2007). The endogenous cannabinoid system plays a critical role in regulating the balance of neurotransmitters such as glutamate and GABA, which function in strengthening and eliminating excitatory synaptic connections in cortical neurocircuitries during adolescence (Chevaleyre, Takahashi, & Castillo, 2006; Schlicker & Kathmann, 2001; Wilson & Nicoll, 2002).

The regulatory role of the endogenous cannabinoid system in GABA and glutamate neurotransmitter release is disrupted by both synthetic cannabinoids (Chevaleyre & Castillo, 2003; Kreitzer & Regehr, 2001; Yoshida et al., 2002) and THC (Hoffman, Oz, Yang, Lichtman, & Lupica, 2007; Mato et al., 2004). Possible mechanisms responsible for this disruption include downregulation (loss of binding sites) and desensitization (uncoupling from G-proteins) of CB_1 receptors. By preventing endocannabinoid-mediated control over the homeostasis of glutamate and GABA, exogenous cannabinoids might dramatically affect the process of maturational refinement of cortical neuronal networks. If this process is disturbed, it could lead to schizophrenia, especially in those already biologically vulnerable. Recently a landmark study demonstrated that alleles of the complement component 4 (C4) gene, located within the major histocompatibility complex (MHC) locus on human chromosome 6 and highly expressed in the brain, mediates synapse elimination during postnatal development in mice. The C4 genes, which are among the strongest genetic association loci in schizophrenia, may therefore help to explain the reduced numbers of synapses in the brains of individuals with schizophrenia (Sekar et al., 2016).

A growing body of animal studies supports the hypothesis that cannabis may have a unique causal contribution to the development of psychosis. The preclinical data generally support a hypothesis that adolescent exposure to cannabinoids might represent a risk factor for the development of psychotic-like symptoms in adulthood since it interferes with maturational events occurring in the adolescent brain. This eventually leads to alterations affecting brain connectivity and functionality similar to those present in patients with schizophrenia. The most likely mechanism underlying these effects involves the disruption of maturational events within the endocannabinoid system during adolescence. Indeed, the impairment in endocannabinoid system maturation might have an impact on the correct neuronal refinement unique to the adolescent brain, leading to altered adult brain functionality and behavior (Rubino & Parolaro, 2014). The current experimental paradigms are generally based on protocols in which animals are treated with THC or synthetic cannabinoids during part of their adolescence, and then behaviors of treated animals are monitored later in life, usually during early adulthood. A full review of the animal literature is beyond the scope of this chapter, but generally, chronic cannabinoid administration in adolescent rats leads to enduring cognitive deficits in adulthood, including working memory deficits and prepulse inhibition abnormalities like those commonly observed in individuals with schizophrenia (Malone & Taylor, 2006; O'Shea, Singh, McGregor, & Mallet, 2004; O'Shea, McGregor, & Mallet, 2006; Schneider, Schomig, & Leweke, 2008).

To summarize, exogenous cannabinoids, including THC, can disrupt the regulatory role of the endocannabinoid system and thus can affect the process of maturational refinement of cortical neuronal networks. Depending on dose, exact time window, duration of exposure, and preexisting vulnerability factors (environmental and genetic), this may ultimately lead to the development of psychosis or schizophrenia. Future research is needed to determine whether THC-induced disruptions in synaptic pruning do indeed underlie the relationship between cannabis use and schizophrenia. Further studies can investigate the contributions of excitatory and inhibitory synapses and the involvement of the endogenous cannabinoid system to the experience-dependent refinement of neural circuitries in the prefrontal cortex during adolescence (Bossong & Niesink, 2010).

CONCLUSIONS

Until we have a more complete understanding of the neurobiology, firmly establishing a causal relationship between cannabis use and psychosis will remain elusive. However, evidence within the literature demonstrates that the relationship between cannabis use and schizophrenia is strong (Table 1). This is a robust association between cannabis use and schizophrenia, and there is moderate to strong evidence that cannabis use precedes onset of schizophrenia (though it is difficult to precisely determine when subtle symptoms of schizophrenia may have emerged). Many of the confounding factors such as childhood trauma, urbanicity, prenatal environment, socioeconomic status, and the use of other drugs such as alcohol and tobacco, which could underlie both cannabis use and schizophrenia, have been adequately controlled for in large studies, though these factors may be synergistic and additive with cannabis use. A dose-response relationship exists in which individuals who use higher amounts (and more potent forms) of cannabis are more likely to develop schizophrenia earlier in life, and more likely to have a more severe course of illness. Plausible biological mechanisms, such as the ability of THC to disrupt the regulatory role of the endocannabinoid system, could explain how cannabis use could lead to the development of schizophrenia. We conclude that, although the current science cannot firmly establish proof of causation, it is prudent for clinicians to recommend that individuals with a family history of psychosis abstain from using cannabis.

TABLE 1 Criteria for Causality: Weighing the Evidence

Criteria	Evidence	Limitations
Association exists between cannabis and schizophrenia	+++	None
Cannabis precedes the onset of schizophrenia	++	Reverse causality possible; difficult to measure subtle mood symptoms that may emerge in childhood, prior to the prodromal phase of schizophrenia, which could predate cannabis use.
Confounders that cause both cannabis use and schizophrenia can be ruled out	+	Impossible for studies to control for every confounding factor; shared genetic disposition leading to both cannabis use and schizophrenia is possible.
Dose-response relationship exists between cannabis use and schizophrenia	++	Reverse causality could indicate that those with worse symptoms use more cannabis.
Biological mechanism for cannabis use causing schizophrenia	+	Incomplete understanding of the neurobiology of schizophrenia.

Key Chapter Points

- Existing literature has established a robust association between cannabis use and schizophrenia, and this association is stronger in frequent cannabis users.
- Though some studies argue for reverse causation (e.g., individuals with subtle mood symptoms that may emerge in childhood, prior to the prodromal phase of schizophrenia, are more likely to use cannabis), evidence for this theory is weak.
- It is possible that cannabis use and schizophrenia have a shared etiology, as individuals with cannabis use disorders and schizophrenia share common etiological factors, such as childhood trauma, low socioeconomic position, urbanicity, adverse prenatal environment, and tobacco use. However, after attempting to control for all of these factors, the relationship between cannabis use and schizophrenia remains significant.
- The link between cannabis and psychosis may be explained by one or more gene-by-environment interactions, in which cannabis use is an environmental trigger of genetic vulnerability to psychotic illness. Genetic variants affecting dopaminergic pathways have been suggested to combine synergistically with cannabis use in the development of psychotic symptoms.

- A biologically plausible pathway may in part explain how cannabis use can cause schizophrenia; it is possible that exogenous cannabinoids, including THC, disrupt the regulatory role of the endocannabinoid system and thus affect the process of maturational refinement of cortical neuronal networks. Depending on dose, exact time window, duration of exposure, and preexisting vulnerability factors (environmental and genetic), this may ultimately lead to the development of psychosis or schizophrenia.

ACKNOWLEDGMENTS

This work was supported by NIDA K01 DA034093 (JMG) and NIDA K24 DA030443 (AEE). These funding sources had no role in writing the manuscript, or the decision to submit the manuscript for publication.

REFERENCES

Alemany, S., Arias, B., Fatjo-Vilas, M., Villa, H., Moya, J., Ibanez, M. I., et al. (2014). Psychosis-inducing effects of cannabis are related to both childhood abuse and COMT genotypes. *Acta Psychiatrica Scandinavica*, *129*, 54–62.

Allardyce, J., Gilmour, H., Atkinson, J., Rapson, T., Bishop, J., & McCreadie, R. G. (2005). Social fragmentation, deprivation and urbanicity: Relation to first-admission rates for psychoses. *The British Journal of Psychiatry: The Journal of Mental Science*, *187*, 401–406.

Allebeck, P., Adamsson, C., Engstrom, A., & Rydberg, U. (1993). Cannabis and schizophrenia: A longitudinal study of cases treated in Stockholm County. *Acta Psychiatrica Scandinavica*, *88*, 21–24.

Alvarez-Jimenez, M., Priede, A., Hetrick, S. E., Bendall, S., Killackey, E. W., Parker, A. G., et al. (2012). Risk factors for relapse following treatment for first episode psychosis: A systematic review and meta-analysis of longitudinal studies. *Schizophrenia Research*, *139*, 116–128.

Andreasson, S., Allebeck, P., Engstrom, A., & Rydberg, U. (1987). Cannabis and schizophrenia. A longitudinal study of Swedish conscripts. *Lancet*, *2*, 1483–1486.

Annerback, E. M., Sahlqvist, L., Svedin, C. G., Wingren, G., & Gustafsson, P. A. (2012). Child physical abuse and concurrence of other types of child abuse in Sweden-Associations with health and risk behaviors. *Child Abuse & Neglect*, *36*, 585–595.

Arendt, M., Rosenberg, R., Foldager, L., Perto, G., & Munk-Jorgensen, P. (2005). Cannabis-induced psychosis and subsequent schizophrenia-spectrum disorders: Follow-up study of 535 incident cases. *The British Journal of Psychiatry: The Journal of Mental Science*, *187*, 510–515.

Arseneault, L., Cannon, M., Poulton, R., Murray, R., Caspi, A., & Moffitt, T. E. (2002). Cannabis use in adolescence and risk for adult psychosis: Longitudinal prospective study. *BMJ*, *325*, 1212–1213.

Auther, A. M., Cadenhead, K. S., Carrion, R. E., Addington, J., Bearden, C. E., Cannon, T. D., et al. (2015). Alcohol confounds relationship between cannabis misuse and psychosis conversion in a high-risk sample. *Acta Psychiatrica Scandinavica*, *132*, 60–68.

Barnes, M. P. (2006). Sativex: Clinical efficacy and tolerability in the treatment of symptoms of multiple sclerosis and neuropathic pain. *Expert Opinion on Pharmacotherapy*, *7*, 607–615.

Bendall, S., Jackson, H. J., Hulbert, C. A., & McGorry, P. D. (2008). Childhood trauma and psychotic disorders: A systematic, critical review of the evidence. *Schizophrenia Bulletin*, *34*, 568–579.

Bossong, M. G., & Niesink, R. J. (2010). Adolescent brain maturation, the endogenous cannabinoid system and the neurobiology of cannabis-induced schizophrenia. *Progress in Neurobiology, 92*, 370–385.

Boydell, J., van Os, J., Caspi, A., Kennedy, N., Giouroukou, E., Fearon, P., et al. (2006). Trends in cannabis use prior to first presentation with schizophrenia, in South-East London between 1965 and 1999. *Psychological Medicine, 36*, 1441–1446.

Buhler, B., Hambrecht, M., Loffler, W., Ander Heiden, W., & Hafner, H. (2002). Precipitation and determination of the onset and course of schizophrenia by substance abuse—A retrospective and prospective study of 232 population-based first illness episodes. *Schizophrenia Research, 54*, 243–251.

Byrne, M., Agerbo, E., Eaton, W. W., & Mortensen, P. B. (2004). Parental socio-economic status and risk of first admission with schizophrenia—A Danish national register based study. *Social Psychiatry and Psychiatric Epidemiology, 39*, 87–96.

Caspi, A., Moffitt, T. E., Cannon, M., McClay, J., Murray, R., Harrington, H., et al. (2005). Moderation of the effect of adolescent-onset cannabis use on adult psychosis by a functional polymorphism in the catechol-O-methyltransferase gene: Longitudinal evidence of a gene X environment interaction. *Biological Psychiatry, 57*, 1117–1127.

Chao, H. M., Kao, H. T., & Porton, B. (2008). BDNF Val66Met variant and age of onset in schizophrenia. *American Journal of Medical Genetics. Part B, Neuropsychiatric Genetics, 147B*, 505–506.

Chavarria-Siles, I., Contreras-Rojas, J., Hare, E., Walss-Bass, C., Quezada, P., Dassori, A., et al. (2008). Cannabinoid receptor 1 gene (CNR1) and susceptibility to a quantitative phenotype for hebephrenic schizophrenia. *American Journal of Medical Genetics. Part B, Neuropsychiatric Genetics, 147*, 279–284.

Chen, L. P., Murad, M. H., Paras, M. L., Colbenson, K. M., Sattler, A. L., Goranson, E. N., et al. (2010). Sexual abuse and lifetime diagnosis of psychiatric disorders: Systematic review and meta-analysis. *Mayo Clinic Proceedings, 85*, 618–629.

Chevaleyre, V., & Castillo, P. E. (2003). Heterosynaptic LTD of hippocampal GABAergic synapses: A novel role of endocannabinoids in regulating excitability. *Neuron, 38*, 461–472.

Chevaleyre, V., Takahashi, K. A., & Castillo, P. E. (2006). Endocannabinoid-mediated synaptic plasticity in the CNS. *Annual Review of Neuroscience, 29*, 37–76.

Cohen-Cory, S. (2002). The developing synapse: Construction and modulation of synaptic structures and circuits. *Science, 298*, 770–776.

Colizzi, M., Iyegbe, C., Powell, J., Ursini, G., Porcelli, A., Bonvino, A., et al. (2015). Interaction between functional genetic variation of DRD2 and cannabis use on risk of psychosis. *Schizophrenia Bulletin, 41*(5), 1171–1182.

Collip, D., Myin-Germeys, I., & Van Os, J. (2008). Does the concept of "sensitization" provide a plausible mechanism for the putative link between the environment and schizophrenia? *Schizophrenia Bulletin, 34*, 220–225.

Compton, M. T., Kelley, M. E., Ramsay, C. E., Pringle, M., Goulding, S. M., Esterberg, M. L., et al. (2009). Association of pre-onset cannabis, alcohol, and tobacco use with age at onset of prodrome and age at onset of psychosis in first-episode patients. *American Journal of Psychiatry, 166*, 1251–1257.

Costas, J., Sanjuan, J., Ramos-Rios, R., Paz, E., Agra, S., Tolosa, A., et al. (2011). Interaction between COMT haplotypes and cannabis in schizophrenia: A case-only study in two samples from Spain. *Schizophrenia Research, 127*, 22–27.

Cougnard, A., Marcelis, M., Myin-Germeys, I., De Graaf, R., Vollebergh, W., Krabbendam, L., et al. (2007). Does normal developmental expression of psychosis combine with environmental risk to cause persistence of psychosis? A psychosis proneness-persistence model. *Psychological Medicine, 37*, 513–527.

Decoster, J., van Os, J., Kenis, G., Henquet, C., Peuskens, J., De Hert, M., et al. (2011). Age at onset of psychotic disorder: Cannabis, BDNF Val66Met, and sex-specific models of gene-environment interaction. *American Journal of Medical Genetics. Part B, Neuropsychiatric Genetics*, *156B*, 363–369.

Derkinderen, P., Valjent, E., Toutant, M., Corvol, J. C., Enslen, H., Ledent, C., et al. (2003). Regulation of extracellular signal-regulated kinase by cannabinoids in hippocampus. *The Journal of Neuroscience: The Official Journal of the Society for Neuroscience*, *23*, 2371–2382.

Di Forti, M., Iyegbe, C., Sallis, H., Kolliakou, A., Falcone, M. A., Paparelli, A., et al. (2012). Confirmation that the AKT1 (rs2494732) genotype influences the risk of psychosis in cannabis users. *Biological Psychiatry*, *72*, 811–816.

Di Forti, M., Marconi, A., Carra, E., Fraietta, S., Trotta, A., Bonomo, M., et al. (2015). Proportion of patients in south London with first-episode psychosis attributable to use of high potency cannabis: A case-control study. *The Lancet Psychiatry*, *2*, 233–238.

Di Forti, M., Morgan, C., Dazzan, P., Pariante, C., Mondelli, V., Marques, T. R., et al. (2009). High-potency cannabis and the risk of psychosis. *The British Journal of Psychiatry: The Journal of Mental Science*, *195*, 488–491.

Di Forti, M., Sallis, H., Allegri, F., Trotta, A., Ferraro, L., Stilo, S. A., et al. (2014). Daily use, especially of high-potency cannabis, drives the earlier onset of psychosis in cannabis users. *Schizophrenia Bulletin*, *40*, 1509–1517.

Di Forti, M., Vassos, E., Lynskey, M., Craig, M., & Murray, R. (2015). Data versus speculation concerning the greater risk of psychosis associated with the use of high potency cannabis. Letter. *Lancet Psychiatry*, *2*(5), 382.

D'Souza, D. C., Perry, E., MacDougall, L., Ammerman, Y., Cooper, T., Wu, Y. T., et al. (2004). The psychotomimetic effects of intravenous delta-9-tetrahydrocannabinol in healthy individuals: Implications for psychosis. *Neuropsychopharmacology: Official Publication of the American College of Neuropsychopharmacology*, *29*, 1558–1572.

D'Souza, D. C., Sewell, R. A., & Ranganathan, M. (2009). Cannabis and psychosis/schizophrenia: Human studies. *European Archives of Psychiatry and Clinical Neuroscience*, *259*, 413–431.

Eaton, W. W. (1974). Residence, social class, and schizophrenia. *Journal of Health and Social Behavior*, *15*, 289–299.

Egan, M. F., Kojima, M., Callicott, J. H., Goldberg, T. E., Kolachana, B. S., Bertolino, A., et al. (2003). The BDNF val66met polymorphism affects activity-dependent secretion of BDNF and human memory and hippocampal function. *Cell*, *112*, 257–269.

Estrada, G., Fatjo-Vilas, M., Munoz, M. J., Pulido, G., Minano, M. J., Toledo, E., et al. (2011). Cannabis use and age at onset of psychosis: Further evidence of interaction with COMT Val158Met polymorphism. *Acta Psychiatrica Scandinavica*, *123*, 485–492.

Fergusson, D. M., Horwood, L. J., & Swain-Campbell, N. R. (2003). Cannabis dependence and psychotic symptoms in young people. *Psychological Medicine*, *33*, 15–21.

Gage, S. H., Hickman, M., Heron, J., Munafo, M. R., Lewis, G., Macleod, J., et al. (2014). Associations of cannabis and cigarette use with psychotic experiences at age 18: Findings from the Avon Longitudinal Study of Parents and Children. *Psychological Medicine*, *44*, 3435–3444.

Giordano, G. N., Ohlsson, H., Sundquist, K., Sundquist, J., & Kendler, K. S. (2015). The association between cannabis abuse and subsequent schizophrenia: A Swedish national co-relative control study. *Psychological Medicine*, *45*, 407–414.

Goldstein, A. L., Henriksen, C. A., Davidov, D. M., Kimber, M., Pitre, N. Y., & Afifi, T. O. (2013). Childhood maltreatment, alcohol use disorders, and treatment utilization in a national sample of emerging adults. *Journal of Studies on Alcohol and Drugs*, *74*, 185–194.

Gonzalez-Pinto, A., Vega, P., Ibanez, B., Mosquera, F., Barbeito, S., Gutierrez, M., et al. (2008). Impact of cannabis and other drugs on age at onset of psychosis. *The Journal of Clinical Psychiatry, 69*, 1210–1216.

Gourion, D., Goldberger, C., Leroy, S., Bourdel, M. C., Olie, J. P., & Krebs, M. O. (2005). Age at onset of schizophrenia: Interaction between brain-derived neurotrophic factor and dopamine D3 receptor gene variants. *Neuroreport, 16*, 1407–1410.

Gurillo, P., Jauhar, S., Murray, R. M., & MacCabe, J. H. (2015). Does tobacco use cause psychosis? Systematic review and meta-analysis. *The Lancet Psychiatry, 2*, 718–725.

Harley, M., Kelleher, I., Clarke, M., Lynch, F., Arseneault, L., Connor, D., et al. (2010). Cannabis use and childhood trauma interact additively to increase the risk of psychotic symptoms in adolescence. *Psychological Medicine, 40*, 1627–1634.

Harrison, G. (1990). Searching for the causes of schizophrenia: The role of migrant studies. *Schizophrenia Bulletin, 16*, 663–671.

Harrison, G., Fouskakis, D., Rasmussen, F., Tynelius, P., Sipos, A., & Gunnell, D. (2003). Association between psychotic disorder and urban place of birth is not mediated by obstetric complications or childhood socio-economic position: A cohort study. *Psychological Medicine, 33*, 723–731.

Heim, C., Newport, D. J., Heit, S., Graham, Y. P., Wilcox, M., Bonsall, R., et al. (2000). Pituitary-adrenal and autonomic responses to stress in women after sexual and physical abuse in childhood. *JAMA, 284*, 592–597.

Henquet, C., Krabbendam, L., Spauwen, J., Kaplan, C., Lieb, R., Wittchen, H. U., et al. (2005). Prospective cohort study of cannabis use, predisposition for psychosis, and psychotic symptoms in young people. *BMJ, 330*, 11.

Henquet, C., Rosa, A., Krabbendam, L., Papiol, S., Fananas, L., Drukker, M., et al. (2006). An experimental study of catechol-o-methyltransferase Val158Met moderation of delta-9-tetrahydrocannabinol-induced effects on psychosis and cognition. *Neuropsychopharmacology: Official Publication of the American College of Neuropsychopharmacology, 31*, 2748–2757.

Hiscock, R., Bauld, L., Amos, A., Fidler, J. A., & Munafo, M. (2012). Socioeconomic status and smoking: A review. *Annals of the New York Academy of Sciences, 1248*, 107–123.

Ho, B. C., Wassink, T. H., Ziebell, S., & Andreasen, N. C. (2011). Cannabinoid receptor 1 gene polymorphisms and marijuana misuse interactions on white matter and cognitive deficits in schizophrenia. *Schizophrenia Research, 128*, 66–75.

Hoffman, A. F., Oz, M., Yang, R., Lichtman, A. H., & Lupica, C. R. (2007). Opposing actions of chronic Delta9-tetrahydrocannabinol and cannabinoid antagonists on hippocampal long-term potentiation. *Learning & Memory, 14*, 63–74.

Houston, J. E., Murphy, J., Adamson, G., Stringer, M., & Shevlin, M. (2008). Childhood sexual abuse, early cannabis use, and psychosis: Testing an interaction model based on the National Comorbidity Survey. *Schizophrenia Bulletin, 34*, 580–585.

Jockers-Scherubl, M. C., Danker-Hopfe, H., Mahlberg, R., Selig, F., Rentzsch, J., Schurer, F., et al. (2004). Brain-derived neurotrophic factor serum concentrations are increased in drug-naive schizophrenic patients with chronic cannabis abuse and multiple substance abuse. *Neuroscience Letters, 371*, 79–83.

Katz, L. C., & Shatz, C. J. (1996). Synaptic activity and the construction of cortical circuits. *Science, 274*, 1133–1138.

Kelleher, I., Keeley, H., Corcoran, P., Ramsay, H., Wasserman, C., Carli, V., et al. (2013). Childhood trauma and psychosis in a prospective cohort study: Cause, effect, and directionality. *American Journal of Psychiatry, 170*, 734–741.

Kelley, M. E., Wan, C. R., Broussard, B., Crisafio, A., Cristofaro, S., Johnson, S., et al. (2016). Marijuana use in the immediate 5-year premorbid period is associated with increased risk of onset of schizophrenia and related psychotic disorders. *Schizophrenia Research*, *171*, 62–67.

Kelly, B. D., O'Callaghan, E., Waddington, J. L., Feeney, L., Browne, S., Scully, P. J., et al. (2010). Schizophrenia and the city: A review of literature and prospective study of psychosis and urbanicity in Ireland. *Schizophrenia Research*, *116*, 75–89.

Konings, M., Stefanis, N., Kuepper, R., de Graaf, R., ten Have, M., van Os, J., et al. (2012). Replication in two independent population-based samples that childhood maltreatment and cannabis use synergistically impact on psychosis risk. *Psychological Medicine*, *42*, 149–159.

Krabbendam, L., & van Os, J. (2005). Schizophrenia and urbanicity: A major environmental influence—Conditional on genetic risk. *Schizophrenia Bulletin*, *31*, 795–799.

Kreitzer, A. C., & Regehr, W. G. (2001). Retrograde inhibition of presynaptic calcium influx by endogenous cannabinoids at excitatory synapses onto Purkinje cells. *Neuron*, *29*, 717–727.

Ksir, C., & Hart, C. L. (2016). Cannabis and psychosis: A critical overview of the relationship. *Current Psychiatry Reports*, *18*, 12.

Kuepper, R., Ceccarini, J., Lataster, J., van Os, J., van Kroonenburgh, M., van Gerven, J. M., et al. (2013). Delta-9-tetrahydrocannabinol-induced dopamine release as a function of psychosis risk: 18F-fallypride positron emission tomography study. *PLoS One*, *8*, e70378.

Kuepper, R., Henquet, C., Lieb, R., Wittchen, H. U., & van Os, J. (2011). Non-replication of interaction between cannabis use and trauma in predicting psychosis. *Schizophrenia Research*, *131*, 262–263.

Kuepper, R., Morrison, P. D., van Os, J., Murray, R. M., Kenis, G., & Henquet, C. (2010). Does dopamine mediate the psychosis-inducing effects of cannabis? A review and integration of findings across disciplines. *Schizophrenia Research*, *121*, 107–117.

Kuepper, R., van Os, J., Lieb, R., Wittchen, H. U., & Henquet, C. (2011). Do cannabis and urbanicity co-participate in causing psychosis? Evidence from a 10-year follow-up cohort study. *Psychological Medicine*, *41*, 2121–2129.

Large, M., Sharma, S., Compton, M. T., Slade, T., & Nielssen, O. (2011). Cannabis use and earlier onset of psychosis: A systematic meta-analysis. *Archives of General Psychiatry*, *68*, 555–561.

Larkin, W., & Read, J. (2008). Childhood trauma and psychosis: Evidence, pathways, and implications. *Journal of Postgraduate Medicine*, *54*, 287–293.

Leeson, V. C., Harrison, I., Ron, M. A., Barnes, T. R., & Joyce, E. M. (2012). The effect of cannabis use and cognitive reserve on age at onset and psychosis outcomes in first-episode schizophrenia. *Schizophrenia Bulletin*, *38*, 873–880.

Luna, B. (2009). Developmental changes in cognitive control through adolescence. *Advances in Child Development and Behavior*, *37*, 233–278.

Malone, D. T., & Taylor, D. A. (2006). The effect of Delta9-tetrahydrocannabinol on sensorimotor gating in socially isolated rats. *Behavioural Brain Research*, *166*, 101–109.

Manrique-Garcia, E., Zammit, S., Dalman, C., Hemmingsson, T., Andreasson, S., & Allebeck, P. (2012). Cannabis, schizophrenia and other non-affective psychoses: 35 years of follow-up of a population-based cohort. *Psychological Medicine*, *42*, 1321–1328.

Marcelis, M., Takei, N., & van Os, J. (1999). Urbanization and risk for schizophrenia: Does the effect operate before or around the time of illness onset? *Psychological Medicine*, *29*, 1197–1203.

March, D., Hatch, S. L., Morgan, C., Kirkbride, J. B., Bresnahan, M., Fearon, P., et al. (2008). Psychosis and place. *Epidemiologic Reviews*, *30*, 84–100.

Marconi, A., Di Forti, M., Lewis, C. M., Murray, R. M., & Vassos, E. (2016). Meta-analysis of the association between the level of cannabis use and risk of psychosis. *Schizophrenia Bulletin*, *42*(5), 1262–1269.

Matheson, S. L., Shepherd, A. M., Pinchbeck, R. M., Laurens, K. R., & Carr, V. J. (2013). Childhood adversity in schizophrenia: A systematic meta-analysis. *Psychological Medicine*, *43*, 225–238.

Mato, S., Chevaleyre, V., Robbe, D., Pazos, A., Castillo, P. E., & Manzoni, O. J. (2004). A single in-vivo exposure to delta 9THC blocks endocannabinoid-mediated synaptic plasticity. *Nature Neuroscience*, *7*, 585–586.

Mauri, M. C., Volonteri, L. S., De Gaspari, I. F., Colasanti, A., Brambilla, M. A., & Cerruti, L. (2006). Substance abuse in first-episode schizophrenic patients: A retrospective study. *Clinical Practice & Epidemiology in Mental Health*, *2*, 4.

McGrath, L. M., Cornelis, M. C., Lee, P. H., Robinson, E. B., Duncan, L. E., Barnett, J. H., et al. (2013). Genetic predictors of risk and resilience in psychiatric disorders: A cross-disorder genome-wide association study of functional impairment in major depressive disorder, bipolar disorder, and schizophrenia. *American Journal of Medical Genetics. Part B, Neuropsychiatric Genetics*, *162B*, 779–788.

McGrath, J. J., Mortensen, P. B., Visscher, P. M., & Wray, N. R. (2013). Where GWAS and epidemiology meet: Opportunities for the simultaneous study of genetic and environmental risk factors in schizophrenia. *Schizophrenia Bulletin*, *39*, 955–959.

Mondelli, V., Pariante, C. M., Navari, S., Aas, M., D'Albenzio, A., Di Forti, M., et al. (2010). Higher cortisol levels are associated with smaller left hippocampal volume in first-episode psychosis. *Schizophrenia Research*, *119*, 75–78.

Murray, R. M., Jones, P., O'Callaghan, E., Takei, N., & Sham, P. (1992). Genes, viruses and neurodevelopmental schizophrenia. *Journal of Psychiatric Research*, *26*, 225–235.

Murray, R. M., O'Callaghan, E., Castle, D. J., & Lewis, S. W. (1992). A neurodevelopmental approach to the classification of schizophrenia. *Schizophrenia Bulletin*, *18*, 319–332.

Myles, N., Newall, H., Compton, M. T., Curtis, J., Nielssen, O., & Large, M. (2012). The age at onset of psychosis and tobacco use: A systematic meta-analysis. *Social Psychiatry and Psychiatric Epidemiology*, *47*, 1243–1250.

Naoe, Y., Shinkai, T., Hori, H., Fukunaka, Y., Utsunomiya, K., Sakata, S., et al. (2007). No association between the brain-derived neurotrophic factor (BDNF) Val66Met polymorphism and schizophrenia in Asian populations: Evidence from a case-control study and meta-analysis. *Neuroscience Letters*, *415*, 108–112.

Numata, S., Ueno, S., Iga, J., Yamauchi, K., Hongwei, S., Ohta, K., et al. (2006). Brain-derived neurotrophic factor (BDNF) Val66Met polymorphism in schizophrenia is associated with age at onset and symptoms. *Neuroscience Letters*, *401*, 1–5.

Onwuameze, O. E., Nam, K. W., Epping, E. A., Wassink, T. H., Ziebell, S., Andreasen, N. C., et al. (2013). MAPK14 and CNR1 gene variant interactions: Effects on brain volume deficits in schizophrenia patients with marijuana misuse. *Psychological Medicine*, *43*, 619–631.

O'Shea, M., McGregor, I. S., & Mallet, P. E. (2006). Repeated cannabinoid exposure during perinatal, adolescent or early adult ages produces similar longlasting deficits in object recognition and reduced social interaction in rats. *Journal of Psychopharmacology*, *20*, 611–621.

O'Shea, M., Singh, M. E., McGregor, I. S., & Mallet, P. E. (2004). Chronic cannabinoid exposure produces lasting memory impairment and increased anxiety in adolescent but not adult rats. *Journal of Psychopharmacology*, *18*, 502–508.

Ozaita, A., Puighermanal, E., & Maldonado, R. (2007). Regulation of PI3K/Akt/GSK-3 pathway by cannabinoids in the brain. *Journal of Neurochemistry*, *102*, 1105–1114.

Pazos, M. R., Nunez, E., Benito, C., Tolon, R. M., & Romero, J. (2005). Functional neuroanatomy of the endocannabinoid system. *Pharmacology, Biochemistry, and Behavior*, *81*, 239–247.

Pedersen, C. B., & Mortensen, P. B. (2001). Evidence of a dose-response relationship between urbanicity during upbringing and schizophrenia risk. *Archives of General Psychiatry*, *58*, 1039–1046.

Pelayo-Teran, J. M., Suarez-Pinilla, P., Chadi, N., & Crespo-Facorro, B. (2012). Gene-environment interactions underlying the effect of cannabis in first episode psychosis. *Current Pharmaceutical Design*, *18*, 5024–5035.

Power, R. A., Verweij, K. J., Zuhair, M., Montgomery, G. W., Henders, A. K., Heath, A. C., et al. (2014). Genetic predisposition to schizophrenia associated with increased use of cannabis. *Molecular Psychiatry*, *19*, 1201–1204.

Purves, D. A. G., Fitzpatrick, D., Hall, W. C., LaMantia, A. S., McNamara, J. O., & White, L. E. (2008). Modification of brain circuits as a result of experience. In W. LE (Ed.), *Neuroscience* (pp. 611–633). Sunderland, MA: Sinauer Associates, Inc.

Read, J., van Os, J., Morrison, A. P., & Ross, C. A. (2005). Childhood trauma, psychosis and schizophrenia: A literature review with theoretical and clinical implications. *Acta Psychiatrica Scandinavica*, *112*, 330–350.

Rogosch, F. A., Oshri, A., & Cicchetti, D. (2010). From child maltreatment to adolescent cannabis abuse and dependence: A developmental cascade model. *Development and Psychopathology*, *22*, 883–897.

Rossler, W., Hengartner, M. P., Angst, J., & Ajdacic-Gross, V. (2012). Linking substance use with symptoms of subclinical psychosis in a community cohort over 30 years. *Addiction*, *107*, 1174–1184.

Rothman, K. J., & Greenland, S. (1998). *Modern epidemiology* (2nd ed.). Philadelphia, PA: Lippincott-Raven.

Rubino, T., & Parolaro, D. (2014). Cannabis abuse in adolescence and the risk of psychosis: A brief review of the preclinical evidence. *Progress in Neuropsychopharmacology and Biological Psychiatry*, *52*, 41–44.

Sara, G. E., Burgess, P. M., Malhi, G. S., Whiteford, H. A., & Hall, W. C. (2014). Stimulant and other substance use disorders in schizophrenia: Prevalence, correlates and impacts in a population sample. *The Australian and New Zealand Journal of Psychiatry*, *48*, 1036–1047.

Schizophrenia Working Group of the Psychiatric Genomics Consortium. (2014). Biological insights from 108 schizophrenia-associated genetic loci. *Nature*, *511*, 421–427.

Schlicker, E., & Kathmann, M. (2001). Modulation of transmitter release via presynaptic cannabinoid receptors. *Trends in Pharmacological Sciences*, *22*, 565–572.

Schneider, M., Schomig, E., & Leweke, F. M. (2008). Acute and chronic cannabinoid treatment differentially affects recognition memory and social behavior in pubertal and adult rats. *Addiction Biology*, *13*, 345–357.

Schwandt, M. L., Heilig, M., Hommer, D. W., George, D. T., & Ramchandani, V. A. (2013). Childhood trauma exposure and alcohol dependence severity in adulthood: Mediation by emotional abuse severity and neuroticism. *Alcoholism: Clinical and Experimental Research*, *37*, 984–992.

Seifert, J., Ossege, S., Emrich, H. M., Schneider, U., & Stuhrmann, M. (2007). No association of CNR1 gene variations with susceptibility to schizophrenia. *Neuroscience Letters*, *426*, 29–33.

Sekar, A., Bialas, A. R., de Rivera, H., Davis, A., Hammond, T. R., Kamitaki, N., et al. (2016). Schizophrenia risk from complex variation of complement component 4. *Nature*, *530*, 177–183.

Semple, D. M., McIntosh, A. M., & Lawrie, S. M. (2005). Cannabis as a risk factor for psychosis: Systematic review. *Journal of Psychopharmacology*, *19*, 187–194.

Sevy, S., Robinson, D. G., Napolitano, B., Patel, R. C., Gunduz-Bruce, H., Miller, R., et al. (2010). Are cannabis use disorders associated with an earlier age at onset of psychosis? A study in first episode schizophrenia. *Schizophrenia Research*, *120*, 101–107.

Shin, S. H., Edwards, E. M., & Heeren, T. (2009). Child abuse and neglect: Relations to adolescent binge drinking in the national longitudinal study of Adolescent Health (AddHealth) Study. *Addictive Behaviors*, *34*, 277–280.

Sideli, L., Fisher, H. L., Murray, R. M., Sallis, H., Russo, M., Stilo, S. A., et al. (2015). Interaction between cannabis consumption and childhood abuse in psychotic disorders: Preliminary findings on the role of different patterns of cannabis use. *Early Intervention in Psychiatry*. http://dx.doi.org/10.1111/eip.12285.

Sorensen, H. J., Mortensen, E. L., Reinisch, J. M., & Mednick, S. A. (2011). A prospective study of smoking in young women and risk of later psychiatric hospitalization. *Nordic Journal of Psychiatry*, *65*, 3–8.

Spauwen, J., Krabbendam, L., Lieb, R., Wittchen, H. U., & van Os, J. (2004). Does urbanicity shift the population expression of psychosis? *Journal of Psychiatric Research*, *38*, 613–618.

Stefanis, N. C., Dragovic, M., Power, B. D., Jablensky, A., Castle, D., & Morgan, V. A. (2013). Age at initiation of cannabis use predicts age at onset of psychosis: The 7- to 8-year trend. *Schizophrenia Bulletin*, *39*, 251–254.

Stompe, T., Ortwein-Swoboda, G., Strobl, R., & Friedmann, A. (2000). The age of onset of schizophrenia and the theory of anticipation. *Psychiatry Research*, *93*, 125–134.

Suarez-Pinilla, P., Roiz-Santianez, R., Ortiz-Garcia de la Foz, V., Guest, P. C., Ayesa-Arriola, R., Cordova-Palomera, A., et al. (2015). Brain structural and clinical changes after first episode psychosis: Focus on cannabinoid receptor 1 polymorphisms. *Psychiatry Research*, *233*, 112–119.

Torrey, E. F., Mortensen, P. B., Pedersen, C. B., Wohlfahrt, J., & Melbye, M. (2001). Risk factors and confounders in the geographical clustering of schizophrenia. *Schizophrenia Research*, *49*, 295–299.

Van Mastrigt, S., Addington, J., & Addington, D. (2004). Substance misuse at presentation to an early psychosis program. *Social Psychiatry and Psychiatric Epidemiology*, *39*, 69–72.

van Os, J., Bak, M., Hanssen, M., Bijl, R. V., de Graaf, R., & Verdoux, H. (2002). Cannabis use and psychosis: A longitudinal population-based study. *American Journal of Epidemiology*, *156*, 319–327.

Van Winkel, R., van Beveren, N. J., Simons, C., & Genetic, R. (2011). Outcome of Psychosis I (2011) AKT1 moderation of cannabis-induced cognitive alterations in psychotic disorder. *Neuropsychopharmacology: Official Publication of the American College of Neuropsychopharmacology*, *36*, 2529–2537.

Veen, N. D., Selten, J. P., van der Tweel, I., Feller, W. G., Hoek, H. W., & Kahn, R. S. (2004). Cannabis use and age at onset of schizophrenia. *American Journal of Psychiatry*, *161*, 501–506.

Vilhena-Churchill, N., & Goldstein, A. L. (2014). Child maltreatment and marijuana problems in young adults: Examining the role of motives and emotion dysregulation. *Child Abuse & Neglect*, *38*, 962–972.

Waddington, J. L., Lane, A., Larkin, C., & O'Callaghan, E. (1999a). The neurodevelopmental basis of schizophrenia: Clinical clues from cerebro-craniofacial dysmorphogenesis, and the roots of a lifetime trajectory of disease. *Biological Psychiatry*, *46*, 31–39.

Waddington, J. L., Lane, A., Scully, P., Meagher, D., Quinn, J., Larkin, C., et al. (1999b). Early cerebro-craniofacial dysmorphogenesis in schizophrenia: A lifetime trajectory model from neurodevelopmental basis to 'neuroprogressive' process. *Journal of Psychiatric Research*, *33*, 477–489.

Weinberger, D. R. (1996). On the plausibility of "the neurodevelopmental hypothesis" of schizophrenia. *Neuropsychopharmacology: Official Publication of the American College of Neuropsychopharmacology*, *14*, 1S–11S.

Weiser, M., Knobler, H. Y., Noy, S., & Kaplan, Z. (2002). Clinical characteristics of adolescents later hospitalized for schizophrenia. *American Journal of Medical Genetics*, *114*, 949–955.

Weiser, M., Reichenberg, A., Grotto, I., Yasvitzky, R., Rabinowitz, J., Lubin, G., et al. (2004). Higher rates of cigarette smoking in male adolescents before the onset of schizophrenia: A historical-prospective cohort study. *American Journal of Psychiatry*, *161*, 1219–1223.

Whitford, T. J., Rennie, C. J., Grieve, S. M., Clark, C. R., Gordon, E., & Williams, L. M. (2007). Brain maturation in adolescence: Concurrent changes in neuroanatomy and neurophysiology. *Human Brain Mapping*, *28*, 228–237.

Wiles, N. J., Zammit, S., Bebbington, P., Singleton, N., Meltzer, H., & Lewis, G. (2006). Self-reported psychotic symptoms in the general population: Results from the longitudinal study of the British National Psychiatric Morbidity Survey. *The British Journal of Psychiatry: The Journal of Mental Science*, *188*, 519–526.

Wilson, R. I., & Nicoll, R. A. (2002). Endocannabinoid signaling in the brain. *Science*, *296*, 678–682.

Yoshida, T., Hashimoto, K., Zimmer, A., Maejima, T., Araishi, K., & Kano, M. (2002). The cannabinoid CB1 receptor mediates retrograde signals for depolarization-induced suppression of inhibition in cerebellar Purkinje cells. *The Journal of Neuroscience: The Official Journal of the Society for Neuroscience*, *22*, 1690–1697.

Zammit, S., Allebeck, P., Andreasson, S., Lundberg, I., & Lewis, G. (2002). Self reported cannabis use as a risk factor for schizophrenia in Swedish conscripts of 1969: Historical cohort study. *BMJ*, *325*, 1199.

Zammit, S., Allebeck, P., Dalman, C., Lundberg, I., Hemmingsson, T., & Lewis, G. (2003). Investigating the association between cigarette smoking and schizophrenia in a cohort study. *American Journal of Psychiatry*, *160*, 2216–2221.

Zammit, S., Lewis, G., Dalman, C., & Allebeck, P. (2010). Examining interactions between risk factors for psychosis. *The British Journal of Psychiatry: The Journal of Mental Science*, *197*, 207–211.

Zammit, S., Owen, M. J., Evans, J., Heron, J., & Lewis, G. (2011). Cannabis, COMT and psychotic experiences. *The British Journal of Psychiatry: The Journal of Mental Science*, *199*, 380–385.

Chapter 11

Cannabis Use as a Determinant of Earlier Age at Onset of Schizophrenia and Related Psychotic Disorders

Claire Ramsay Wan*, Beth Broussard[†]

**Cambridge Health Alliance, Somerville, MA, United States [†]Lenox Hill Hospital, New York, NY, United States*

CANNABIS USE IN SCHIZOPHRENIA

Cannabis use and cannabis use disorders are highly prevalent among individuals with schizophrenia-spectrum disorders. A meta-analysis of studies of patients with psychotic disorders reported a median prevalence rate of 16% for cannabis use disorder and 27% for lifetime cannabis use (Koskinen, Löhönen, Koponen, Isohanni, & Miettunen, 2010). Among patients with first-episode psychosis (FEP), the prevalence is even greater, with a median rate of current comorbid cannabis use disorder of 29% and a lifetime rate of 44%. While it is plausible that we overestimate cannabis use by comparing clinical samples (e.g., people who are seeking medical care) to general population samples, these data suggest that the rates of cannabis use among those with a schizophrenia-spectrum disorder are much higher than in the general population (Ramsay Wan & Compton, 2016).

WHAT IS CONSIDERED THE ONSET OF PSYCHOSIS?

Although the usual age at onset is between 20 and 30 years in males and between 24 and 34 years in females (Compton & Broussard, 2009), establishing an individual's age at onset of a psychotic (AOP) disorder can be challenging, as psychosis can have an insidious onset. Many studies investigating AOP use the patient's first hospitalization or first antipsychotic medication usage as a marker for the onset of illness. Yet patients commonly delay seeking treatment by weeks, months, or even years after developing symptoms such as hallucinations or delusions. In fact, the median duration of untreated psychosis (DUP) in the United States has been found to be 74 weeks (Addington et al., 2015).

The Complex Connection between Cannabis and Schizophrenia. http://dx.doi.org/10.1016/B978-0-12-804791-0.00011-2

Some studies use standardized interview guides to gather retrospective data on the development of positive symptoms (hallucinations and delusions) and estimate the date of onset based on the patient's report, and occasionally, additional input from family members or other accounts (Barnes, Mutsatsa, Hutton, Watt, & Joyce, 2006; Compton et al., 2009; González-Pinto et al., 2008). A few researchers gathered retrospective information on the onset of prodromal symptoms, which are nonspecific mood and cognitive disturbances that commonly precede the onset of frank hallucinations or delusions (Goldberger et al., 2010; Leeson, Harrison, Ron, Barnes, & Joyce, 2012). It is reasonable to argue that, for patients who have prodromal symptoms preceding psychosis, these initial changes are the first true clinical signs of their developing illness, though they are not yet recognizable as part of a psychotic disorder. Finally, some researchers gathered prospective rather than retrospective data: adolescents and young adults with attenuated psychotic symptoms or a family history of psychosis and deteriorating function are considered to be at ultra-high risk of developing psychosis and are actively recruited and followed by research groups that study the early course of psychotic disorders (Cannon et al., 2014; Dragt et al., 2012; Kristensen & Cadenhead, 2007; Phillips, Curry, & Yung, 2002). In most of these samples, only a minority of participants go on to develop a major mental illness. Nonetheless, these studies give invaluable information on the early course of psychosis.

WHY IS AGE AT ONSET OF SCHIZOPHRENIA IMPORTANT?

The typical age at onset of schizophrenia—during young adulthood when important milestones related to independence, relationships, education, and career are commonly achieved—makes it a particularly devastating period to experience any type of mental illness. Although young adults experiencing a first episode may achieve these milestones, disruptions caused by the illness set difficult obstacles in one's path. An important prognostic indicator, earlier age at onset has been found to correlate with a number of poor outcomes such as longer delay to treatment (Addington et al., 2015), reduced response to antipsychotic treatment (Carbon & Correll, 2014), and greater symptom severity (Compton, 2010, Crumlish et al., 2008). In a systematic review, Clemmensen, Vernal, and Steinhausen (2012) found that 60.1% of individuals with early-onset schizophrenia experienced a "poor" outcome, based on global or study-specific functioning scales. Long-term outcomes of patients with onset in adulthood are more encouraging. In one review of long-term follow-up studies, 50% of people with AOP in adulthood experienced a "good" outcome (Harrison et al., 2001), as opposed to the 15.4% found by Clemmensen and colleagues.

CANNABIS USE IS A PREDICTOR OF EARLIER AGE AT ONSET OF PSYCHOSIS

As the extent of the literature on cannabis use and AOP has increased, so too has the quality of research in this area. Early publications simply reported that among patients with psychosis, substance users were generally younger, sparking questions about whether psychoactive substances hasten the onset of these illnesses. Later reports identified cannabis as a substance of interest and collected increasingly relevant data on this topic. Some of the most recent studies allow researchers to isolate the effects of premorbid cannabis use and other substance use separately, quantify the duration and patterns of cannabis use, collect genetic and other data on potential interactions, control for other predictors of AOP, and collect high-quality data on stages of onset of illness. As a detailed review of every report in the literature on this topic would quickly become redundant, this chapter will instead summarize findings from a meta-analysis and examine six questions of particular interest regarding the relationship between cannabis use and AOP.

Large, Sharma, Compton, Slade, and Nielssen (2011) published a systematic review and meta-analysis on cannabis use and AOP. From the 83 identified studies, using strict inclusion criteria for rigorous methodology, they found that the AOP was 2.7 years younger (standardized mean difference = −0.414) among cannabis users than nonsubstance users. For more broadly defined substance users compared with nonsubstance users, the AOP was 2.0 years younger (standardized mean difference = −0.315), but alcohol use was not associated with any difference. The authors did not find statistically significant differences in effect sizes by sex or based on the severity of use. While a meta-regression showed that the effect of substance use on AOP was reduced in studies including patients over 45 years of age, cannabis use still made an independent contribution to AOP in these studies. The proportion of patients with a schizophrenia-spectrum disorder (as compared to affective psychoses) was not a statistically significant predictor, nor was the inclusion of other (non-cannabis) substance use in the control group. Additional analyses tested the effect of methodological factors on individual study outcomes and the validity of the meta-analysis itself. Regarding the validity of the meta-analysis, several tests to assess for publication bias were negative. On meta-regression, methodology factors, such as whether the data were collected from a sample with FEP or those with chronic illness, whether the onset of illness was defined as the first psychotic symptom or first treatment, whether systematic methods were used to assess for substance use and AOP, and how other substance use was handled in analyses did not significantly alter the results.

The meta-analysis described provides high-quality evidence of a pattern across the literature: that cannabis use is clearly associated with an earlier AOP. Yet, many explanations for this data are plausible. As interest in this phenomenon has increased, researchers have collected data addressing some related questions, each of which will be addressed in this chapter.

1. What is the temporal relationship between cannabis use and AOP?
2. Is the relationship between cannabis use and an earlier AOP different for nonaffective versus affective psychotic disorders?
3. Is there a dose-response relationship between cannabis use and AOP?
4. Is cannabis use associated with an earlier age at onset of prodromal symptoms?
5. Is cannabis use associated with a shorter prodrome?
6. Is the relationship between cannabis use and AOP moderated by sex, other substance use, genetic factors, or any other known variables?

WHAT IS THE TEMPORAL RELATIONSHIP BETWEEN CANNABIS USE AND AGE AT ONSET?

As will be discussed later in the chapter and extensively reviewed in Chapter 10, the well-described association between cannabis use and an earlier AOP supports the increasingly accepted theory that cannabis use in adolescence is a component cause of psychotic disorders. Recent research has examined whether cannabis use in fact precedes illness onset, both in regards to the onset of psychotic symptoms and the even earlier onset of nonspecific prodromal symptoms. Some studies have focused specifically on cannabis use prior to the onset of psychosis (Compton et al., 2009; Decoster et al., 2011; Di Forti et al., 2014; Galvez-Buccollini et al., 2012; Kelley et al., 2016), greatly reducing the likelihood that the phenomenon described in the literature is due to reverse causality.

Cannabis Use Usually Precedes Onset of Psychosis

Unsurprisingly, studies of patients with recent-onset schizophrenia typically find that most of those who use cannabis began doing so prior to their relatively recent onset of psychosis. For instance, among 201 patients with FEP and a comorbid cannabis use disorder from Melbourne, Australia, 88% developed their cannabis use disorder prior to psychotic symptoms (Schimmelmann et al., 2011). In another sample of 49 patients with FEP and comorbid cannabis use disorders, 74% had developed the cannabis use disorder more than one year before (Sevy et al., 2010). Studies of patients with chronic schizophrenia that retrospectively determine AOP and cannabis use also report that cannabis use precedes psychosis in most cases. In a sample of 121 males with chronic schizophrenia who had used cannabis, cannabis use preceded the onset of positive symptoms in 73% (Goldberger et al., 2010). Another report from a large sample of patients with schizophrenia-spectrum ($n=676$) and bipolar affective disorders with psychotic features (n=90) found that >95% of those who used cannabis started before the onset of illness.

In fact, cannabis use often precedes psychotic symptoms by several years. A recent study of 57 individuals with a primary psychotic disorder and a history of cannabis use prior to the onset of psychosis reported that participants used cannabis for 7 years, on average, before developing psychosis (Galvez-Buccollini et al., 2012). The average age at first cannabis use was 15, whereas the average for AOP was 22. Remarkably similar findings (a 6–8-year gap between average age of first cannabis use in mid-adolescence and AOP in the early 20s) have been reported in at least three other studies with diverse inclusion criteria and data collection methods (Leeson et al., 2012; Öngür et al., 2009; Stefanis et al., 2013).

Cannabis Use Sometimes Precedes the Onset of Prodromal Symptoms

Given that most schizophrenia prodromes last fewer than 6–8 years, the data presented above suggest that cannabis use precedes the prodromal period. However, there are few reports in the literature that specifically address the timing of cannabis use relative to prodromal symptoms. In a sample of 121 males with chronic schizophrenia who had used cannabis, one thorough retrospective study found that cannabis use occurred before prodromal symptoms in 35% (Goldberger et al., 2010). In a study of patients with FEP, Leeson et al. (2012) reported that the average age at prodrome onset among cannabis users was 21 years, well above the average age at initial cannabis use (15 years). The authors noted that they only included 53 of the 71 cannabis users in this analysis, excluding those who could not state "with confidence" when cannabis use began.

In a study of individuals (n=245, age range 16–35) considered at ultra-high risk for developing psychosis (i.e., a putative prodromal group), Dragt et al. (2012) investigated the temporal relationship between cannabis use and various nonspecific, possibly prodromal symptoms such as anxiety, social withdrawal, depressed mood, and decreased concentration. A few attenuated psychotic symptoms such as brief hallucinations were also included. They reported that in the majority of patients, cannabis use started more than one year prior to each symptom. However, cannabis use started after certain symptoms for a sizeable minority. The proportion of participants who reported initiating cannabis use afterwards varied by symptoms: 25% reported starting cannabis use after developing anxiety, 16% after experiencing derealization, and only 4% reported initiating cannabis use after having hallucinations. In all, only 37 of 245 patients transitioned to psychosis over the course of this longitudinal study. Therefore, this sample may not be representative of individuals with psychotic disorders or those who are actually prodromal.

IS THE RELATIONSHIP BETWEEN CANNABIS USE AND EARLIER ONSET OF PSYCHOSIS DIFFERENT FOR NONAFFECTIVE VERSUS AFFECTIVE PSYCHOTIC DISORDERS?

In light of the theory that cannabis use is a component cause of schizophrenia and of research into whether cannabis use also contributes to the development of affective psychotic disorders, one question of interest is whether cannabis use is associated with a greater decrease in AOP in those with nonaffective versus affective psychotic disorders. Many studies demonstrating an association between cannabis use and an earlier AOP have restricted samples to those with nonaffective disorders (Barnes et al., 2006; Bersani, Orlandi, Kotzalidis, & Pancheri, 2002; Compton et al., 2009; Decoster et al., 2011; Dekker et al., 2012; Di Forti et al., 2014; Kelley et al., 2016), but other studies have reported similar findings in samples including both affective and nonaffective psychoses (De Hert et al., 2011; González-Pinto et al., 2008).

When comparing results in those with affective and nonaffective psychoses, two studies have found quite different results. In a study of 766 patients with schizophrenia-spectrum (n=676) and bipolar affective disorders with psychotic features (n=90) recruited from inpatient and outpatient settings, De Hert et al. (2011) found that cannabis use was associated with earlier age at first hospitalization in both affective and nonaffective groups. While cannabis use was more common among patients with a schizophrenia-spectrum disorder, the difference in AOP associated with cannabis use was larger among those with bipolar disorder. Öngür et al. (2009) also investigated cannabis use and other factors affecting AOP in a sample of patients with schizophrenia (n=80), schizoaffective disorder (n=61), and bipolar disorder with psychotic features (n=92). Using a multiple linear regression and controlling for sex, diagnostic category, and alcohol use disorder, the authors reported that cannabis use disorder was associated with an almost 3-year decrease in AOP across the sample. Diagnostic category was not associated with statistically significant differences in AOP and no statistically significant interaction between cannabis use disorder and diagnostic category on AOP was found. More research is required to determine whether the association between cannabis use and AOP is similar or different in those with nonaffective and affective psychotic disorders.

IS THERE A DOSE-RESPONSE RELATIONSHIP BETWEEN CANNABIS USE AND AGE AT ONSET?

Several studies have investigated whether the association between cannabis use and AOP follows a pattern that is consistent with a dose-response to cannabis. Some have used severity of misuse as a metric for determining dose-response. In a study of 131 patients hospitalized for FEP (including those with both nonaffective and affective psychotic disorders), AOP in patients with cannabis use, abuse, and dependence was 7, 8.5, and 12 years younger, respectively, as compared to those with no cannabis use (González-Pinto et al., 2008). This effect was observed independent of sex and other drug use [hazard ratio (HR) = 2.6, $P < .001$]. To correct for the potential of bias caused by increased rates of cannabis use in younger individuals, the authors repeated the analyses in a subset of younger patients, with similar results (HR = 1.9, $P = .018$).

Another study investigated the impact of the frequency and patterns of escalation in use, prior to the onset of symptoms, in 109 patients hospitalized for FEP (Compton et al., 2009). Participants reported their age at onset of cannabis use at three rates of frequency: occasionally, weekly, and daily. When maximal use any time prior to illness onset was used in a categorical manner, a dose-response was not observed. However, a faster progression to daily use was associated with an earlier AOP in Cox regressions, after controlling for gender (HR = 1.997). Conversely, weekly and daily alcohol use was associated with a later AOP. These data suggest that a pattern of rapid escalation in cannabis use is associated with an earlier AOP. Findings were replicated in a similar sample of 247 FEP patients (Kelley et al., 2016).

Two forms of dose-response were observed by Decoster et al. (2011) in a sample of 585 patients with schizophrenia. Among cannabis users (those who had used cannabis more than five times prior to the age at onset) both the age at first cannabis use and the frequency of the most intensive level of use prior to psychosis onset were trichotomized. In multivariate models that controlled for sex and other drug use, both intensity of use and age at first use of cannabis were statistically significant predictors of AOP, with HRs of 1.29 (95% CI 1.03–1.60, $P = .023$) for intensity of use and 1.50 (95% CI 1.24–1.82, $P < .001$) for age at first use. When the authors conducted the Cox regression model adjusting for sex and other drug use, age at first use rather than intensity of use was more strongly associated with an earlier AOP (HR for age at first use = 1.56, 95% CI 1.26–1.93, $P < .001$; HR for intensity of use was not statistically significant).

Perhaps the most thorough investigation into cannabis use intensity and AOP was conducted by Di Forti et al. (2014) in a report on 410 patients with first-episode nonaffective or affective psychosis (Di Forti et al., 2014). Cannabis use was defined by lifetime exposure (ever using cannabis) and further characterized by frequency during the patient's most consistent pattern of use as well as the potency of the substance used, with hash type (average of 4% Δ^9-tetrahydrocannabinol [THC]) considered low potency and skunk type (average of 16% THC) considered high potency. In addition, the researchers dichotomized age at first use by <15 years old or ≥ 15 years old, noting that early adolescence represents a period of potential critical exposure for increasing risk of developing psychosis. Using Cox regression, the authors found that cannabis use was a statistically significant predictor of AOP, with mean and median age differences of 3.2 and 2.9 years, respectively. This effect was slightly attenuated once sex was added to the model (HR = 1.39 compared to an unadjusted HR of 1.42), and remained statistically significant. When the authors added frequency of use and potency of the cannabis used to the model, both were statistically significant predictors of an earlier AOP. Finally, in a Kaplan-Meier survival curve analysis, individuals who started using cannabis before the age of 15 were more likely to have developed psychosis at any age, but on Cox regression this effect was not statistically significant after adjusting for frequency and potency of cannabis use. Tobacco and other illicit drug use were not associated with AOP.

Each of the five studies described above report one or more findings that suggest that the association between cannabis use and an earlier AOP follows a dose-response pattern. While the studies use varied measures for quantifying the amount or length of cannabis exposure and thus do not replicate one another, taken together, they clearly indicate a dose-response between cannabis exposure and an earlier AOP.

IS CANNABIS USE ASSOCIATED WITH AN EARLIER AGE AT ONSET OF PRODROMAL SYMPTOMS?

Schizophrenia-spectrum disorders are chronic illnesses that often begin insidiously. While the onset of frank psychotic experiences meeting diagnostic criteria is usually considered the onset of the illness, many people have had nonspecific symptoms and reduced functioning for months or years preceding this point. This period is retrospectively defined as their prodrome. Some researchers posit that this should be considered the true onset of the illness, rather than later symptoms that are more clearly recognizable as belonging to a psychotic disorder. While less research is available on the impact of cannabis use on the age at onset of prodromal symptoms, some studies have collected data to determine whether cannabis use impacts this earlier manifestation. Two studies found a correlation between cannabis use and age at onset of the prodrome, but two other studies had mixed or negative findings.

As mentioned earlier, Compton et al. (2009) addressed this question of premorbid cannabis use and AOP among patients hospitalized for FEP in their report. Age at onset of the prodrome was determined by consensus using the Symptom Onset in Schizophrenia inventory and information collected from the affected individual, medical records, and 1–2 family members or friends when available. Similar to their findings regarding cannabis use and AOP, they reported that a faster progression to daily cannabis use was a predictor of age at onset of prodromal symptoms (HR = 2.065). While frequency of maximal premorbid cannabis use as a categorical variable was not a predictor of the age at onset of prodromal symptoms, rapid escalation of use to a daily frequency was associated with an earlier prodrome.

Cannabis use was also associated with an earlier age of prodromal symptoms in a sample of 99 patients, aged 16–60 years, with a first episode of a schizophrenia-spectrum disorder and no history of alcohol misuse or substance use other than cannabis (Leeson et al., 2012). Researchers collected data on the age at first cannabis use and the amount and maximal frequency of use over the participant's lifetime as well as in the 12 and 3 months prior to study participation. Age at onset of prodrome was determined by retrospective self-report using the Nottingham Onset Scale. Cannabis users had an earlier age at onset of prodrome (21 years) than non-cannabis users (26 years, $P=.003$). Furthermore, age at first cannabis use correlated significantly with age at prodrome onset ($r=.47$, $P<.001$).

In an incidence sample of 133 patients (18–54 years of age) with first-episode nonaffective psychosis from The Hague, Netherlands, Veen et al. (2004) did not find an association between cannabis use and age at onset of social or occupational dysfunction, which could be considered a proxy for prodromal symptoms. Data on the onset of symptoms were collected via self-report using the Comprehensive Assessment of Symptoms and History and were supplemented by and reconciled with data from the medical record and an informant, when available, using the Instrument for the Retrospective Assessment of the Onset of Schizophrenia. In univariate testing, both sex and cannabis use were associated with an earlier age at symptom onset. Multivariate testing was constricted by a very small sample of women ($n=36$), only six of whom reported using cannabis. Thus, in the analyses the authors used dummy variables to test for the effects of sex and cannabis separately, with males who used cannabis as a reference group. In multiple linear regressions, sex was a predictor of age at onset of social or occupational dysfunction after controlling for cannabis use, but cannabis use did not remain statistically significant when controlling for sex ($\beta=2.4$, 95% CI $=-1.4$ to 6.3, $P=.20$).

In a study of 190 patients with chronic schizophrenia, Goldberger et al. (2010) compared AOP and age at prodrome onset between those who had used cannabis at least four times ($n=121$) and those who had not ($n=69$). They did not find a statistically significant difference in AOP (21.4 ± 4.9 years compared with 22.1 ± 5.4, $P=.14$) or age at onset of prodromal symptoms (16.7 ± 5.3 years compared with 16.7 ± 5.5, $P=.82$). However, further analyses in a subgroup considered "cannabis-sensitive" did reveal differences in AOP. Overall, this study did not replicate the findings from other samples that cannabis users had an earlier age at onset of the prodrome (Goldberger et al., 2010).

Two studies published by a group from the Netherlands found that cannabis use was associated with an earlier age at onset of potential prodromal symptoms in samples of young adults considered to be at ultra-high risk for psychosis. A team headed by Dragt investigated the association between cannabis use and deteriorating functioning in a group of 68 such individuals (Dragt et al., 2010). The sample included individuals aged 12–35 years with schizotypal personality disorder or a family history of psychosis who recently had a substantial decline in functioning. They reported on baseline data from a longitudinal study that monitored these individuals for the development of psychotic symptoms (17% conversion rate). The age at onset of possible prodromal symptoms was determined from self-report using the Interview for the Retrospective Assessment of the Onset of Schizophrenia. Among the cannabis users, age at initiation of cannabis use and age at onset of a cannabis use disorder were both correlated with age at onset of a cluster of putative prodromal symptoms ($\rho = 0.48$, $P = .003$, and $\rho = 0.67$, $P = .001$, respectively). One critique is that many of these nonspecific symptoms, such as anxiety, paranoia, and derealization, may also be caused by cannabis use and/or dependence. The authors did not report whether cannabis use predicted conversion to psychosis and the conversion group was too small for subgroup analyses.

In a larger sample ($n = 245$) including individuals recruited from sites in several other European countries, Dragt et al. (2012) repeated and extended their analyses on cannabis use and age at onset of prodromal symptoms. Participants aged 16–35 years who had attenuated or brief and limited psychotic symptoms, or genetic risk and a recent decline in functioning, were recruited into the European Prediction of Psychosis Study and followed to determine whether the symptoms resolved, progressed to psychosis, or evolved into another illness. Information about symptomatology, premorbid functioning, and cannabis use and dependence were collected at baseline, and age at onset of possible prodromal symptoms was again determined using the Interview for the Retrospective Assessment of the Onset of Schizophrenia. The authors controlled for potential confounders including sex, premorbid functioning, and alcohol use disorders in the analyses. They found that a younger age at onset of cannabis use was significantly related to younger ages at onset of multiple symptoms of a possible prodrome, including anxiety, avoiding contact, depressed mood, derealization, and weakness of thinking and concentration (with correlations ranging from 0.33 to 0.58, all $P < .004$). After again adjusting for potential confounders, younger age at onset of cannabis dependence was significantly related to younger age at onset of anxiety, avoiding contact, impairment of memory, and weakness of thinking and concentration (correlations ranging from 0.57 to 0.84, all $P < .003$). Neither cannabis use nor dependence predicted transition to psychosis.

IS CANNABIS USE ASSOCIATED WITH A SHORTER PRODROME?

The prodromal period presents an opportunity for early intervention, and researchers are currently investigating methods to identify patients in this early stage, as well as potential therapies designed to delay or reduce the rate of conversion to psychosis. Discontinuing any ongoing cannabis use may have a positive effect at this stage. While the literature contains many studies addressing cannabis use/misuse and AOP, and a handful of studies addressing the relationship with age at onset of prodrome, to our knowledge, no groups have investigated whether the transition from prodrome to psychosis is more rapid in those who use cannabis. However, a few studies have reported that cannabis or substance use is a risk factor for conversion to psychosis over a set follow-up period. In a 2.5-year prospective study following 291 individuals at risk for developing psychosis, Cannon et al. (2014) found that a history of substance use was a predictor of conversion to psychosis but did not report predictors of length of prodrome. A second and smaller longitudinal study of 48 "at risk" patients reported that both cannabis abuse/dependence and nicotine use at baseline were associated with increased risk of developing psychotic symptoms over the course of at least one year of follow-up (Kristensen & Cadenhead, 2007). However, two other studies with similar samples failed to replicate this finding (Dragt et al., 2012; Phillips et al., 2002). To date, we do not have enough information to determine the impact of cannabis use on this stage of illness development.

IS THE RELATIONSHIP BETWEEN CANNABIS USE AND AGE AT ONSET MODERATED BY SEX, OTHER SUBSTANCE USE, GENETIC FACTORS, OR ANY OTHER KNOWN VARIABLES?

Sex

A common critique of the literature reporting an association between cannabis use and an earlier AOP is that both variables are associated with male sex. Some researchers argue that sex and other demographic variables might be enough to explain the phenomenon. However, many of the studies reviewed in this chapter control for sex in their analyses regarding cannabis and AOP (Compton et al., 2009; Decoster et al., 2011; Dekker et al., 2012; Di Forti et al., 2014; Dragt et al., 2012; Galvez-Buccollini et al., 2012; González-Pinto et al., 2008; Kelley et al., 2016; Tosato et al., 2013; Veen et al., 2004). A few studies on cannabis and AOP did not find sex differences within their samples, however, and therefore did not control for it in further analyses (Stefanis et al., 2013; Sugranyes et al., 2009).

Others posit that the potential for cannabis use affecting the development of psychotic symptoms may in fact be attenuated by sex, either through a longer period of potential premorbid exposure in women, or by directly interacting with chemical signals that are altered by estrogen (Decoster et al., 2011). Four studies formally tested for statistical interactions between cannabis and sex in their analyses, with mixed results. Compton et al. (2009) found that progression to daily cannabis use was associated with a larger increased relative risk for onset of psychosis in female patients (HR = 5.154) than in male patients (HR = 3.359, P = .054). Second, a sex-by-cannabis-by-brain-derived neurotrophic factor (BDNF) gene interaction was reported by Decoster et al. (2011), as described in more detail below. Two other studies failed to find a sex interaction with cannabis use. However, in their sample of 766 patients with affective and nonaffective psychoses, De Hert et al. investigated 2- and 3-way interaction terms between cannabis use, sex, and diagnostic category on AOP (De Hert et al., 2011). While each of the independent variables contributed to variability in the model, the only statistically significant interaction term was between sex and diagnosis, such that the sex difference in AOP was different in affective than in nonaffective psychotic disorders. Similarly, in their study of 410 patients with FEP described earlier, Di Forti et al. (2014) noted that while both sex and cannabis use status were predictors of AOP, an interaction term between these two variables was not statistically significant in the Cox regression model. They asserted that the apparent sex interaction may simply be an artifact of an earlier AOP in men and greater levels of cannabis use. These conflicting results warrant investigation in future studies.

Other Substances

While cannabis is the most heavily used substance in most samples, many studies note that some cannabis users concurrently misuse alcohol or other drugs, which may also influence the development of psychotic disorders. For instance, in a report on 123 patients with FEP, Barnett et al. (2007) found that not only cannabis but also cocaine and ecstasy were associated with an earlier AOP. In many samples, only a small minority of cannabis users have concurrent other substance use (Compton & Ramsay, 2009; Sevy et al., 2010; Veen et al., 2004). However, there are exceptions, such as studies from Melbourne, Australia, and Cambridge, UK, where more than half of those with a cannabis use disorder also misused other substances (Barnett et al., 2007; Schimmelmann et al., 2011). The data presented below suggest that cannabis is associated with an earlier AOP regardless of other drug use. Indeed, alcohol use appears to be associated with a later AOP in some studies.

Three studies reported an association between cannabis use and AOP while entirely excluding patients with a history of other drug use. In a small sample of 57 patients with a schizophrenia-spectrum disorder, a history of cannabis use, and no history of using other drugs (with the exception of prior alcohol use or dependence), Galvez-Buccollini et al. (2012) reported correlations between age at onset of cannabis use and AOP, as well as age at first hospitalization ($\beta=0.4$, 95% CI=0.1–0.7, $P=.004$ and $\beta=0.4$, 95% CI=0.1–0.8, $P=.008$, respectively). These analyses adjusted for potential confounders including sex, age, lifetime diagnosis of alcohol abuse or dependence, and family history of schizophrenia. Similarly, in a sample of 99 patients with FEP with no alcohol misuse and no other substance use lasting more than one month, described earlier, Leeson et al. (2012) demonstrated a decreased age at onset of psychosis by more than 5 years among cannabis users as compared to nonusers ($P=.002$). The study by Schimmelmann et al. (2011) is another example in which the study sample was restricted to comparisons between cannabis-only-users and nonusers. Of 625 patients with FEP, 468 used cannabis, but a large number (267) were excluded due to having another substance use disorder. The remaining 201 with cannabis use and no other substance use disorder were compared to non-cannabis users. After controlling for sex, a comparison of all cannabis users to nonusers did not yield a statistically significant difference in AOP (21.8 years compared with 21.1 years of age). However, those who started using cannabis at 14 years or younger ($n=37$) had an AOP of 19.4, which was statistically distinguishable from AOP of nonusers ($P=.024$). Within the cannabis use group, again controlling for sex, age at onset of cannabis predicted AOP (linear regression analysis: $\beta=-0.49$, $P<0.001$), explaining approximately 25% of the variance in AOP.

Öngür et al. (2009) investigated cannabis use and other factors affecting AOP in a sample of 233 patients with nonaffective and affective psychosis (Öngür et al., 2009). Using multiple linear regressions, they investigated possible interactions between abuse of or dependence to cannabis, alcohol, and other drugs. An interesting statistically significant interaction between cannabis and alcohol abuse/dependence was observed, with post hoc analyses suggesting that cannabis abuse/dependence was associated with a 4.8-year decrease in age at onset in the absence of alcohol abuse/dependence but only a 1.2-year decrease with comorbid alcohol abuse/dependence ($P=.05$). Thus, comorbid alcohol misuse appeared to decrease the effect of cannabis misuse. In this sample, AOP was not significantly associated with the use of other drugs and no statistically significant interaction between cannabis and other drug abuse/dependence impacting AOP was found ($\beta=1.18$, $P=.35$). Compton et al. (2009) noted similar findings in their sample of 109 patients hospitalized for FEP; weekly or daily alcohol use was associated with a later, not earlier, AOP.

Gene-by-Environment Interactions

Recent advances in genetics research have generated tremendous excitement and frustration in the field of psychiatry, where the pathogenesis of many illnesses continues to be poorly understood. While it is clear that schizophrenia is caused in part by genetic factors, no single genetic variant identified to date explains a large portion of the risk. Rather, researchers now posit that a constellation of genetic variants creates a "perfect storm," coming together in some individuals to increase the risk of schizophrenia. At the same time, data consistently show that environmental hazards such as cannabis use in adolescence, childhood trauma, and being an immigrant all increase the risk of illness by a small amount. Only a small proportion of individuals exposed to such hazards develop psychosis. The field has now turned to inquiries regarding potential gene-by-environment interactions that may explain why cannabis use is associated with increased risk of schizophrenia, even though only a small proportion of individuals who use cannabis develop psychosis (van Os, Rutten, & Poulton, 2008).

In France, Goldberger et al. conducted an interesting study, in which they sought to differentiate between subgroups that were cannabis-sensitive and nonsensitive among 190 male patients with schizophrenia, excluding those with a history of cocaine, amphetamine, ecstasy, or LSD (lysergic acid diethylamide) use (Goldberger et al., 2010). Data on age at onset of symptoms, lifetime substance use disorders, and family history of psychosis were collected through semi-structured interviews using the Diagnostic Interview for Genetic Studies for DSM-IV Disorders with patients and the Family Interview for Genetic Studies with family members. Participants were considered cannabis-sensitive if their psychotic symptoms emerged within one month after starting cannabis use or markedly increased with cannabis use, or if their symptoms were repeatedly exacerbated following cannabis use. While 121 of the 190 patients had used cannabis, 44 were deemed to be cannabis-sensitive. When all cannabis users were compared with nonusers, no differences in AOP (21.4 ± 4.9 years compared with 22.1 ± 5.4, $P = .14$) or age at onset of prodromal symptoms (16.7 ± 5.3 years compared with 16.7 ± 5.5, $P = .82$) were found. However, cannabis-sensitive patients had an earlier AOP than nonsensitive patients (mean difference of 2.6 years, $P = .005$). In addition, cannabis-sensitive patients were more likely to have a family history of nonaffective psychotic disorders than nonsensitive patients. The authors suggested that this cannabis-sensitive subgroup may have specific genetic vulnerabilities to cannabis exposure that could explain why cannabis is a risk factor for schizophrenia, even though a large portion of the exposed population does not develop psychosis.

Catechol-O-Methyltransferase

One genetic variant that is of special interest in research on schizophrenia is found in the catechol-O-methyltransferase (COMT) gene, which codes for a protein that degrades catecholamines, including dopamine, epinephrine, and norepinephrine. A functional polymorphism causing a valine substitution for methionine at codon 158 (Val158Met) changes thermostability of this protein, resulting in an increased activity and altered dopamine levels in several areas of the brain, including the prefrontal cortex. Alone, this polymorphism may be associated with a slightly increased risk of schizophrenia (Ira, Zanoni, Ruggeri, Dazzan, & Tosato, 2013), and it is a gene of great interest in investigations of possible gene-by-environment interactions that may increase the risk of this illness. Caspi et al. (2005) reported such an interaction in a birth cohort study of 1037 children (data reported on 806 participants) who were followed to the age of 26 years. Specifically, as discussed in Chapter 10, they found that adolescent cannabis use was associated with the development of schizophreniform disorder and adult psychotic experiences in individuals with the Val158Met polymorphism (Met/Val or Val/Val) but not in those who were Met homozygotes. In addition, the Val158Met substitution was associated with poorer performance on neuropsychological tasks that measure working memory and attentional modulation, as well as smaller volumes of the hippocampus, amygdala-uncus, and middle temporal gyrus in the temporal area and smaller gray matter density in the left anterior cingulate cortex and the anterior cingulate cortex in the frontal lobe. Ira et al. (2013) asserted that these changes may represent related endophenotypes for schizophrenia, and increase risk for the illness when combined with other genetic or environmental factors.

Pelayo-Terán et al. (2008) found that COMT Val homozygotes had an earlier AOP and greater negative symptoms than Met carriers, leading them to posit that this polymorphism may lower dopaminergic transmission in the prefrontal cortex, playing a causal role in negative symptoms and the onset of illness. The same group then investigated whether cannabis use moderates this relationship (Pelayo-Terán et al., 2010). The study recruited 169 patients with first-episode nonaffective psychosis from inpatient and outpatient settings. Cannabis use status was dichotomized into regular versus sporadic or no use. The authors included sex in the analyses, citing its association with age at onset and the effects of estrogen on COMT expression (Xie et al., 1999). A multivariate ANCOVA was performed with AOP as the outcome and sex, cannabis use, and COMT genotype as fixed variables. Sex and cannabis use contributed independently to AOP, but the COMT genotype did not. However, a statistically significant interaction was observed between cannabis use and the COMT gene on AOP ($F=3.816$; $P=.024$; Eta Squared$=0.045$), such that among nonusers, the Met homozygous group had a later AOP than the Met/Val and Val/Val groups. The authors suggested that cannabis use may reduce a delay in AOP that is otherwise associated with the Met/Met genotype.

A different interaction between the COMT Val allele and cannabis use on AOP was reported by Estrada et al. (2011), albeit in a very small study of 80 inpatients with recent-onset schizophrenia-spectrum disorders and 77 inpatients with conduct and affective disorders as a comparison group. Age at first cannabis use and frequency (daily, weekly, or monthly) of use were collected through semi-structured interviews. The authors conducted linear regressions, testing age at first cannabis use and frequency of use, combined with sex or Val158Met genotype. Age at first cannabis use was positively correlated with age at first psychiatric admission. A statistically significant interaction between lifetime cannabis use and Val158Met genotype was observed such that the Val allele was associated with an earlier AOP among cannabis users, with earliest AOP in the Val/Val group, followed by the Met/Val group, and then the Met/Met group ($\beta=2.72$, SE$=1.30$, $P=.04$). A marginally statistically significant interaction was also observed between age at first cannabis use and Val158Met genotype ($\beta=0.93$, SE$=0.46$, $P=.05$), with the Val allele again associated with an earlier AOP. The authors conducted additional analyses showing that this interaction was not present in the comparison group with conduct and affective disorders, although cannabis use was by itself associated with an earlier age of first hospital admission.

In both studies, cannabis use was associated with an earlier AOP, but the interaction with the Val158Met genotype was different in the two studies. Pelayo-Terán et al. (2010) found that the Val allele was associated with an earlier AOP among nonusers, whereas Estrada et al. (2011) found that the Val allele was associated with an earlier AOP in cannabis users. More data are necessary to build a consensus in this area of research.

Brain-Derived Neurotrophic Factor

The BDNF gene is another gene that is proposed to be of particular importance in the pathogenesis of schizophrenia. BDNF encodes a protein that plays a crucial role in intracellular trafficking, including modulation of synaptic plasticity, neurotransmitter release, and dendritic complexity. Abnormalities in BDNF expression at critical points of development may have differential effects on the structure, formation, and functioning of neural circuits involving dopamine D_3 receptors and serotonergic signaling, both of which are clearly implicated in the pathogenesis of schizophrenia (Notaras, Hill, & Van den Buuse, 2015). In some studies the polymorphism causing a valine substitution for methionine at codon 66 of the BDNF protein (Val66Met) has been reported to result in decreased activity-dependent BDNF secretion and appears to modulate AOP (Decoster et al., 2011; Notaras et al., 2015).

In a study of 585 patients with schizophrenia, Decoster et al. (2011) investigated the effects of sex, cannabis use, and the BDNF gene on AOP. Male sex and cannabis use were both associated with an earlier AOP, with an average age at first hospital admission 3.7 years earlier in males ($P<.001$) and 2.7 years earlier in cannabis users ($P<.001$), both of which remained statistically significant in multivariate analyses controlling for one another and for other drug use. A weak association between BDNF Val66Met genotype and AOP was found, with Met carriers on average displaying onset 1.2 years earlier than Val/Val genotypes ($P=.050$). BDNF genotype remained statistically significant in Cox regression analyses controlling for sex, cannabis use, and other drug use, with a hazard ratio of 1.23 (95% CI: 1.03–1.46, $P=.021$). Of particular interest, Decoster et al. (2011) found no BDNF genotype-by-cannabis use interaction for AOP, but did find a statistically significant interaction between sex, cannabis use, and BDNF genotype. AOP decreased more than 7 years in females with both cannabis use and the BDNF Met genotype, but in females with the Val/Val genotype, no association was found between cannabis use and AOP (Decoster et al., 2011).

CHALLENGES IN DOCUMENTING AGE AT ONSET OF ILLNESS AND PREMORBID CANNABIS USE

The literature described above must be understood in the context of the many challenges in documenting AOP and premorbid cannabis use. One such challenge is the commonly lengthy DUP. As mentioned earlier, the median DUP in the United States is approximately 74 weeks (Addington et al., 2015). This delay of treatment leads to first professional contact well after the onset of frank psychotic symptoms in most patients. Therefore, data collected is nearly always retrospective and thus subject to inherent limitations. The majority of the studies included used age at first hospital admission as a proxy for AOP. While this is a widely used practice, it likely overestimates the age at onset in many individuals and creates the potential for bias.

Prospective studies also present challenges for documenting AOP. Identification of prodromal symptoms in adolescents and young adults can prove difficult and the number of false positives is high. Criteria for clinical high-risk or prodromal samples include presentation of attenuated psychotic symptoms, brief and self-limiting psychotic symptoms, or significant functioning decrease in those with a genetic risk. In high-risk samples, the mean risk of transition to psychosis is 29% in 31 months of follow-up from first presentation (Fusar-Poli et al., 2012). When examining the premorbid period, functioning has been found to be predictive of conversion to psychosis (Lyngberg et al., 2015), with the North American Prodrome Longitudinal Study (NAPLS) finding early adolescent social dysfunction to be a significant predictor (Tarbox et al., 2013). While correct identification of high-risk individuals is important, impaired functioning in adolescence can be seen as a predictor of broader mental health issues.

Following the previously described meta-analysis by Large et al. (2011), a related team conducted a second meta-analysis with special attention to methodologies of studies and factors that may impact the findings regarding cannabis use and AOP (Myles, Newall, Nielssen, & Large, 2012). This group found no meaningful differences in the results when they restricted the studies to those with higher-quality study designs. Specifically, they did not find any major impact by restricting the analyses to studies that met quality criteria such as using systematic measures for substance use, systematic measures for psychiatric diagnosis, and using the age at first positive symptom rather than age at initial treatment for illness onset. While research on the impact of cannabis use on AOP is subject to many sources of potential bias and error as a result of retrospective data collection and self-report, it is reassuring that the findings remain largely unchanged when comparing studies with different methodologies and taking study design factors into account.

WHAT IS THE DIRECTION OF THE ASSOCIATION BETWEEN CANNABIS USE AND AGE AT ONSET OF PSYCHOSIS?

While AOP is an important prognostic indicator for the course of schizophrenia-spectrum disorders and merits consideration of its own accord, much of the interest regarding the association between cannabis use and AOP relates to the now well-accepted theory that cannabis use is a component cause of schizophrenia (see Chapter 10). However, several hypotheses could explain the association between cannabis use and AOP, just as they could explain the association between cannabis use and risk of psychotic disorders:

1. Cannabis use and earlier onset of psychosis may both be the consequence of a shared diathesis.

2. Individuals with an emerging earlier onset of psychotic disorder may be more likely to use cannabis, perhaps in an attempt to self-medicate.
3. Cannabis use may hasten the onset of psychosis, either by raising the risk of its development or by accelerating the process that was already under way.

The first hypothesis proposes that an etiological factor predisposing individuals to schizophrenia and to cannabis use is a shared one. In this model of causation, there is no relationship between the two. Studies have provided mixed support for the shared diathesis model. A review of population-based, longitudinal studies by Smit, Bolier, and Cujpers (2004) concluded that further research is needed to arrive at firm conclusions and that results held stronger support for alternative theories, such as cannabis use as a risk factor for onset of schizophrenia, and it having a stronger effect in those predisposed to schizophrenia. A 14-year, longitudinal population study in the Netherlands (Ferdinand et al., 2005) found that antecedent cannabis use predicted future psychotic symptoms and vice versa, providing further support for a shared diathesis. Henquet et al. (2005) found that although cannabis use at baseline increased the risk of psychotic symptoms at follow-up, it had the strongest effect in those predisposed to psychosis. However, predisposition to psychosis at baseline did not predict later cannabis use—evidence against the shared diathesis model.

Regarding the hypothesis that patients use cannabis for self-medicating, either to relieve untreated psychosis or the subtle symptoms that precede development of a full psychotic disorder, at least some research supports this theory, such as a prospective study of cannabis-naïve children, which reported that those who had at least one brief/transient psychotic-like symptom were more likely to use cannabis (Ferdinand et al., 2005). However, the temporal relationship between cannabis use and psychosis described earlier, where cannabis use often precedes the illness onset by 6–7 years, does not support this theory. While nonspecific or prodromal symptoms do typically emerge before frank psychosis, this period usually can be measured in weeks or, at most, months. Also, when directly asked, individuals with psychotic disorders report using cannabis for similar reasons as unaffected peers: for social reasons, to experience euphoria, and simply to have something to do (Ramsay Wan & Compton, 2016).

The predominant hypothesis regarding the association between cannabis use and AOP is that cannabis use asserts a toxic effect and thereby precipitates the onset of schizophrenia in patients who may already be at risk for the illness. This theory is supported by the temporal relationship between the two (cannabis use typically precedes illness onset) and a dose-response relationship between cannabis use and AOP that has been documented in various ways, as described above. In addition, the reports regarding possible gene-by-cannabis use interactions on AOP are consistent with our best understanding of the constellation of factors that cause this complex illness to emerge. The neurobiological mechanisms underlying this increased vulnerability to psychosis are discussed in greater detail in Chapter 10.

CONCLUSIONS

Though some questions related to the relationship between cannabis use and AOP need further investigation to confidently answer, we do know that most patients with recent-onset schizophrenia typically initiate cannabis use years prior to the onset of psychotic symptom and most likely before the prodromal period. In addition, all the studies reviewed suggest that there is a dose-response relationship between cannabis use and an earlier AOP. The current predominant hypothesis for the association between AOP and cannabis use is that cannabis use hastens the onset of schizophrenia in patients who may already be at risk by asserting a toxic effect. Data presented in this chapter also suggests that cannabis use continues to be associated with an earlier AOP regardless of other substance use, and alcohol use appears to be associated with a later AOP.

Even though the studies reviewed in this chapter have employed varying methodologies, a second meta-analysis of Large's work (Myles et al., 2012) revealed that the findings remained mostly unchanged by restricting the analysis only to studies that met quality criteria. No significant differences were found. Continued investigation to further understand the relationship between cannabis use and AOP is necessary given the correlation between AOP and a number of poor outcomes, in addition to the high prevalence of cannabis use among individuals with schizophrenia-spectrum disorders.

Key Chapter Points

1. An important prognostic indicator, age at onset has been found to correlate with a number of poor outcomes.
2. A meta-analysis by Large et al. (2011) provides high-quality evidence of a pattern across the literature: that cannabis use is clearly associated with an earlier AOP.
3. Cannabis use initiation often precedes psychotic symptoms by several years. A 6–7-year age gap between average age of first cannabis use and AOP has been found in a number of studies.
4. A clear dose-response relationship between cannabis exposure and an earlier AOP has been illustrated in multiple studies.
5. Cannabis use is associated with an earlier AOP regardless of other drug use. Indeed, alcohol use appears to be associated with a later AOP in some studies.
6. The predominant hypothesis regarding the association between cannabis use and AOP is that cannabis use asserts a toxic effect and thereby precipitates the onset of schizophrenia in patients who may already be at risk for the illness.
7. Questions that remained unanswered, warranting further investigation, pertain to the relationship between cannabis use and an earlier AOP in nonaffective versus affective psychotic disorders, a possible correlation between cannabis use and earlier age at onset of prodromal symptoms, whether cannabis use is associated with a shorter prodrome, and possible moderators of the relationship between cannabis use and AOP.

REFERENCES

Addington, J., Heinssen, R. K., Robinson, D. G., Schooler, N. R., Marcy, P., Brunette, M. F., et al. (2015). Duration of untreated psychosis in community treatment settings in the United States. *Psychiatric Services*, *66*(7), 753–756. http://dx.doi.org/10.1176/appi.ps.201400124.

Barnes, T., Mutsatsa, S. H., Hutton, S., Watt, H. C., & Joyce, E. M. (2006). Comorbid substance use and age at onset of schizophrenia. *British Journal of Psychiatry*, *188*, 237–242. Available at: http://sro.sussex.ac.uk/14504/.

Barnett, J. H., Werners, U., Secher, S. M., Hill, K. E., Brazil, R., Masson, K., et al. (2007). Substance use in a population-based clinic sample of people with first-episode psychosis. *British Journal of Psychiatry*, 515–520.

Bersani, G., Orlandi, V., Kotzalidis, G. D., & Pancheri, P. (2002). Cannabis and schizophrenia: Impact on onset, course, psychopathology and outcomes. *European Archives of Psychiatry and Clinical Neuroscience*, *252*, 86–92.

Cannon, T. D., Cadenhead, K., Cornblatt, B., Woods, S. W., Addington, J., Walker, E., et al. (2014). Prediction of psychosis in youth at high clinical risk. *Archives of General Psychiatry*, *65*, 28–37.

Carbon, M., & Correll, C. U. (2014). Clinical predictors of therapeutic response to antipsychotics in schizophrenia. *Dialogues in Clinical Neuroscience*, *16*(4), 505–524.

Caspi, A., Moffitt, T. E., Cannon, M., McClay, J., Murray, R., Harrington, H., et al. (2005). Moderation of the effect of adolescent-onset cannabis use on adult psychosis by a functional polymorphism in the catechol-O-methyltransferase gene: Longitudinal evidence of a gene X environment interaction. *Biological Psychiatry*, *57*, 1117–1127.

Clemmensen, L., Vernal, D. L., & Steinhausen, H. C. (2012). A systematic review of the long-term outcomes of early onset schizophrenia. *BMC Psychiatry*, *12*, 150. http://dx.doi.org/10.1186/1471-244X-12-150.

Compton, M. T. (2010). Age at onset and mode of onset of psychosis: Two Prognostic Indicators in the early course of schizophrenia. *Medscape*, 2010. July 15.

Compton, M. T., & Broussard, B. (2009). *The first episode of psychosis: A guide for patients and their families*. New York, NY: Oxford University Press.

Compton, M. T., Kelley, M. E., Ramsay, C. E., Pringle, M., Goulding, S. M., Esterberg, M. L., et al. (2009). Association of pre-onset cannabis, alcohol, and tobacco use with age at onset of prodrome and age at onset of psychosis in first-episode patients. *American Journal of Psychiatry*, *166*, 1251–1257.

Compton, M. T., & Ramsay, C. E. (2009). The impact of pre-onset cannabis use on age at onset of prodromal and psychotic symptoms. *Primary Psychiatry*, *16*, 35–43.

Crumlish, N., Whitty, P., Clarke, M., Browne, S., Kamali, M., Gervin, M., et al. (2008). Beyond the critical period: Longitudinal study of 8-year outcome in first-episode non-affective psychosis. *British Journal of Psychiatry*, *194*(1), 18–24. http://dx.doi.org/10.1192/bjp.bp.107.048942.

De Hert, M., Wampers, M., Jendricko, T., Franic, T., Vidovic, D., De Vriendt, N., et al. (2011). Effects of cannabis use on age at onset in schizophrenia and bipolar disorder. *Schizophrenia Research*, *126*, 270–276. http://dx.doi.org/10.1016/j.schres.2010.07.003.

Decoster, J., Van Os, J., Kenis, G., Henquet, C., Peuskens, J., De Hert, M., et al. (2011). Age at onset of psychotic disorder: Cannabis, BDNF Val66Met, and sex-specific models of gene-environment interaction. *American Journal of Medical Genetics Part B, Neuropsychiatric Genetics*, *156*, 363–369.

Dekker, N., Meijer, J., Koeter, M., Van Den Brink, W., Van Beveren, N., Kahn, R. S., et al. (2012). Age at onset of non-affective psychosis in relation to cannabis use, other drug use and gender.

Psychological Medicine, 1–9. Available at: http://www.ncbi.nlm.nih.gov/entrez/query.fcgi?cmd=Retrieve&db=PubMed&dopt=Citation&list_uids=22452790.

Di Forti, M., et al. (2014). Daily use, especially of high-potency cannabis, drives the earlier onset of psychosis in cannabis users. *Schizophrenia Bulletin*, *40*, 1509–1517.

Dragt, S., Nieman, D. H., Becker, H. E., Van De Fliert, R., Dingemans, P. M., De Haan, L., et al. (2010). Age of onset of cannabis use is associated with age of onset of high-risk symptoms for psychosis. *Canadian Journal of Psychiatry*, *55*, 165–171. Available at: http://www.ncbi.nlm.nih.gov/pubmed/20370967.

Dragt, S., Nieman, D. H., Schultze-Lutter, F., Van Der Meer, F., Becker, H., De Haan, L., et al. (2012). Cannabis use and age at onset of symptoms in subjects at clinical high risk for psychosis. *Acta Psychiatrica Scandinavica*, *125*, 45–53.

Estrada, G., Fatjó-Vilas, M., Muñoz, M. J., Pulido, G., Miñano, M. J., Toledo, E., et al. (2011). Cannabis use and age at onset of psychosis: Further evidence of interaction with COMT Val158Met polymorphism. *Acta Psychiatrica Scandinavica*, *123*, 485–492. Available at https://www.ncbi.nlm.nih.gov/pubmed/21231925.

Ferdinand, R. F., Sondeijker, F., van der Ende, J., Selten, J. P., Huizink, A., & Verhulst, F. C. (2005). Cannabis use predicts future psychotic symptoms, and vice versa. *Addiction*, *100*, 612–618. Available at: http://www.ncbi.nlm.nih.gov/entrez/query.fcgi?cmd=Retrieve&db=PubMed&dopt=Citation&list_uids=15847618/nhttp://onlinelibrary.wiley.com/store/10.1111/j.1360-0443.2005.01070.x/asset/j.1360-0443.2005.01070.x.pdf?v=1&t=hlovivsj&s=2b07e34a3058dac1fe42478bcde6f1.

Fusar-Poli, P., Bonoldi, I., Yung, A. R., Borgwardt, S., Kempton, M. J., Valmaggia, L., et al. (2012). Predicting psychosis: Meta-analysis of transition outcomes in individuals at high clinical risk. *Archives of General Psychiatry*, *69*(3), 220–229. http://dx.doi.org/10.1001/archgenpsychiatry.2011.1472.

Galvez-Buccollini, J. A., Proal, A. C., Tomaselli, V., Trachtenberg, M., Coconcea, C., Chun, J., et al. (2012). Association between age at onset of psychosis and age at onset of cannabis use in non-affective psychosis. *Schizophrenia Research*, *139*, 157–160. http://dx.doi.org/10.1016/j.schres.2012.06.007.

Goldberger, C., Dervaux, A., Gourion, D., Bourdel, M., Laqueille, X., Krebs, M., et al. (2010). Variable individual sensitivity to cannabis in patients with schizophrenia. *International Journal of Neuropsychopharmacology*, *13*, 1145–1154.

González-Pinto, A., Vega, P., Ibáñez, B., Mosquera, F., Barbeito, S., Gutiérrez, M., et al. (2008). Impact of cannabis and other drugs on age at onset of psychosis. *Journal of Clinical Psychiatry*, *69*, 1210–1216.

Harrison, G., Hopper, K., Craig, T., Laska, E., Siegel, C., Wanderling, J., et al. (2001). Recovery from psychotic illness: A 15- and 25-year international follow-up study. *British Journal of Psychiatry*, *178*(6), 506–517.

Henquet, C., Krabbendam, L., Spauwen, J., Kaplan, C., Lieb, R., Wittchen, H. U., et al. (2005). Prospective cohort study of cannabis use, predisposition for psychosis, and psychotic symptoms in young people. *BMJ*, *330*, 11.

Ira, E., Zanoni, M., Ruggeri, M., Dazzan, P., & Tosato, S. (2013). COMT, neuropsychological function and brain structure in schizophrenia: A systematic review and neurobiological interpretation. *Journal of Psychiatry and Neuroscience*, *38*, 366–380.

Kelley, M. E., Ramsay Wan, C., Broussard, B., Crisafio, A., Cristofaro, S., Johnson, S., et al. (2016). Marijuana use in the immediate 5-year premorbid period is associated with increased risk of onset of schizophrenia and related psychotic disorders. *Schizophrenia Research*, *171*, 62–67.

Koskinen, J., Löhönen, J., Koponen, H., Isohanni, M., & Miettunen, J. (2010). Rate of cannabis use disorders in clinical samples of patients with schizophrenia: A meta-analysis. *Schizophrenia Bulletin, 36*, 1115–1130.

Kristensen, K., & Cadenhead, K. S. (2007). Cannabis abuse and risk for psychosis in a prodromal sample. *Psychiatry Research, 151*, 151–154.

Large, M., Sharma, S., Compton, M. T., Slade, T., & Nielssen, O. (2011). Cannabis use and earlier onset of psychosis: A systematic meta-analysis. *Archives of General Psychiatry, 68*, 555–561. Available at: http://www.ncbi.nlm.nih.gov/pubmed/21300939.

Leeson, V. C., Harrison, I., Ron, M. A., Barnes, T. R. E., & Joyce, E. M. (2012). The effect of cannabis use and cognitive reserve on age at onset and psychosis outcomes in first-episode schizophrenia. *Schizophrenia Bulletin, 38*, 873–880.

Lyngberg, K., Buchy, L., Liu, L., Perkins, D., Woods, S., & Addington, J. (2015). Patterns of premorbid functioning in individuals at clinical high risk of psychosis. *Schizophrenia Research, 169*(1), 209–213. http://dx.doi.org/10.1016/j.schres.2015.11.004.

Myles, N., Newall, H., Nielssen, O., & Large, M. (2012). The association between cannabis use and earlier age at onset of schizophrenia and other psychoses: Meta-analysis of possible confounding factors. *Current Pharmaceutical Design, 18*, 5055–5069.

Notaras, M., Hill, R., & Van den Buuse, M. (2015). A role for the BDNF gene Val66Met polymorphism in schizophrenia? A comprehensive review. *Neuroscience and Biobehavioral Reviews, 51*, 15–30. http://dx.doi.org/10.1016/j.neubiorev.2014.12.016.

Öngür, D., Lin, L., & Cohen, B. M. (2009). Clinical characteristics influencing age at onset in psychotic disorders. *Comprehensive Psychiatry, 50*, 13–19. http://dx.doi.org/10.1016/j.comppsych.2008.06.002.

Pelayo-Terán, J. M., Crespo-Facorro, B., Carrasco-Marín, E., Pérez-Iglesias, R., Mata, I., Arranz, M. J., et al. (2008). Catechol-*O*-methyltransferase Val158Met polymorphism and clinical characteristics in first episode non-affective psychosis. *American Journal of Medical Genetics Part B, Neuropsychiatric Genetics, 147*, 550–556.

Pelayo-Terán, J. M., Pérez-Iglesias, R., Mata, I., Carrasco-Marín, E., Vázquez-Barquero, J. L., & Crespo-Facorro, B. (2010). Catechol-*O*-methyltransferase (COMT) Val158Met variations and cannabis use in first-episode non-affective psychosis: Clinical-onset implications. *Psychiatry Research, 179*, 291–296. http://dx.doi.org/10.1016/j.psychres.2009.08.022.

Phillips, L., Curry, C., & Yung, A. (2002). Cannabis use is not associated with the development of psychosis in an "ultra" high-risk group. *The Australian and New Zealand Journal of Psychiatry, 36*, 800–806. Available at: http://onlinelibrary.wiley.com/doi/10.1046/j.1440-1614.2002.01089.x/full.

Ramsay Wan, C., & Compton, M. T. (2016). Marijuana use and psychosis: From reefer madness to marijuana use as a component cause. In M. T. Compton (Ed.), *Marijuana and mental health* (pp. 119–148). Arlington, VA: American Psychiatric Association Publishing.

Schimmelmann, B. G., Conus, P., Cotton, S. M., Kupferschmid, S., Karow, A., Schultze-Lutter, F., et al. (2011). Cannabis use disorder and age at onset of psychosis—A study in first-episode patients. *Schizophrenia Research, 129*, 52–56. http://dx.doi.org/10.1016/j.schres.2011.03.023.

Sevy, S., Robinson, D. G., Napolitano, B., Patel, R. C., Gunduz-Bruce, H., Miller, R., et al. (2010). Are cannabis use disorders associated with an earlier age at onset of psychosis? A study in first episode schizophrenia. *Schizophrenia Research, 120*, 101–107. http://dx.doi.org/10.1016/j.schres.2010.03.037.

Smit, F., Bolier, L., & Cujpers, P. (2004). Cannabis use and the risk of later schizophrenia: A review. *Addiction, 99*, 425–430.

Stefanis, N. C., Dragovic, M., Power, B. D., Jablensky, A., Castle, D., & Morgan, V. A. (2013). Age at initiation of cannabis use predicts age at onset of psychosis: The 7- to 8-year trend. *Schizophrenia Bulletin, 39*, 251–254.

Sugranyes, G., Flamarique, I., Parellada, E., Baeza, I., Goti, J., Fernandez-Egea, E., et al. (2009). Cannabis use and age of diagnosis of schizophrenia. *European Psychiatry, 24*, 282–286. http://dx.doi.org/10.1016/j.eurpsy.2009.01.002.

Tarbox, S. I., Addington, J., Cadenhead, K. S., Cannon, T. D., Cornblatt, B. A., Perkins, D. O., et al. (2013). Premorbid functional development and conversion to psychosis in clinical high-risk youths. *Development and Psychopathology, 25*(4 pt 1), 1171–1186. http://dx.doi.org/10.1017/S0954579413000448.

Tosato, S., Lasalvia, A., Bonetto, C., Mazzoncini, R., Cristofalo, D., De Santi, K., et al. (2013). The impact of cannabis use on age of onset and clinical characteristics in first-episode psychotic patients. Data from the Psychosis Incident Cohort Outcome Study (PICOS). *Journal of Psychiatric Research, 47*, 438–444. http://dx.doi.org/10.1016/j.jpsychires.2012.11.009.

van Os, J., Rutten, B. P. F., & Poulton, R. (2008). Gene-environment interactions in schizophrenia: Review of epidemiological findings and future directions. *Schizophrenia Bulletin, 34*, 1066–1082. Available at: https://www.ncbi.nlm.nih.gov/pmc/articles/PMC2632485/.

Veen, N. D., Selten, J. P., van der Tweel, I., Feller, W. G., Hoek, H. W., & Kahn, R. S. (2004). Cannabis use and age at onset of schizophrenia. *American Journal of Psychiatry, 161*, 501–506. Available at: http://www.ncbi.nlm.nih.gov/entrez/query.fcgi?cmd=Retrieve&db=PubMed&dopt=Citation&list_uids=14992976/nhttp://ajp.psychiatryonline.org/article.aspx?articleid=176684.

Xie, T., Ho, S. L., & Ramsden, D. (1999). Characterization and implications of estrogenic down-regulation of human catechol-O-methyltransferase gene transcription. *Molecular Pharmacology, 56*(1), 31–38.

Chapter 12

The Prevalence and Effects of Cannabis Use Among Individuals With Schizophrenia and Related Psychotic Disorders

Ana Fresán*, Rebeca Robles-García*, Carlos-Alfonso Tovilla-Zarate†
**Instituto Nacional de Psiquiatría Ramón de la Fuente Muñíz, Mexico City, Mexico †Universidad Juárez Autónoma de Tabasco, Comalcalco, Mexico*

INTRODUCTION

Recreational cannabis use has been decriminalized or legalized in several countries, and more recently in several states in the United States (US), with varying restrictions on its use. Cannabis use is legal in the Netherlands, and in other countries such as Italy, Switzerland, Germany, and Spain, its consumption is not penalized or at least does not involve major legal ramifications. There are also many countries where cannabis use remains a crime, with penalties ranging from mandatory substance use treatment to incarceration (Mead, 2014). Increasing cannabis legalization, combined with the fact that cannabis is already one of the most commonly used substances among individuals with schizophrenia and related psychotic disorders, warrants in-depth analysis of the effects of cannabis use among vulnerable populations.

The deleterious effects of cannabis use may not manifest in all individuals with psychosis to the same extent, but many do experience serious negative complications, which can impact multiple clinical domains, including the course of illness, psychosocial functioning, and social integration. This chapter presents an analysis of the prevalence of cannabis use in individuals with psychotic disorders, as well as the most common clinical effects and psychosocial complications in this population as described in the scientific literature to date.

PREVALENCE OF CANNABIS USE IN INDIVIDUALS WITH SCHIZOPHRENIA AND RELATED PSYCHOTIC DISORDERS

Cannabis is the world's most widely used illicit drug, with a long history in human society. There seems to have been recent increases in cannabis use, and it

The Complex Connection between Cannabis and Schizophrenia. http://dx.doi.org/10.1016/B978-0-12-804791-0.00012-4

is estimated that 2.8%–4.5% of the population between 15 and 64 years worldwide uses cannabis, comprising an estimated 125–203 million of the estimated 149–271 million substance users globally (Degenhardt & Hall, 2012). The annual prevalence of cannabis use continues to be highest in Oceania (10.7%), the Americas (8.4%), and Africa (7.5%), while Europe and Asia report the lowest prevalence rates (4.5% and 1.9%, respectively) (United Nations Office on Drugs and Crime, 2015). According to the United Nations World Drug Report 2015, from 2009 to 2013, there was little change in illicit drug use rates and their health consequences globally. In the specific case of cannabis, the number of people requiring treatment for its use is increasing, as many users meet criteria for cannabis use disorders, suggesting that cannabis may be increasingly harmful to public health (United Nations Office on Drugs and Crime, 2015).

Cannabis is also the most commonly used illicit substance among those with schizophrenia and related disorders. This has been a subject of interest ever since the French physician Moreau de Tours described its harmful effects in 1845 (Moreau de Tours, 1845). Since the 1960s, when cannabis use increased substantially, many epidemiological and clinical studies have been undertaken to determine the precise link between cannabis use and psychosis.

Epidemiological Studies

Epidemiological studies focus on the association between cannabis use and the development of psychosis. The fact that cannabis can induce psychotic-like symptoms have been reported for more than 50 years, with the first study, published in 1958, describing psychotic-like symptoms in healthy volunteers after a single use of cannabis (Ames, 1958).

To our knowledge, the first epidemiological evidence of cannabis use as a risk factor for the onset of schizophrenia began with a longitudinal study reported by Andreasson, Allebeck, Engström, and Rydberg (1987). With data from 50,465 participants over a period of 15 years, they found that high users of cannabis (defined as having used cannabis on more than 50 occasions) were six times more likely to have schizophrenia than nonusers (OR 6.0, 95% CI 4.0–8.9). This result remained significant even after controlling for other psychiatric illnesses and social background. In a reanalysis of this dataset, in which data were available on self-reported use of cannabis and other drugs (50.087 subjects), cannabis use was associated with an increased risk of developing schizophrenia in subjects who had ever used cannabis (OR 1.2, 95% CI 1.1–1.4), subjects who had only used cannabis and no other drugs (OR 1.3, 95% CI 1.1–1.5), and most prominently in subjects who had used cannabis on more than 50 occasions (OR 6.7, 95% CI 2.1–21.7) (Zammit, Allebeck, Andreasson, Lundberg, & Lewis, 2002). In a more recent report of a 3-year longitudinal study in a sample of 34,653 people within the United States, Davis, Compton, Wang, Levin, and Blanco (2013) found that the risk for schizotypal personality disorder increased with certain patterns of cannabis use: the odds ratio in cannabis users compared to nonusers was 2.03 (95% CI 1.70–2.42), which increased in subjects who regularly used cannabis (OR 2.88, 95% CI 2.38–3.48), and was even higher among those with cannabis dependence (OR 7.97, 95% CI 6.0–10.60).

A number of studies have investigated whether exposure to cannabis during adolescence is a risk factor for the future development of a psychotic disorder. In a longitudinal study including 759 adolescents from Dunedin, New Zealand, Arseneault et al. (2002) found that cannabis users assessed at ages 15 and 18 had more psychotic symptoms (DSM-IV criteria for schizophreniform disorder) than those who never used cannabis or had only used it once or twice (OR 1.95, 95% CI 0.76–5.01). In another study conducted in a representative Greek cohort of 3500 19 year olds, cannabis use was positively associated with both the positive and negative dimensions of psychosis; in addition, first use of cannabis earlier than age 16 was associated with a much stronger effect on both psychosis dimensions, regardless of lifetime frequency of use (Stefanis et al., 2004). Consistent with these findings, a study in Germany sought to determine whether cannabis use in adolescence increased risk for psychosis. Kuepper et al. (2011) conducted follow-ups at three different time points (baseline, 3.5 years, 8.4 years, and 10 years) in a sample of 1923 adolescents, and observed that the risk of a psychotic experience increased among cannabis users (OR 1.5, 95% IC 1.1–3.1). In a large-scale Finnish study including 6298 15- and 16-year-old adolescents, Miettunen et al. (2008) examined cannabis use and prodromal psychotic symptoms. The study assessed cannabis use and screened for prodromal symptoms (e.g., feelings that something strange or inexplicable is taking place in oneself or in the environment, feelings that one is being followed or influenced in a special way, and experience of thoughts running wild or difficulty in controlling the speed of thoughts). They found that adolescents who had tried cannabis (5.6% of the total sample) were more likely to display three or more prodromal symptoms even after controlling for confounders including previous behavioral symptoms (OR = 2.23; 95% CI 1.70–2.94).

These represent only a portion of the significant epidemiologic studies conducted to determine whether cannabis use increases the risk of psychosis. Many other important epidemiologic studies have been conducted around the world, which support the notion of a causal link between cannabis use and increased risk of psychosis, though controversy and uncertainty remains in the field.

Although this link has been robust and consistent, many questions remain regarding measurement, confounding variables, and even a noncausal link between cannabis use and psychosis. Obviously, not all individuals using cannabis develop a psychotic disorder, and not all of those who display psychotic symptoms have a history of cannabis use. Cannabis use may therefore interact with other factors that lead to the development of a psychotic disorder (Le Bec, Fatséas, Denis, Lavie, & Auriacombe, 2009).

Clinical Studies

Although epidemiological studies have often focused on the development of psychosis in cannabis users, clinical studies commonly focus on cannabis use in individuals with psychosis. Findings from clinical studies have been conflicting, and have failed to determine the exact prevalence of cannabis use in people with psychosis for several reasons. First, many studies reporting cannabis use prevalence in individuals with psychotic disorders have used different methodologies with inherent limitations related to each method. Second, since psychotic disorders are relatively low-prevalence illnesses, most research has been conducted using small sample sizes and with restricted criteria, providing results that may be relevant only to specific clinical populations. Third, many studies using the terms cannabis "abuse" or "dependence" fail to provide detailed data on cannabis use, limiting the meaningfulness of these results. Thus, to date, the most reliable estimates of cannabis use prevalence in people with psychosis are expressed in meta-analysis reports, which provide the opportunity to statistically combine and evaluate the results of multiple comparable studies. Since several studies have reported a decline in cannabis use after first-episode psychosis (FEP) (Koskinen, Löhönen, Koponen, Isohanni, & Miettunen, 2010), meta-analysis results of the prevalence of cannabis use among FEP patients and chronic-treatment samples will be discussed separately in this chapter.

A recent meta-analysis of the prevalence of cannabis use in FEP, with a total of 35 samples and combined sample size of 6321 subjects, reported that 33.7% (95% CI=29–38%) of patients were using cannabis at the time of their FEP (Myles, Myles, & Large, 2016). This study also showed geographical differences in prevalence rates, with the highest rate of cannabis use reported in Australia (Table 1).

TABLE 1 Geographical Region Analysis for Meta-Analysis of Cannabis Use Prevalence in FEP (Myles et al., 2016)

Region	Number of Samples	Rate of Cannabis Use[a]	Effect Size With 95% CI	Between Group Heterogeneity
Australia	3	0.51	−0.60 to 0.71	$Q=15.5$, $df(Q)=4$ $P=.004$
Europe	15	0.37	−0.79 to −0.22	
United Kingdom	8	0.30	−1.21 to −0.42	
North America	7	0.29	−1.29 to −0.45	

[a]*Rates expressed in proportion of population.*

Another meta-analysis of 53 studies (n=9792) reported the prevalence of cannabis use among samples of patients in treatment (Green, Young, & Kavanagh, 2005). Inpatient and community patient studies were included if the prevalence of cannabis use or misuse among patients with psychosis could be calculated. The total number of reported cannabis users in studies that contained current use data was divided by the total sample size of the respective studies to calculate weighted averages. Estimates involved calculating current, 12-month, and lifetime prevalence rates. The average prevalence rate of cannabis use reported was 42.9% for current use, 44.7% for 12-month use, and 53.5% for lifetime use.

This meta-analysis also included geographical areas (studies were conducted in Australasia, Europe, North America, and the United Kingdom) and determined their influence on the odds of cannabis use or misuse. None of the regions were associated with increased odds of current cannabis use, but for lifetime use, the odds were highest for Australasia (OR=1.70; 95% CI 1.38–2.10) and North America (OR=1.53; 95% CI 1.14–2.05).

Determinants of and Motivations for Cannabis Use

The widespread use of cannabis among individuals with psychotic disorders has been explained in at least four different ways: cannabis use as a component cause of psychosis; the self-medication hypothesis; similar motivations for use as the general population; and social determinants including stigma and rejection by peers, which may increase patients' risk of becoming part of drug users' networks, which could be more accepting of them (Lamb, 1982). The first explanation is based on the scientific literature described earlier in this chapter and in other chapters; although it is clear that cannabis use alone is neither essential nor sufficient to lead to the onset of psychotic symptoms. Moreover, the association could reflect cannabis use among individuals experiencing prodromal symptoms of schizophrenia or other psychotic disorders (Radhakrishnan, Wilkinson, & D'Souza, 2014), opening up the possibility of understanding their cannabis use as a self-medication of early experiences, and/or relating to the social determinants of the psychotic disorder. The self-medication hypothesis posits that substance use is a consequence of patients' attempts to decrease the symptoms accompanying the disease or to alleviate underlying distressing emotional states (Khantzian, 1997). This could be extended to include the management of medication side effects for those in treatment. A number of studies support this explanation of cannabis use in patients with schizophrenia and other psychotic disorders.

For example, in a study by Addington and Duchak (1997) that examined the reasons for cannabis use in outpatients with cannabis dependence, 40% reported that the use of the drug was "to decrease voices." Regarding the use of cannabis to alleviate underlying distressing emotional states (or secondary dysphoria related to the disease), Green, Kavanagh, and Young (2004) demonstrated, based on a case–control study comparing patients with healthy controls, that patients with psychosis were more likely to use cannabis to reduce anxiety, depression, or boredom. Schofield et al. (2006) studied cannabis use as a way to reduce the side effects of medication. In patients who had recently used cannabis, almost 50% of those having inner unrest/agitation or difficulty sleeping, and at least 25% of those experiencing side effects such as sleepiness/sedation, muscle tension, and lack of emotions, reported cannabis-associated reductions in these symptoms. However, Dekker, Linszen, and De Haan (2009) concluded, after their review of 14 studies examining self-reported reasons for cannabis use and self-reported effects of cannabis use in patients with psychotic disorders, that only 12.9% of patients reported reasons for use related to the alleviation of psychotic symptoms or the side effects of medication. Interestingly, according to a study by Fowler, Carr, Carter, and Lewin (1998), the small proportion of patients who used it for illness- and medication-related reasons tended to be cannabis misusers rather than users.

Contrary to the self-medication hypothesis, the vast majority of studies examining self-reported reasons for cannabis use in patients with psychotic disorders report that, despite an awareness of negative effects of cannabis on positive symptoms, patients use it for social reasons (i.e., being friendly, improving interactions with others, "going along with the group," or "as something to do with friends") (61.7%) and/or to enhance positive affect (i.e., feelings of happiness, "to increase pleasure," "to get high," or "to satisfy curiosity") and relax (42.1%) (Dekker et al., 2009). These findings highlight the importance of taking into account the common effects of cannabis as well the motivations for cannabis use in the general population in order to understand its use by individuals with schizophrenia and other psychotic disorders. Indeed, according to a study by Green et al. (2004), two main reasons for cannabis use among persons with and without psychotic disorders are related to social activities (37.8% for psychosis and 28.9% for non-psychosis), and positive mood alteration (35.6% for psychosis and 42.1% for non-psychosis).

In sum, cannabis use among patients with psychotic disorders might be best understood as the result of multiple determinants and motivations, which may interact in a dynamic and synergistic way at different stages of the disorder. Cannabis may precipitate and exacerbate psychotic symptoms, and then be used, at least from the person's perspective, to alleviate some of those symptoms and their social consequences, and even the side effects of medication. But it may also be associated with positive experiences common to cannabis users both with psychotic disorders and in the general population. This insight may be useful in the treatment of patients with both psychotic and cannabis use disorders.

CLINICAL EFFECTS OF CANNABIS USE IN THE CONTEXT OF SCHIZOPHRENIA AND RELATED PSYCHOTIC DISORDERS

Positive and Negative Symptoms

The symptomatology of psychotic disorders can generally be divided into "positive" and "negative" symptom clusters. Examples of positive symptoms include auditory hallucinations and delusions, while examples of negative symptoms include amotivation, affective flattening, and social deficits. Together with negative symptoms, "cognitive deficits" tend to cause the most disability in schizophrenia and other psychotic disorders, though positive symptoms often draw the most attention from treatment providers and the public alike. The association between cannabis use and more prominent positive psychotic symptoms and a generally more severe course of illness has been shown in the scientific literature. The impact of cannabis use has been reported in terms of acute effects (e.g., precipitation of an acute episode or exacerbation of existing psychotic symptoms) and longer-term effects (e.g., progressive and insidious pattern of deteriorating symptoms and functioning), which have been replicated in patients during their FEP as well as in chronic patients (Castle, 2013). A recent meta-analysis confirmed a small-to-moderate effect of substance use on more severe positive symptoms, although the confounding effects of other variables in this association, such as male gender and younger age, could not be ruled out. This association might also be explained by reverse causation, whereby patients with more prominent symptoms use cannabis to ameliorate symptoms and distress (Large, Mullin, Gupta, Harris, & Nielssen, 2014). Although no firm conclusions can be drawn from this data in psychotic disorders, the psychoactive effects of cannabis that are mediated by Δ^9-tetrahydrocannabinol (THC) include a range of transient positive symptoms, similar to those observed in patients with psychotic disorders.

Psychotomimetic effects of THC, such as paranoia, suspiciousness, conceptual disorganization, and perceptual alterations, have consistently been reported in healthy individuals. It is possible that certain individuals may be more prone to the psychotomimetic effects of cannabis, or may even have some risk factors for psychosis that make them more vulnerable to the effects of, or more likely to start using, cannabis (van Winkel & GROUP Investigators, 2015). One might be tempted to infer from this that individuals with psychosis are obviously more susceptible to the effects of cannabis and therefore to expect worse positive symptoms even at low cannabis doses, but more studies are needed to determine whether cannabis use definitively affects positive symptom manifestation in individuals with psychotic disorders.

Information on the effects of cannabis use on negative symptoms is even more limited. A recent study showed that some negative symptoms, specifically apathy and lack of motivation, lead to the onset of subclinical depressive symptoms, which in turn are associated with an increased risk of cannabis use as a self-medication strategy to alleviate a negative emotional state (González-Ortega et al., 2015). In addition, chronic, heavy cannabis use has been associated with an "amotivational syndrome," in which the individual exhibits a gradual detachment from the world and a progressive loss of general reactivity. It closely resembles a clinical picture of negative symptoms, since it is characterized by apathy, lack of motivation, social withdrawal, lack of interest, anhedonia, and impaired general achievement. However, since the amotivational syndrome has been described in individuals with chronic cannabis use rather than those with psychotic disorders, the concept is not directly relevant to the impact of cannabis on negative symptoms in persons with psychotic disorders (Castle, 2013; Rovai et al., 2013).

Cognitive Impairment

On the whole, researchers have reported inconsistent results on the relationship between cognitive impairment in schizophrenia and cannabis use. Before any firm conclusions can be drawn, it is important to bear in mind the fact that cognition is a multifaceted entity, directly influenced by a range of biological and environmental factors. Emerging evidence related to cognitive performance and cannabis use in individuals with schizophrenia suggests that cannabis users in general have better cognitive performance than patients who do not engage in comorbid cannabis use. Researchers have suggested that persons with psychosis who use cannabis may have higher premorbid IQ and are thus better able to procure an illegal substance while evading the law (Radhakrishnan et al., 2014). However, IQ is not the only variable associated with this ability; other factors, such as personality features and availability of the drug where people live, may have a stronger association with cannabis use than IQ. Such hypotheses warrant further study. In addition, the higher premorbid IQ is not immune to the effects of continued cannabis use, which, in general, tends to result in a decline in overall cognitive performance. Other authors have argued that there may be no differences in cognitive performance between cannabis users and nonusers with psychosis, and that any probable effect associated with cannabis use— better or worse cognitive performance—can be better explained by other factors related to cannabis use, such as age, age of illness onset, and socioeconomic status (Power et al., 2015). For example, during adolescence, when the onset of cannabis use as well as schizophrenia and other psychotic disorders generally occurs, the brain is not fully developed and is more vulnerable to structural and functional abnormalities. At this stage of life, those who are at high risk for psychosis may also be at risk for additional microstructural alterations resulting from cannabis use, and indeed the clinical manifestations of its use in terms of cognitive decline have been observed across almost all domains of cognitive functioning, and have been associated with an earlier onset and higher frequency of cannabis use (Damjanović et al., 2015).

The inconsistent findings about the relationships between cannabis use and cognition in individuals with psychotic disorders may reflect several methodological differences across studies. One of these differences involves how each research group defines cannabis use: present/absent, none/mild/moderate/severe, or user/nonuser/dependent, among others. Moreover, sample size and cognitive assessments vary across studies. Despite this, the link between cognitive deficits and continued cannabis use is not new and cannot be overlooked. In general, cannabis use can cause short-term disturbances in memory, attention, working memory, IQ, and executive functions, some of which may return to their premorbid state after the acute effects of cannabis have worn off. However, cognitive effects may not abate after cessation for chronic and heavy users, particularly for heavy users with a psychotic disorder. Whether cognitive impairments are permanent is an important research question. Although no scientific consensus has yet been reached, studies have shown variable periods of time until full recovery, ranging from a week to almost three months after cannabis cessation (Crean, Crane, & Mason, 2011). This cognitive recovery may occur in those who are not heavy cannabis users or who are not prone to psychosis, although cannabis cessation may not lead to complete cognitive recovery in those with psychotic disorders, likely at least partially due to aspects of the underlying brain disease process.

Treatment Adherence and Relapses

Adherence to antipsychotic medication is a critical factor in achieving symptomatic remission and functional recovery in individuals with schizophrenia and other psychotic disorders. The disease itself, poor insight into the illness, adverse effects of medications, and lack of information about the need for treatment maintenance are among the many reasons why patients stop antipsychotic medications (Wade, Tai, Awenat, & Haddock, 2017). It has been estimated that nearly 50% of patients with schizophrenia discontinue treatment within the first year and up to 74% discontinue treatment at 18 months, and that up to 75% of patients have a clinical relapse between 6 and 24 months after stopping antipsychotic medications (Stroup et al., 2009). Often each subsequent relapse is more serious and requires more effort to recover clinical stability. Thus, relapses associated with lack of medication adherence negatively impact the patient (e.g., social, familial, academic, and occupational performance), society (e.g., particularly the patient's relatives and the surrounding environment), and mental health systems (e.g., greater use of direct and indirect resources).

Some authors have suggested that premorbid cannabis use is not directly related to treatment adherence, time to remission, or likelihood of relapse in persons with FEP in the first year of treatment (Colizzi et al., 2016). Nevertheless, relapse and poor medication adherence are two of the most important negative clinical outcomes repeatedly associated with cannabis use in psychosis (Barbeito et al., 2013). Complex interactions between clinical variables may have direct bearing on these apparently contradictory findings. Persons with comorbid cannabis use are generally informed that stopping cannabis is the most desirable course of action during antipsychotic treatment. When coping with the need for specialized treatment for psychosis, individuals have to make crucial decisions. One, in the case of those with premorbid cannabis use, is whether to continue or cease cannabis use, and another is whether or not to take medication for psychotic symptoms. These two decisions may interact over time during the evolution of the illness, and the person's decisions may change based on his or her personal beliefs about cannabis use and antipsychotic medication (Faridi, Joober, & Malla, 2012). During the FEP, a cannabis user may conclude that all the symptoms are related to cannabis and therefore decide to quit using cannabis and not start or continue antipsychotic medication. Other individuals may decide to start antipsychotic medication but, as a result of adverse antipsychotic effects, may decide to use cannabis to ameliorate them. Similar alternative decision-making possibilities are numerous and mental health service providers should be aware that, depending on an array of factors, such as the individual's motivations and knowledge, cannabis use may promote nonadherence to medications and future relapses. Providers should therefore encourage patients and their families to become involved in decision-making around treatment, and provide education and counseling about the disorder, aimed at reducing cannabis use and/or encouraging continued treatment adherence despite persistent cannabis use (Faridi et al., 2012).

Studies on the Clinical Effects of Cannabis Use Cessation

Little is known about the differences between persons with psychotic disorders who continue to use cannabis compared to those who stop using it during treatment. Several reports have confirmed that almost 50% of patients with FEP either give up or reduce their cannabis use (Faridi et al., 2012), although patients meeting criteria for cannabis use disorder are more likely to continue to use it, compared with those who not do not meet criteria. The precise reasons why persons with psychosis decide to stop or continue using cannabis are unclear. To date, only two research studies on individuals with FEP have assessed the differences between those who continue to use cannabis after their first admission for psychosis and those who cease cannabis use during treatment. A study by Baeza et al. (2009), which included children and adolescents with FEP, showed that those who quit cannabis use had better short-term outcomes, with a significant reduction in the intensity of psychopathology after a 6-month follow-up. On the other hand, a study by Faridi et al. (2012), which included adolescents and young FEP patients, did not show any difference in the severity of psychopathology in a 12-month follow-up between continuous cannabis users and those who stopped using cannabis during treatment. In the latter report, the authors stated that the differences between these disparate study findings may be partly explained by the different follow-up periods, as well as the different ages of patients; the effects of stopping cannabis in a younger population may not be reflected in an older population. Moreover, other variables such as treatment adherence may influence the differing outcomes related to cannabis use, similar to the earlier example described in this chapter, where treatment adherence may act as a mediator between cannabis use and remission.

Less information is available concerning cannabis cessation in individuals with more chronic schizophrenia, though it seems that cannabis withdrawal may be a complicating factor in this population. The first systematic research report on withdrawal symptoms in persons with psychotic disorders included 120 patients with regular cannabis use. In this study, the most frequently reported withdrawal symptoms were craving (59.2%), anxiety (52.57%), feeling bored (47.5%), and sadness (45.8%). The authors concluded that these symptoms may be associated with clinically significant behavioral changes and warrant further study (Boggs et al., 2013).

PSYCHOSOCIAL EFFECTS OF CANNABIS USE IN PERSONS WITH SCHIZOPHRENIA AND RELATED PSYCHOTIC DISORDERS

Substance use has been associated with a wide range of negative outcomes in individuals with psychotic disorders, including in the psychosocial sphere. Indeed, clinical and psychosocial effects of drug use in persons with psychosis appear to be interrelated in a complex, bidirectional way. For example, higher rates of psychiatric relapse and hospital admission may be explained by the well-documented increase in the proportion of homeless individuals with schizophrenia who use drugs (Caton et al., 1994). From this point of view, it is possible to consider that certain psychosocial variables (such as interpersonal and family problems) can be understood not just as outcomes but as predictors of both a worse clinical course of psychosis and of drug use.

Psychosocial impairment is a defining feature of both schizophrenia and substance abuse. The question most relevant to this discussion is whether cannabis use and psychosis have a synergic effect on the interpersonal functional problems usually observed in individuals with schizophrenia. The development of effective treatment plans requires a thorough understanding not only of the risk factors for the development of a comorbid substance use disorder, but also of the consequences of substance use. To elucidate the consequences of drug use (particularly cannabis) on psychosocial functioning in persons with schizophrenia or other psychotic disorders, this section presents an analysis of the scientific literature in the following key areas: interpersonal (family, friends, work, etc.), social exclusion, aggression and violence, and suicide. It is important to note that many studies on the psychosocial effects of substance use on persons with psychotic disorders do not distinguish between specific substances, so general terms such as "drug use" or "substance use" may be used rather than "cannabis use" in this discussion. However, as cannabis use is the most common illicit drug used by individuals with psychotic disorders, it is highly likely that any general findings related to substance use also apply specifically to cannabis use.

Interpersonal Consequences

It is now well established that interpersonal problems, mainly with family, are more common in persons with serious mental illness who also misuse alcohol and drugs (Dixon, McNary, & Lehman, 1995), with cannabis being the illicit drug most frequently used by individuals with schizophrenia, and cocaine and opioids being the least frequently used drugs in this population (Lehman, Myers, Corty, & Thompson, 1994). Moreover, a study by Salyers and Mueser (2001) comparing three groups (No/Low Alcohol, Alcohol Only, and Drug Use—with or without alcohol) of patients with schizophrenia on their persistence of substance use over time in the categories, found that those in the Drug Use group reported the greatest problems in interpersonal and family relationships. Moreover, individuals in the Drug Use group were rated as having the greatest amount of friction in their households compared with both the No/Low Alcohol and Alcohol Only groups. However, although it may seem counterintuitive, there is also evidence of better social functioning associated with substance misuse in individuals with schizophrenia (Tsuang, Simpson, & Kronfol, 1982). This could be because a certain level of social skills is needed to acquire and use drugs and alcohol (Cohen & Klein, 1970) and/or because drugs and alcohol facilitate certain aspects of social functioning. In short, scientific literature reveals an emerging and complex pattern in the relationship between social adjustment and substance use in persons with schizophrenia. While premorbid substance use tends to be associated with better social and leisure functioning, individuals with consistent drug and alcohol use are more likely to have interpersonal difficulties, mainly within their families.

Social Exclusion

Social exclusion and rejection are unfortunately common in both individuals with schizophrenia and those with substance use disorders, at least partly due to negative stereotypes about them, such as propensity for violence (Phelan, Link, Stueve, & Pescosolido, 2000). These misconceptions have major ramifications on the early identification, diagnosis, and treatment of psychosis. Moreover, a comorbid substance use disorder among individuals with schizophrenia increases their risk for homelessness (Caton et al., 1994), which in turns predisposes them to institutionalization, incarceration (Appleby & Desai, 1987), and earlier mortality (Hibbs, Benner, & Klugman, 1994). This compounding of adverse life events in vulnerable populations has been termed structural violence (Kelly, 2005), and clearly impels the eradication of stigma toward both mental illnesses and drug use, as well as the ending of discrimination against those with these disorders.

Aggression and Violence

There is a multidirectional relationship between aggression and violence, cannabis use, and psychotic disorders. On the one hand, compared with people without mental disorders, those with schizophrenia and other psychotic disorders are more likely to be victims of both threatened and actual physical assaults, while people with cannabis use disorders experience more attempted physical assaults (Silver, Arseneault, Langley, Caspi, & Moffitt, 2005). Therefore, individuals with both psychosis and cannabis use should be regarded as more vulnerable to aggression and violence. On the other hand, several studies have shown that people with mental illnesses engage in violent acts more often than those without, particularly when their disorders involve alcohol or drug use. This section focuses on the evidence related to drug use, especially cannabis use, among those with psychotic disorders who commit aggressive or violent acts against others. However, it is important not to ignore victimization among persons with mental illnesses, or to reinforce the belief that those with mental illnesses are inherently dangerous.

Contrary to the widely held belief that cannabis is a harmless drug that may even be beneficial for relaxation, its use has been associated with violent behavioral patterns among the general population, not only among those with schizophrenia or related psychotic disorders (Niveau & Dang, 2003). Indeed, substance use and misuse has been associated with homicide by offenders without mental illnesses (Shaw et al., 2006), as well as with homicides in the context of schizophrenia (Joyal, Putkonen, Paavola, & Tiihonen, 2004). In addition, and contrary to the thinking from decades ago (Coid, 1983), the rates of homicide committed by people diagnosed with schizophrenia correlate strongly with total homicide rates ($r=0.86$) (Large et al., 2014). It therefore seems that violent acts by both members of the general population and those with schizophrenia and other psychotic disorders have certain common etiological factors, such as social disadvantage, access to weapons, and substance use. Thus, although schizophrenia alone has been shown to increase the risk of engaging in violence to some degree, this risk is significantly magnified by drug use and misuse (Wallace, Mullen, & Burgess, 2004). There is evidence, for example, that a high percentage of people with psychotic disorders who committed homicide also had a substance use disorder, very often involving cannabis (Appleby, Kapur, & Shaw, 2006). Moreover, the high correlation between substance misuse and schizophrenia, when combined with the vulnerability of persons with psychosis to social factors that elevate risk of violence, may explain the increased likelihood of aggression in people with psychotic disorders. Thus, since comorbidity between psychosis and drug use could be particularly relevant to violence and homicide, early identification and treatment of both psychosis and substance use including cannabis use should be important public health and safety policy priorities.

Suicide

Schizophrenia and other psychotic disorders are strongly associated with risk of suicide (Inskip, Harris, & Barraclough, 1998), and drug misuse is among the better established risk factors for suicide in schizophrenia. According to a systematic review by Hawton, Sutton, Haw, Sinclair, and Deeks (2005), people with schizophrenia and drug misuse have a three-times-greater risk of suicide than

TABLE 2 Factors With Robust Evidence of Increased Risk Of Suicide in Schizophrenia (Hawton et al., 2005)

Risk Factor	OR (95% CI)
Fear of mental disintegration	12.1 (1.89–81.3)
Previous suicide attempts	4.09 (2.79–6.01)
Recent loss	4.03 (1.37–11.8)
Poor adherence to treatment	3.75 (2.20–6.37)
Drug misuse	3.21 (1.99–5.17)
Previous depressive disorders	3.03 (2.06–4.46)
Agitation or motor restlessness	2.61 (1.54–4.41)

those who do not use drugs, making drug use an important risk factor for suicide in individuals with schizophrenia (after fear of mental disintegration, recent loss, previous suicide attempts, and poor adherence to treatment) (Table 2). Since cannabis is the most common illicit drug used by those with psychotic disorders, it is implicated in suicide risk for this population. Surprisingly, alcohol use did not appear to be a risk factor for suicide in persons with schizophrenia, although it is clearly a major risk factor for suicide in the general population (Murphy, 2000). In sum, it is clear that suicide prevention among individuals with schizophrenia is likely to be achieved by improving adherence to antipsychotic treatment, effective management of affective symptoms, and special care of patients after losses, together with the identification of and attention to comorbid substance use disorders.

CONCLUSIONS AND RESEARCH NEEDS

Given the established relationship between cannabis use and schizophrenia in the scientific literature, there is an urgent need for research on effective strategies to prevent the use of this drug, mainly targeting young people and highlighting the possible consequences of cannabis use in precipitating and/or exacerbating psychotic symptoms. For those who have already developed a psychotic disorder, treatment interventions should encourage individuals to discontinue or at least reduce cannabis use; research is required to create or identify effective strategies to achieve this goal. Finally, research should be undertaken to develop and evaluate strategies for translating science into effective policies and evidence-based programs in healthcare systems around the world.

Key Chapter Points

- Cannabis is by far the most frequently used illicit drug among individuals with schizophrenia.
- Persons with schizophrenia or other psychotic disorders have various motivations for using cannabis, including the enhancement of positive feelings, relief of dysphoria, social pressure, and reasons related to the illness itself and side effects of medications.
- Among individuals with psychotic disorders, cannabis use can precipitate or exacerbate symptoms of psychosis, leading to relapses and an increased need for inpatient treatment, and eventually resulting in a range of poor clinical and psychosocial outcomes.
- Those with comorbid psychotic and cannabis use disorders should be encouraged to continue antipsychotic medication despite continued cannabis use.
- Social isolation among individuals with schizophrenia may create a vicious cycle by increasing the risk of being part of drug users' networks, which may lead to sustained substance use, in turn potentially contributing to interpersonal problems, rejection, and alienation.
- Cannabis use among persons with psychotic disorders may increase the risk for significant negative effects, including violence and suicide, providing a strong rationale for routinely incorporating evaluation and management of cannabis use—and all drug use—into standard clinical practice.

REFERENCES

Addington, J., & Duchak, V. (1997). Reasons for substance use in schizophrenia. *Acta Psychiatrica Scandinavica*, *96*, 329–333.

Ames, F. (1958). A clinical and metabolic study of acute intoxication with *Cannabis sativa* and its role in the model psychoses. *Journal of Mental Science*, *104*, 972–999.

Andreasson, S., Allebeck, P., Engström, A., & Rydberg, U. (1987). Cannabis and schizophrenia a longitudinal study of Swedish conscripts. *Lancet*, *2*, 1483–1486.

Appleby, L., & Desai, P. (1987). Residential instability: A perspective on system imbalance. *American Journal of Orthopsychiatry*, *57*, 515–524.

Appleby, L., Kapur, N., & Shaw, J. (2006). *Avoidable deaths: Five year report of the national confidential inquiry into suicide and homicide by people with mental illness*. www.bbmh.manchester.ac.uk/cmhs/research/centreforsuicideprevention/nci/reports/,ed.

Arseneault, L., Cannon, M., Poulton, R., Murray, R., Caspi, A., & Moffitt, T. (2002). Cannabis use in adolescence and risk for adult psychosis: Longitudinal prospective study. *BMJ*, *325*, 1212–1213.

Baeza, I., Graell, M., Moreno, D., Castro-Fornieles, J., Parellada, M., González-Pinto, A., et al. (2009). Cannabis use in children and adolescents with first episode psychosis: Influence on psychopathology and short-term outcome (CAFEPS study). *Schizophrenia Research*, *113*, 129–137.

Barbeito, S., Vega, P., Ruiz de Azúa, S., Saenz, M., Martinez-Cengotitabengoa, M., González-Ortega, I., et al. (2013). Cannabis use and involuntary admission may mediate long-term adherence in first-episode psychosis patients: A prospective longitudinal study. *BMC Psychiatry*, *13*, 1–9.

Boggs, D., Kelly, D., Liu, F., Linthicum, J., Turner, H., Schroeder, J., et al. (2013). Cannabis withdrawal in chronic cannabis users with schizophrenia. *Journal of Psychiatric Research*, *47*, 240–245.

Castle, D. (2013). Cannabis and psychosis: What causes what? *F1000 Medicine Reports*, *5*, 1.

Caton, C., Shrout, P., Eagle, P., Opler, L., Felix, A., & Domínguez, B. (1994). Risk factors for homelessness among schizophrenic men: A case control study. *American Journal of Public Health*, *84*, 265–270.

Cohen, M., & Klein, D. (1970). Drug abuse in a young psychiatric population. *American Journal of Orthopsychiatry*, *40*, 448–455.

Coid, J. (1983). The epidemiology of abnormal homicide and murder followed by suicide. *Psychological Medicine*, *13*, 855–860.

Colizzi, M., Carra, E., Fraietta, S., Lally, J., Quattrone, D., Bonaccorso, S., et al. (2016). Substance use, medication adherence and outcome one year following a first episode of psychosis. *Schizophrenia Research*, *170*, 311–317.

Crean, R., Crane, N., & Mason, B. (2011). An evidence based review of acute and long-term effects of cannabis use on executive cognitive functions. *Journal of Addiction Medicine*, *5*, 1–8.

Damjanović, A., Pantović, M., Damjanović, A., Dunjić-Kostić, B., Ivković, M., Milovanović, S., et al. (2015). Cannabis and psychosis revisited. *Psychiatria Danubina*, *27*, 97–100.

Davis, G., Compton, M., Wang, S., Levin, F., & Blanco, C. (2013). Association between cannabis use, psychosis, and schizotypal personality disorder: Findings from the National Epidemiologic Survey on Alcohol and Related Conditions. *Schizophrenia Research*, *151*, 197–202.

Degenhardt, L., & Hall, W. (2012). Extent of illicit drug use and dependence, and their contribution to the global burden of disease. *Lancet*, *379*, 55–70.

Dekker, N., Linszen, D., & De Haan, L. (2009). Reasons for cannabis use and effects of cannabis use as reported by patients with psychotic disorders. *Psychopathology*, *42*, 350–360.

Dixon, L., McNary, S., & Lehman, A. (1995). Substance abuse and family relationships of persons with severe mental illness. *American Journal of Psychiatry*, *152*, 456–458.

Faridi, K., Joober, R., & Malla, A. (2012). Medication adherence mediates the impact of sustained cannabis use on symptom levels in first-episode psychosis. *Schizophrenia Research*, *141*, 78–82.

Fowler, I., Carr, V., Carter, N., & Lewin, T. (1998). Patterns of current and lifetime substance use in schizophrenia. *Schizophrenia Bulletin*, *24*, 443–455.

González-Ortega, I., Alberich, S., Echeburúa, E., Aizpuru, F., Millán, E., Vieta, E., et al. (2015). Subclinical depressive symptoms and continued cannabis use: Predictors of negative outcomes in first episode psychosis. *PLoS One*, *10*, e0123707.

Green, B., Kavanagh, D., & Young, R. (2004). Reasons for cannabis use in men with and without psychosis. *Drug and Alcohol Review*, *23*, 445–453.

Green, B., Young, R., & Kavanagh, D. (2005). Cannabis use and misuse prevalence among people with psychosis. *British Journal of Psychiatry*, *187*, 306–313.

Hawton, K., Sutton, L., Haw, C., Sinclair, J., & Deeks, J. (2005). Schizophrenia and suicide: Systematic review of risk factors. *British Journal of Psychiatry*, *187*, 9–20.

Hibbs, J., Benner, L., & Klugman, L. (1994). Mortality in cohort of homeless adults in Philadelphia. *New England Journal of Medicine*, *331*, 304–309.

Inskip, H., Harris, E., & Barraclough, B. (1998). Lifetime risk of suicide for affective disorder, alcoholism and schizophrenia. *British Journal of Psychiatry*, *172*, 35–37.

Joyal, C., Putkonen, A., Paavola, P., & Tiihonen, J. (2004). Characteristics and circumstances of homicidal acts committed by offenders with schizophrenia. *Psychological Medicine*, *34*, 433–442.

Kelly, B. (2005). Structural violence and schizophrenia. *Social Science and Medicine*, *61*, 721–730.

Khantzian, E. (1997). The self-medication hypothesis of substance use disorders: A reconsideration and recent applications. *Harvard Review of Psychiatry*, *4*, 231–244.

Koskinen, J., Löhönen, J., Koponen, H., Isohanni, M., & Miettunen, J. (2010). Rate of cannabis use disorders in clinical samples of patients with schizophrenia: A meta-analysis. *Schizophrenia Bulletin*, *36*, 1115–1130.

Kuepper, R., van Os, J., Lieb, R., Wittchen, H., Höfler, M., & Henquet, C. (2011). Continued cannabis use and risk of incidence and persistence of psychotic symptoms: 10 year follow-up cohort study. *BMJ*, *342*, 1–8.

Lamb, H. (1982). Young adult chronic patients: The new drifters. *Hospital & Community Psychiatry*, *33*, 465–468.

Large, M., Mullin, K., Gupta, P., Harris, A., & Nielssen, O. (2014). Systematic meta-analysis of outcomes associated with psychosis and co-morbid substance use. *The Australian and New Zealand Journal of Psychiatry*, *48*, 418–432.

Le Bec, P., Fatséas, M., Denis, C., Lavie, E., & Auriacombe, M. (2009). Cannabis and psychosis: Search of a causal link through a critical and systematic review. *Encephale*, *35*, 377–385.

Lehman, A., Myers, P., Corty, E., & Thompson, J. (1994). Prevalence and patters of dual diagnosis among psychiatric inpatients. *Comprehensive Psychiatry*, *35*, 106–112.

Mead, A. (2014). International control of cannabis. In R. Pertwee (Ed.), *Handbook of cannabis* (pp. 44–64). Oxford: Oxford University Press.

Miettunen, J., Törmänen, S., Murray, G., Jones, P., Mäki, P., Ebeling, H., et al. (2008). Association of cannabis use with prodromal symptoms of psychosis in adolescence. *British Journal of Psychiatry*, *192*, 470–471.

Moreau de Tours, J. (1845). *Du Hachish et de L'aliènation Mentale. Ètudes Psychologiques*. Paris: Masson.

Murphy, G. (2000). Psychiatric aspects of suicidal behaviour: Substance abuse. In K. Hawton & K. Van Heeringen (Eds.), *The international handbook of suicide and attempted suicide* (pp. 135–146). Chichester: Wiley.

Myles, H., Myles, N., & Large, M. (2016). Cannabis use in first episode psychosis: Meta-analysis of prevalence, and the time course of initiation and continued use. *The Australian and New Zealand Journal of Psychiatry*, *50*, 208–219.

Niveau, G., & Dang, C. (2003). Cannabis and violent crime. *Medicine, Science and the Law*, *43*, 115–121.

Phelan, J., Link, B., Stueve, A., & Pescosolido, B. (2000). Public conceptions of mental illness in 1950 and 1996: What is mental illness and is it to be feared? *Journal of Health and Social Behavior*, *41*, 188–207.

Power, B., Dragovic, M., Badcock, J., Morgan, V., Castle, D., Jablensky, A., et al. (2015). No additive effect of cannabis on cognition in schizophrenia. *Schizophrenia Research*, *168*, 245–251.

Radhakrishnan, R., Wilkinson, S., & D'Souza, D. (2014). Gone to pot—A review of the association between cannabis and psychosis. *Frontiers in Psychiatry*, *5*, 54.

Rovai, L., Maremmani, A., Pacini, M., Pani, P., Rugani, F., Lamanna, F., et al. (2013). Negative dimension in psychiatry. Amotivational syndrome as a paradigm of negative symptoms in substance abuse. *Rivista di Psichiatria*, *48*, 1–9.

Salyers, M., & Mueser, K. (2001). Social functioning, psychopathology and medication side effects in relation to substance use and abuse in schizophrenia. *Schizophrenia Research*, *48*, 109–123.

Schofield, D., Tennant, C., Nash, L., Degenhardt, L., Cornish, A., Hobbs, C., et al. (2006). Reasons for cannabis use in psychosis. *Australian & New Zealand Journal of Psychiatry*, *40*, 570–574.

Shaw, J., Hunt, I., Flynn, S., Amos, T., Meehan, J., Robinson, J., et al. (2006). The role of alcohol and drugs in homicides in England and Wales. *Addiction*, *101*, 1117–1124.

Silver, E., Arseneault, L., Langley, J., Caspi, A., & Moffitt, T. (2005). Mental disorder and violent victimization in a total birth cohort. *American Journal of Public Health*, *95*, 2015–2021.

Stefanis, N., Delespaul, P., Henquet, C., Bakoula, C., Stefanis, C., & Van Os, J. (2004). Early adolescent cannabis exposure and positive and negative dimensions of psychosis. *Addiction*, *10*, 1333–1341.

Stroup, S., Lieberman, J., McEvoy, J., Davis, S., Swartz, M., Keefe, R., et al. (2009). Results of phase 3 of the CATIE schizophrenia trial. *Schizophrenia Research*, *107*, 1–12.

Tsuang, M., Simpson, J., & Kronfol, Z. (1982). Subtypes of drug abuse with psychosis. *Archives of General Psychiatry*, *39*, 141–147.

United Nations Office on Drugs and Crime. (2015). *World Drug Report 2015*. Vienna: United Nations Publication.

van Winkel, R., & GROUP Investigators. (2015). Further evidence that cannabis moderates familial correlation of psychosis-related experiences. *PLoS One*, *10*, e0137625.

Wade, M., Tai, S., Awenat, Y., & Haddock, G. (2017). A systematic review of service-user reasons for adherence and nonadherence to neuroleptic medication in psychosis. *Clinical Psychology Review*, *51*, 75–95.

Wallace, C., Mullen, P., & Burgess, P. (2004). Criminal offending in schizophrenia over a 25-year period marked by deinstitutionalization and increasing prevalence of comorbid substance use disorders. *American Journal of Psychiatry*, *161*, 716–727.

Zammit, S., Allebeck, P., Andreasson, S., Lundberg, I., & Lewis, G. (2002). Self reported cannabis use as a risk factor for schizophrenia in Swedish conscripts of 1969: Historical cohort study. *BMJ*, *325*, 1199.

Chapter 13

The Treatment of Cannabis Use Disorder Among Individuals With a Psychotic Disorder

Peter Bosanac*, Ana Lusicic†, David J. Castle*
**St Vincent's Hospital, Melbourne, VIC, Australia*
†Orygen Youth Health, Melbourne, VIC, Australia

INTRODUCTION

Cannabis, one of the most commonly used illicit drugs in the world, is often a complicating factor in managing psychotic disorders, and it is well known that, irrespective of extent of use, it can markedly and adversely affect the course of illness (Wynne & Castle, 2012). Moreover, cannabis use has been deemed to be the most preventable risk factor in the development of psychotic disorders (UK Schizophrenia Commission, 2012). Neurobiological mechanisms that underlie the effects of cannabis, as described in other chapters, are linked to the endocannabinoid system, which plays a key role in the modulation of cognition, reward processing (impaired in schizophrenia and related disorders), emotion, and sleep. This understanding is highly relevant to informing treatments for cannabis use disorder generally and in individuals with a psychotic disorder in particular.

Treatment approaches to cannabis use disorder among those with psychotic disorders require a specific focus on screening, assessment (including the nature, extent, and reasons for use, as well as readiness to change), and developing models of care—particularly integrated treatment, involving concurrent mental health and addiction interventions—to facilitate evidence-based treatments (NICE, 2015). Notwithstanding, there is only a limited psychological and pharmacological evidence base from which to recommend specific interventions in this clinical population (Wynne & Castle, 2012). This chapter provides an overview of treatment approaches, including models of care and screening, as well as psychological and pharmacological strategies.

The Complex Connection between Cannabis and Schizophrenia. http://dx.doi.org/10.1016/B978-0-12-804791-0.00013-6

ASSESSMENT

The management of cannabis use among individuals with psychotic disorders requires comprehensive direct and collateral assessment (obtaining consent to speak with family, caregivers, or significant others), as well as education about the potential harms of cannabis use in this context. Aspects of the assessment of cannabis use among individuals with a psychotic disorder are summarized in Table 1. Assessment includes screening for and identification of cannabis use, which is often underreported, as well as evaluation of the nature, extent, and reasons for cannabis (and other substance) use. This must be done in tandem with assessment of the individual's readiness to change. Readiness to change can be evaluated via a transtheoretical model approach, assessing precontemplation (no intention to change cannabis use in the next 6 months), contemplation (future change intended within 1–6 months), preparation for action (imminent change intended, within 1 month), action (change in cannabis use for less than 6 months), and maintenance (change in cannabis use for at least 6 months). An example of a tool to assess readiness for change, and in turn align with motivational interviewing (MI) approaches, is the University of Rhode Island Change Assessment (URICA) scale (McConnaughy, Prochaska, & Velicer, 1983).

The initial assessment requires a direct history from the patient about psychotic symptoms and cannabis use, as well as a longitudinal history that includes exacerbating and mitigating factors. Positive psychotic symptoms include: firmly held beliefs that are based on an abnormal inference of external

TABLE 1 Aspects of the Assessment of Cannabis Use Among Individuals With a Psychotic Disorder

- History of psychotic symptoms and cannabis use, including longitudinal history of concurrent course, circumstances, exacerbating and mitigating factors, as well as available supports
- History of use, previous management, impact of cannabis use on psychotic and other symptoms
- Screening for dependence (e.g., using the Severity of Dependence Scale)
- Readiness to change
- Collateral history and urine or serum toxicology to optimize the veracity of history
- Clinical and social correlates, including cigarette smoking, alcohol use, homelessness, legal and financial problems, unresponsiveness to conventional treatment, and frequency of relapses
- Mental status examination for psychotic phenomena (delusions, disorganized thoughts, hallucinations), as well as ideation about self-harm and associated immediacy
- Features of cannabis intoxication (e.g., lethargy, impaired motor coordination, slurred speech, and postural hypotension)

reality, that are not modified by evidence to the contrary, and are not consistent with the person's cultural and educational background (delusions); and perceptual experiences involving any of the sensory modalities, but not occurring externally to the mind (hallucinations). Negative symptoms of schizophrenia encompass impaired emotional engagement (anhedonia), reduced speech (alogia), and diminished motivation (avolition), among others.

In persons with psychotic disorders, it is highly informative to understand the interaction between substance use and specific symptoms. Specifically, understanding the individual's reasons for using cannabis requires exploration of: enhancement of positive mood and intoxication; tolerating negative emotions; social and interpersonal mediators such as conformity and acceptance; dampening of psychotic symptoms or medication side effects; and memories, thoughts, and perceptions influencing expectations about the direct and indirect effects of cannabis use. Individuals with psychotic disorders tend to endorse alleviation of negative affect as the predominant reason for use (see Spencer, Castle, & Michie, 2002). This is important, as understanding reasons for use can aid clinical engagement and inform treatment. Mental status examination exploring positive and negative symptoms, as well as features of cannabis intoxication (e.g., lethargy, impaired motor coordination, slurred speech, and postural hypotension) and cannabis withdrawal (e.g., sleep disturbance, fatigue, cravings, anorexia, depressive and anxiety symptoms, irritability, restlessness, and agitation) is essential, as well as ideation about self-harm or suicide (Alvarez-Jimenez et al., 2012; Potvin, Stip, & Roy, 2003) (which can vary as circumstances or patterns of cannabis use change), potential for violence (Abram & Teplin, 1991), and homicidal ideation.

The longitudinal assessment and management of risk to self and/or others in individuals with a psychotic disorder and comorbid cannabis use are central components of care. An individual's symptoms associated with previous self-harm or a suicide attempt can serve as a guide to assess the level of suicide risk in their specific cases. This is in tandem with establishing maximal rapport via an empathic and nonjudgmental approach and the utilization of initial open-ended questions. If the individual being assessed is not able to answer questions directly or cooperate with the assessment, then correlates of undisclosed suicidal ideation or a plan include anger, despair, despondence, agitation, and guardedness. Clinical judgment alone is not sufficient to predict immediate risk, and there are no evidence-based risk factors that are unequivocally predictive, particularly in the absence of collateral history (BMJ, 2012). Nonetheless, the presence of a suicide plan and immediate access to lethal means herald immediacy of risk. Immediate management in these circumstances may require invoking local Mental Health Act legislation and involvement of emergency services (e.g., police, ambulance) if there are threats of, or explorations of methods to enact, suicide (BMJ, 2012). In terms of violence, there are associations between aggression and positive symptoms of schizophrenia and substance use (Walsh, Buchanan, & Fahy, 2002); associated factors include poor impulse control, hostility, compulsory treatment for first-episode psychosis, impaired insight and poor treatment adherence, a forensic history, and past suicide attempts (Witt, van Dom, & Fazel, 2013).

MODELS OF CARE

In terms of optimal models of care, integrated treatment (concurrent treatment of psychotic and cannabis use disorders) to facilitate evidence-based treatments (NICE, 2015) appears to be more effective than serial (treatment of cannabis use disorder followed by a mental health intervention, or vice versa) or parallel (concurrent treatment by different services) approaches (NICE, 2015; Wynne & Castle, 2012). Integrated treatment emphasizes the importance of treating both substance use and the mental illness at the same time (Mueser et al., 2003; Wynne & Castle, 2012) and recognizes that responsibility for integrating service lies with the service provider rather than the patient (Drake, Mueser, Brunette, & McHugo, 2004, Drake, O'Neal, & Wallach, 2008). Moreover, integrated treatment has been well received in terms of patient satisfaction (Schulte et al., 2011). Yet, integrated programs, which are predominantly outpatient as opposed to residential-based, vary widely with regard to the type, number, duration, intensity, breadth, and setting of interventions (De Witte, Crunelle, Sabbe, Moggi, & Dom, 2014).

Integrated treatment usually encompasses, *inter alia*, motivational interviewing (MI), cognitive-behavioral therapy (CBT), relapse prevention, case management (including assertive community treatment when necessary), and family interventions. MI endeavors to increase motivation for change by focusing on supported decision-making (personal choice, responsibility, and consciousness) about the risks and advantages of continued cannabis use. CBT addresses antecedent automatic thoughts, cognitive schema, and ensuing emotional and behavioral responses that underpin and reinforce continued cannabis use in the setting of psychotic disorders, by identifying, challenging, and modifying distorted cognitions at the interface of comorbidity; and in turn developing/learning more adaptive skills to cope with stressors and specific problems. Case management entails access to and coordination of the continuum of available care. Family interventions, albeit diverse, aim to develop a family's (or support network's) knowledge about the nexus of cannabis dependence and the psychotic disorder, as well as optimizing communication between the patient, family members, and clinical services (De Witte et al., 2014).

Evaluation of the effectiveness of treatment and models of care for cannabis use in individuals with psychotic disorders requires the assessment of: medical comorbidity, relapse rates (gauged by symptomatic exacerbation and modification of clinical management, as well as inpatient care), substance misuse, general medical health, global and social functioning (e.g., vocational and housing), quality of life, satisfaction with care, and access to and engagement with mental health and other services (NICE, 2015). A systematic review involving 14 studies of the effectiveness of outpatient integrated treatment in this clinical population, despite the heterogeneity of study designs, found that behavioral treatment and specific interventions such as MI and family interventions, as well as integration of multiple interventions, were effective. There was no difference with the assertiveness of case management, suggesting that a lower caseload was not essential to better outcomes (De Witte et al., 2014).

In addition to an evidence-based integrated care plan, coordination of care is essential, so as to ensure that patients with cannabis use and psychotic disorders do not "fall through gaps" in the continuum of care; for example, at the time of discharge from acute or inpatient care to the community or, at the interface of specialist and primary care settings. Case coordination (with key and consistent personnel coordinating care and communication in regard to the individual with cannabis use and a psychotic disorder across treatment settings and services) is vital in optimizing engagement, improving overall health outcomes, and addressing risks (e.g., relapse of cannabis use and/or symptoms of psychosis, deliberate self-harm/suicidality).

The predominant increased risks of cannabis use to physical health include cardiac, respiratory, and vascular disorders, as well as various malignancies. In cannabis users, these risks arise largely from smoking cannabis, which is often done in the context of smoking tobacco (Hashibe et al., 2006; Tashkin, 2013; Thomas, Kloner, & Rezkalla, 2014). Starkly, three cannabis "joints" per day equate to the pulmonary impact of smoking 20 cigarettes in the same period. Thus, baseline and regular physical health/cardiometabolic monitoring is required in individuals with comorbid cannabis use and psychotic disorders, noting that schizophrenia and antipsychotic treatment by themselves carry heightened cardiometabolic risk and likelihood of premature death (De Hert et al., 2011).

SCREENING

Screening is fundamental in diagnosing comorbid psychotic and cannabis use disorders, and in facilitating evidence-based and individualized treatment, as undetected substance use can confuse interpretation of important signs and symptoms of psychosis, possibly leading to overreliance on medication (Drake, Altereman, & Rosenberg, 1993). Barriers to screening for cannabis use include an underappreciation of the prevalence and potential implications on psychosis, a lack of awareness regarding different approaches to screening and detection, and an absence of systematic screening processes within mental health services (Wynne & Castle, 2012).

Successful screening programs require staff education on evidence-based approaches to screening and treatment, routine inclusion of questions about past and current substance use within assessment documentation, and use of brief self-report screening instruments. Although common tools include the Alcohol Smoking and Substance Involvement Screening Test (ASSIST) and Drug Abuse Screening Test (DAST), there is no "gold standard" (Wynne & Castle, 2012).

The nature and extent of cannabis use in individuals with psychotic disorders can be evaluated with relatively brief tools such as the Addiction Severity Index (ASI), Severity of Dependence Scale (SDS), and Cannabis and Substance Use Assessment Schedule (CASUAS), rather than arduous diagnostic interview instruments. In addition, reasons for cannabis use can be further evaluated by instruments such as the Reasons for Substance Use in Schizophrenia scale (ReSUS) or the Substance Use Scale for Psychosis (SUSP), in tandem with readiness for change, which will assist in selecting the most appropriate and practicable management plan (for references, see Wynne & Castle, 2012).

Clinical correlates of cannabis use in individuals with psychotic disorders include cigarette smoking, alcohol use, homelessness, legal and financial problems, unresponsiveness to conventional treatment, and frequent relapses (Wynne & Castle, 2012). These broader medical and social parameters need to be appropriately addressed in a comprehensive package of care for individuals with psychotic disorders who also use cannabis.

PSYCHOLOGICAL INTERVENTIONS

A number of specific psychological interventions to reduce substance use have been tested among individuals with psychotic illnesses. In some of these interventions, cannabis has been the main or exclusive focus, while in others, cannabis has been included as one of a number of substances. Also, most randomized controlled trials (RCTs) of psychological treatments for substance use disorders in individuals with psychotic disorders have evaluated the outcomes of global substance use (Baker, Hides, & Lubman, 2010). It is fair to say that work in this area is notoriously difficult, not least in terms of patient engagement and retention. We provide here an overview of some of the larger and more influential studies in the literature.

A 12-month RCT involving 130 individuals with a psychotic disorder and comorbid substance use (61.3% used cannabis at least weekly) compared 10 weekly, individual sessions of manual-guided MI/CBT to treatment as usual. The MI/CBT intervention group showed a larger improvement in depressive symptomatology at 6 months, but both groups had similarly reduced substance use, with a nonsignificant differential impact on cannabis use (Baker et al., 2006). Another RCT randomized 47 patients with first-episode psychosis to 10 individual sessions of psychoeducation or to psychoeducation, MI, and CBT combined, but did not demonstrate significant differences in regard to severity of psychotic symptoms, cannabis use, or general functioning at the end of the intervention or at 6-month follow-up (Edwards et al., 2006). This study was limited by the lack of assessment of adherence and fidelity with the treatment model, as well as a low participation rate and categorical as opposed to continuous outcome measures.

An RCT of two manual-guided sessions of MI or a two-session manualized standard psychiatric interview of 44 individuals with psychotic disorders, in which cocaine or cannabis were the primary substances used by 57% and 29.5% of participants, respectively, found a 92.1% reduction in cannabis use and a 100% reduction in secondary substance use in the standard interview group, compared to no reduction in the MI group (Martino, Carroll, Nich, & Rounsaville, 2006). However, the study was limited by only one day of substance use being required in the 2 months preceding participation and non-blindedness of follow-up (Baker et al., 2010). A year-long RCT, including 62 participants with psychotic disorders and comorbid cannabis use (at least three cannabis "joints" per week in the preceding month), and assessing 4–6 sessions of individual MI (30–60 minutes in duration) based on written guidelines compared with treatment as usual, found a greater reduction in cannabis use in the MI group, albeit this was not sustained at 12 months (Bonsack et al., 2011).

An RCT of phase-specific psychological therapy in 110 individuals with cannabis use following a first episode of psychosis failed to demonstrate that brief (up to 12 sessions over 4.5 months) or extended (24 sessions over 9 months) psychological intervention (MI and CBT) in accordance with a treatment manual (phase 1, motivation building; phase 2, plan for change) was superior to standard care (intensive case management and crisis response) in terms of quantity and frequency of cannabis use (Barrowclough et al., 2014). A one-year RCT in 88 patients with affective and nonaffective psychotic disorders and comorbid cannabis use disorder found that a group-based psychological intervention, compared with treatment as usual, led to better quality of life, but there was no improvement in cannabis use, psychotic symptoms, global functioning, insight, or attitude to treatment. The psychological intervention consisted of anxiety management, MI, and CBT, delivered over six phases (entry, commitment, goal setting, challenges, relapse prevention and lifestyle, and maintenance), while treatment as usual involved regular review and antipsychotic medication (Madigan et al., 2013). A 6-month RCT of MI and CBT addressing cannabis-related problems in 103 patients with a psychotic disorder in conjunction with treatment as usual, compared with treatment as usual alone, did not lead to a reduction in the frequency of cannabis use, but possibly impacted the amount of cannabis used. The latter reduction occurred in subgroups within the study population, including those who were nonabstinent at baseline, younger, and unemployed (Hjorthoj, Fohlmann, Larsen, & Nordentoft, 2013). Limitations of the study included referral of patients rather than consecutively recruited ones (who may have been among the more willing to address their cannabis use) and the absence of assessment of readiness to change their substance use.

There has been minimal research evaluating family-based interventions in individuals with psychotic and comorbid cannabis use disorders. An RCT exploring this on 108 patients with serious mental illness, 83 of whom had schizophrenia and related disorders and substance use (alcohol and other unspecified substances) disorders, found that both brief (8–12 weeks) and longer-term (9–12 months) family education (problem solving and communication training) resulted in better mental health, substance use (not specified as cannabis use), and functional outcomes. But the retention of families was moderate, which illustrates the need for the modification of family interventions to more effectively engage families (Mueser et al., 2013).

In summary, psychological interventions of up to 3 months, including MI and CBT, appear to be efficacious in the reduction of cannabis use in at least some individuals with psychotic disorders, but not necessarily more effective than control comparators such as treatment as usual, self-help booklets, and psychoeducation. On the other hand, studies evaluating outcome of psychological interventions over longer periods demonstrate cannabis use reverting to that of preexisting levels, irrespective of the specific psychological intervention utilized, or its duration (Baker et al., 2002; Barrowclough et al., 2014; Edwards et al., 2006; Hjorthoj et al., 2013; Madigan et al., 2013).

Combination of Antipsychotic Medication and Psychological Interventions

A recent Cochrane systematic review of eight RCTs comprising 530 participants (McLoughlin et al., 2014) concluded that the concomitant treatment of psychotic symptoms with antipsychotic medication and psychological intervention may lead to reduced cannabis use. However, the small number of studies and participants preclude conclusive support for any specific antipsychotic medication in individuals with concurrent cannabis use and psychotic disorders. Also, MI/CBT for cannabis use with and without coexisting mental health problems has limited long-term effectiveness, and future interventions must aim for an improvement in the magnitude of therapeutic change. In addition, better designed RCTs with longer follow-up are required (Baker et al., 2010).

Aspects of the treatment of cannabis use disorder among individuals with a psychotic disorder are summarized in Table 2.

MEDICATIONS

Considerations of pharmacological treatment of cannabis use disorders followed the discovery of potential neurobiological targets and a better understanding of the neurocircuitry underpinning addiction. Acute effects of cannabis intoxication and withdrawal, as well as the neurobiological processes underlying development of compulsive use, are thought to be largely mediated by CB_1 receptors and the dopaminergic system, in addition to other neurotransmitter systems.

Clinically, acute cannabis intoxication can occur with a number of psychiatric and physical symptoms, including anxiety, psychosis, and impaired coordination and cognitive performance. Treatment is usually targeted at the associated psychiatric and medical manifestations (Crippa et al., 2012). Uncomplicated cannabis intoxication usually resolves within 4–6 hours, but occasionally requires symptomatic treatment, and/or the use of benzodiazepines (if the presence of significant anxiety necessitates this), while ensuring safety

TABLE 2 Aspects of the Treatment of Cannabis Use Disorder Among Individuals With a Psychotic Disorder

- Treatment of psychotic symptoms with antipsychotic medication (including clozapine, risperidone, olanzapine, and ziprasidone) and psychological intervention may lead to reduced cannabis use
- Small number of studies and participants preclude conclusive support for greater efficacy of any specific antipsychotic medication in individuals with concurrent cannabis use and psychotic disorders
- MI (increasing motivation for change by focusing on supported decision-making—personal choice, responsibility, and consciousness—about the risks and advantages of continued cannabis use)
- CBT (addressing antecedent automatic thoughts, cognitive schema, and ensuing emotional and behavioral responses that underpin and reinforce continued cannabis use in the setting of psychotic disorders; identifying, challenging, and modifying distorted cognitions; and developing/learning more adaptive skills to cope with stressors and specific problems)

during the process of detoxification. Cannabis-precipitated psychosis is one of the potential complications of acute intoxication (see Chapter 8). Treatment usually follows conventional guidelines for acute psychosis, and there is scant data to guide the specific choice of agent. In one of the few double-blind, randomized studies specifically of cannabis-precipitated psychosis (Berk, Brook, & Trandafir, 1999), olanzapine and haloperidol were compared without finding significant outcome differences between groups on the Brief Psychiatric Rating Scale (BPRS), Clinical Global Impression (CGI) severity scale, or CGI improvement scale. However, patients in the haloperidol group experienced significantly more extrapyramidal symptoms. It is well known that the acute effects of cannabis can be associated with an exacerbation and persistence of psychotic symptoms in individuals with established psychotic disorders (Grech, Van Os, Jones, Lewis, & Murray, 2005; Kuepper et al., 2011). Treatment usually follows established guidelines for the management of the underlying disorder, albeit there is evidence that some antipsychotics (notably clozapine) have a greater propensity to ameliorate substance use and thus may be preferred in cannabis users with psychosis (see below). But of course, any such decisions require a careful individualized weighing of risks and benefits.

A withdrawal syndrome in cannabis-dependent patients usually evolves during the first week of abstinence, with resolution within a few weeks. Management is symptomatic and based on clinical monitoring and symptom control. In individuals with an established psychotic disorder, withdrawal symptoms are roughly comparable to symptoms experienced in nonpsychotic individuals (Boggs et al., 2013), but can be associated with significant behavioral disturbance and might mimic the core psychotic symptoms; this creates a clinical challenge. Thus, withdrawal symptoms can be mistaken for psychotic relapse and inappropriately treated. A cautious approach is required in this particular patient group regarding difficult pharmacological decisions in the short- and long-term management of cannabis dependence (Koola et al., 2013).

A number of studies have shown that cannabis use disorder can affect treatment effectiveness, adherence to treatment, and outcomes in patients with psychotic disorders; thus, identifying effective pharmacological strategies is of paramount importance. There is a paucity of research exploring pharmacological options in individuals with cannabis use disorder, especially comorbid with psychotic disorders. Currently most of the insights into potential choice of medication in managing cannabis dependence come from understanding of the neurobiological effects of cannabis on the brain, and several treatment studies focusing solely on the management of cannabis use. We provide here a brief overview of pharmacological agents that have been studied in individuals with cannabis use disorder. Where relevant, we make specific reference to patients with a psychotic disorder.

CB_1 Receptor Agonists

Levin et al. (2011) have performed a double-blind, placebo-controlled, 12-week study of dronabinol, a synthetic CB_1 receptor agonist, for the treatment of cannabis dependence, and showed marginal improvement in treatment retention and withdrawal symptoms. Use of dronabinol in schizophrenia is controversial, though there have been case reports of dronabinol improving psychotic symptoms in patients with schizophrenia who historically experienced improvement when smoking cannabis (Schwarcz, Karajgi, & McCarthy, 2009). Cannabinoid replacement therapy is a novel approach to the treatment of cannabis dependence (Allsop et al., 2014) that shows promising results; however, its use in psychotic patients has not yet been studied and remains controversial.

CB_1 Receptor Antagonists

Rimonabant, a CB_1 receptor antagonist, has been shown to attenuate the acute physiological effects of cannabis use in male smokers (Huestis, Boyd, Heishman, & Preston, 2007). We are, however, aware of no studies evaluating the effects of rimonabant in individuals with cannabis dependence and a psychotic disorder, and the psychiatric side effects of depressed mood and anxiety encountered with rimonabant use has led to it being withdrawn from markets worldwide (Christensen, Kristensen, Bartels, Bliddal, & Astrup, 2007). However, other CB_1 receptor antagonists are in development and might have a role to play in the management of cannabis dependence.

Opioid Antagonists

The endogenous opioid system has been implicated in the development of substance dependence and is known to interact with the endocannabinoid system. Animal studies have indicated that opioid antagonists block the acute effects of cannabis use (Haney, 2007). An exploratory investigation using pretreatment with naltrexone followed by cannabis use observed that naltrexone potentiated the acute effects of cannabis (Cooper & Haney, 2010). However, a follow-up study (Haney, Ramesh, Glass, Pavlicova, & Cooper, 2015) showed that maintenance treatment with naltrexone reduced the intake of cannabis, indicating complex interactions between the opioid and endocannabinoid systems. Data on the treatment potential of naltrexone in cannabis dependence remains scarce and further research is required. To date, there are no published trials on naltrexone maintenance treatment in individuals with a psychotic disorder and comorbid cannabis use.

Dopamine Agonists

Entacapone, a catechol-*O*-methyltransferase (COMT) inhibitor that increases the synaptic availability of dopamine, and in turn, reduces unpleasant effects of cravings and potentially drug-seeking behavior, has been of interest in the treatment of cannabis use disorder. A small, open-label trial in cannabis-dependent individuals decreased cravings after 12 weeks of use (Shafa et al., 2009). Clearly, further studies are required, including specifically among those with a psychotic disorder.

Dopamine Antagonists/Antipsychotic Medications

The dopaminergic system is a key player in the development of addiction, mediated at least in part via effects on the mesolimbic reward circuitry. Thus, the use of dopamine receptor antagonists could reduce pleasurable effects of substance use, as well as inhibit neurobiological mechanisms associated with the generation of cravings and drug seeking. However, dopamine blockade can also be associated with an increase in the dysphoria and affective blunting that are known to drive substance use among individuals with psychotic disorders (Spencer et al., 2002). To further complicate matters, dopamine blockade (using haloperidol) does not eliminate all the psychotomimetic effects of cannabis (D'Souza et al., 2008).

Various antipsychotic agents, compatible with their diverse pharmacology (i.e., the varied receptor binding profiles of these agents), appear to have differing effects on cannabis craving and persistence of use. Machielsen et al. (2012) explored differences in cravings for cannabis in patients with nonaffective psychotic disorders receiving risperidone, olanzapine, and clozapine. Patients in the risperidone group reported higher cravings compared to patients receiving clozapine or olanzapine. There was no significant difference in craving scores between the olanzapine and clozapine groups. This finding might reflect pharmacodynamic variation across these agents in affinity to the dopamine D_2 receptor, D_2 receptor dissociation rate, and ratio of D_1/D_2 receptor occupancy. The greater muscarinic and histaminergic binding of olanzapine and clozapine, compared to risperidone, may also be relevant. However, not all findings in the field have been consistent, with van Nimwegen, de Haan, van Beveren, and Linszen (2008) showing no significant differences in cravings in a 6-week, double-blind, randomized trial of olanzapine and risperidone. Akerele and Levin (2007), in a 14-week, double-blind study comparing the efficacy of risperidone versus olanzapine on cannabis and cocaine users with psychotic disorders, observed higher cannabis cravings in the risperidone group compared to the olanzapine group in the last 6 weeks of the study. However, Sevy et al. (2011) concluded that olanzapine and risperidone had similar effects on the treatment of psychotic symptoms in patients with comorbid cannabis use disorder.

Clozapine has been an interesting agent in the treatment of dual diagnosis patients. Green, Burgess, Dawson, Zimmet, and Strous (2003) found that patients treated with clozapine were more likely to abstain from alcohol or cannabis use, compared to patients treated with risperidone over a one-year follow-up period. Brunette et al. (2011) followed 31 patients with schizophrenia, who were randomly switched to clozapine or remained on their usual antipsychotic medication. The clozapine group showed a reduction in cannabis use. A 12-month, randomized study comparing ziprasidone and clozapine showed reduced cannabis use in both groups. Clozapine treatment had a better effect on positive symptoms, but at the expense of more side effects and poorer adherence with treatment (Schnell, Koethe, Krasnianski, et al., 2014). The small number of participants and high dropout rates were limitations of the study. In a small fMRI (functional magnetic resonance imaging) study, Machielsen, Veltman, van den Brink, and de Haan (2014) explored the effects of clozapine and risperidone on brain activity: during a cannabis-word Stroop task (an objective assessment of attentional bias and distraction to salient drug-related stimuli) and correlations with subjective cravings. The clozapine group showed a larger reduction in cannabis cravings associated with a bilateral decrease in activation of the insula during the cannabis-word Stroop task, compared to the risperidone group. The interpretation of findings of the study was affected by the limited inclusion of participants with schizophrenia without cannabis use disorder.

Quetiapine has been of interest in the treatment of cannabis dependence for several reasons, not least of which is its psychopharmacology being in many ways similar to clozapine, as well as its established place as a "come down" drug among substance users. Mariani et al. (2014) assessed various doses of quetiapine treatment in cannabis-dependent patients during an 8-week, open-label pilot outpatient trial. Preliminary findings indicated a reduction in cannabis use, and that a dose of 300 mg was optimal. Interestingly, there are no published trials of quetiapine use in patients with psychotic disorders with comorbid cannabis use disorder. A small series of eight patients with psychotic and cannabis use disorders suggested benefit in augmenting usual antipsychotic treatment with quetiapine (Potvin, Stip, & Roy, 2004). The dopamine partial agonist, aripiprazole, has shown interesting effects on cravings in patients with alcohol, nicotine, cocaine, and amphetamine use (Brunetti et al., 2012), but we are aware of no studies of this agent in cannabis-dependent individuals with psychotic disorders, apart from a single study among persons deemed at "high risk" for a psychotic disorder, in whom there was no effect on cannabis consumption (Rolland, Geoffroy, Jardri, & Cottencin, 2013).

Long-Acting Injectable Antipsychotics

Long-acting antipsychotic agents, delivered by intramuscular injection every 2–4 weeks, are known to enhance adherence and reduce relapse in individuals with schizophrenia. They are often deployed clinically in people with substance use comorbidity. However, few rigorous clinical trials have been conducted in this clinical context. Rubio et al. (2006) studied the effects of zuclopenthixol and risperidone long-acting injectable agents on cocaine, alcohol, cannabis, and opiate users. Long-acting risperidone, at an average dose of 47.2 mg every 15 days with 3.4 mg oral augmentation, was superior to long-acting zuclopenthixol at 200 mg given every 3 weeks, in reducing positive urine samples. Again, further research is required, notably comparing the effects of oral and long-acting injectable antipsychotic agents on cannabis users with schizophrenia.

Mood Stabilizers

Lithium carbonate use in the management of cannabis use disorder has been considered, based on results from preclinical trials in which lithium was shown to inhibit the development of withdrawal symptoms in rats receiving synthetic cannabinoids (Cui et al., 2001). Winstock, Lea, and Copeland (2009), in their small, open-label study of 20 treatment-seeking participants without specified psychotic disorders, used 500 mg of lithium carbonate twice a day and reported a diminution in cannabis withdrawal symptoms. The authors are not aware of any studies on the effects of lithium on individuals with psychotic disorders specifically focused on cannabis use disorder. The anticonvulsant mood stabilizer, divalproex, was examined by Levin et al. (2004) in a randomized, placebo-controlled, double-blind study, but there was no difference between the groups in terms of effect on cannabis use.

Anxiolytics and Antidepressants

A recent Cochrane review of pharmacological treatments for cannabis use disorder included 14 RCTs involving medication versus placebo (Marshall, Gowing, Ali, & Le Foll, 2014). Active comparators to placebo were the selective serotonin reuptake inhibitors, fluoxetine and escitalopram; antidepressants with noradrenergic and serotonergic effects, including nefazodone, mirtazapine, and venlafaxine; and the dopamine and noradrenaline reuptake inhibitor, bupropion. Among other medications included in the review were: the previously mentioned mood stabilizers (divalproex sodium and lithium); the anticonvulsant, gabapentin; the $5HT_{1A}$ partial agonist anxiolytic, buspirone; the selective noradrenaline reuptake inhibitor, atomoxetine; and the glutamatergic modulator, N-acetylcysteine. The quality of the reviewed studies was limited, *inter alia*, by small sample sizes, inconsistency in study design, and the risk of attrition bias. Antidepressants reviewed in the study were found to be "probably of little value in treatment of cannabis dependence." The authors are not aware of these agents having been specifically employed in cannabis use associated with psychotic disorders.

CONCLUSIONS

The course of psychotic disorders may be negatively affected by comorbid cannabis use. Initial screening and evaluation of previous and concurrent cannabis use (and other substances) and recommendation of reduced intake/cessation in individuals with a psychotic disorder is paramount, in tandem with psychoeducation and delivery of comprehensive care, encompassing MI and CBT. While there are no specific psychotropic medications for cannabis use disorders *per se* in individuals with psychotic disorders, a number of second-generation antipsychotic medications may be beneficial.

Given the described deleterious impacts of cannabis use in this population, and the need for effective, evidence-based treatments, further evaluation of specific psychological, pharmacological, and integrated models of care (across primary and specialist care settings) is required in individuals with concurrent cannabis use and psychotic disorders. More specifically, research that evaluates the therapeutic and cost effectiveness of these approaches, including short-term interventions compared with longer interventions, as well as duration of treatment effects (Baker et al., 2010), is needed.

Key Chapter Points

- Cannabis use in individuals with psychotic disorders adversely affects symptoms, functioning, treatment adherence, and physical health, as well as the risk of suicide and violence.
- Individuals with psychotic disorders have a high rate of underreporting substance use, including on questionnaires, as compared with laboratory assays. Treatment approaches to cannabis use among individuals with psychotic disorders require specific focus on screening, assessment (including the nature, extent, and reasons for use, as well as readiness to change), as well as developing models of care, particularly integrated treatment.

- The management of cannabis use among individuals with psychotic disorders requires comprehensive direct assessment and collateral history, as well as education about the potential harms of cannabis use in this context, longitudinal assessment and management of risk to self and/or others, and monitoring of cardiometabolic risk factors.
- The concomitant treatment of psychotic symptoms with antipsychotic medication and psychological intervention may lead to reduced cannabis use, but the small number of studies and participants preclude conclusive support for any specific antipsychotic medication in individuals with concurrent cannabis use and a psychotic disorder. Although MI and CBT for cannabis use in individuals with psychotic disorders seem to have limited effectiveness in the short term, better designed RCTs with longer follow-up are required.
- Further evaluation of specific psychological, pharmacological, and integrated models of care (across primary and specialist care settings), as well as of differentially beneficial components of care for specific demographic groups, is required in individuals with concurrent cannabis use and psychotic disorders. More specifically, research is needed that evaluates the therapeutic and cost effectiveness of these approaches, including short- and long-term interventions, as well as duration of treatment effects.

REFERENCES

Abram, K. M., & Teplin, L. A. (1991). Co-occurring disorders among mentally ill jail detainees. Implications for public policy. *American Psychologist*, *46*(10), 1036–1045.

Akerele, E., & Levin, F. R. (2007). Comparison of olanzapine to risperidone in substance-abusing individuals with schizophrenia. *The American Journal on Addictions*, *16*(4), 260–268.

Allsop, D. J., Copeland, J., Linzteris, N., Montebello, M., Sadler, C., Rivas, G., et al. (2014). Nabiximols as an agonist replacement therapy during cannabis withdrawal: A randomized clinical trial. *JAMA Psychiatry*, *71*(3), 281–291.

Alvarez-Jimenez, M., Priede, A., Hetrick, S. E., Bendall, S., Killackey, E., Parker, A. G., et al. (2012). Risk factors for relapse following treatment for first episode psychosis: A systematic review and meta-analysis of longitudinal studies. *Schizophrenia Research*, *139*, 116–128.

Baker, A., Bucci, S., Lewin, T. J., Kay-Lambkin, F., Constable, P. M., & Carr, V. J. (2006). Cognitive-behavioural therapy for substance use disorders in people with psychotic disorders: Randomised controlled trial. *The British Journal of Psychiatry*, *188*(5), 439–448.

Baker, A., Hides, L., & Lubman, D. I. (2010). Treatment of cannabis use among people with psychotic or depressive disorders: A systematic review. *The Journal of Clinical Psychiatry*, *71*(3), 247–254.

Baker, A., Lewin, T., Reichler, H., Clancy, R., Carr, V., Garrett, R., et al. (2002). Motivational interviewing among psychiatric in-patients with substance use disorders. *Acta Psychiatrica Scandinavica*, *106*(3), 233–240.

Barrowclough, C., Marshall, M., Gregg, L., Fitzsimmons, M., Tommenson, B., Warburton, J., et al. (2014). A phase-specific psychological therapy for people with problematic cannabis use following a first episode of psychosis: A randomized controlled trial. *Psychological Medicine*, *44*, 2749–2761.

Berk, M., Brook, S., & Trandafir, A. I. (1999). A comparison of olanzapine with haloperidol in cannabis-induced psychotic disorder: A double-blind randomized controlled trial. *International Clinical Psychopharmacology*, *14*(3), 177–180.

Boggs, D. L., Kelly, D. L., Liu, F., Linthicum, J. A., Turner, H., Schroeder, J. R., et al. (2013). Cannabis withdrawal in chronic cannabis users with schizophrenia. *Journal of Psychiatric Research, 47*(2), 240–245.

Bonsack, C., Manetti, S. G., Favrod, J., Montagrin, Y., Besson, J., Bovet, P., et al. (2011). Motivational intervention to reduce cannabis use in young people with psychosis: A randomized controlled trial. *Psychotherapy and Psychosomatics, 80*, 287–297.

British Medical Journal. (2012). *Suicide risk management-diagnosis-step-by-step-best practice*http://bestpractice.bmj.com/best-practice/monograph/1016/diagnosis/step-by-step.html.

Brunette, M. F., Dawson, R., O'Keefe, M. A., Narasimhan, M., Noordsy, D. L., Wojcik, J., et al. (2011). A randomized trial of clozapine vs. other antipsychotics for cannabis use disorder in patients with schizophrenia. *Journal of Dual Diagnosis, 7*(1–2), 50–63.

Brunetti, M., Di Tizio, L., Dezi, S., Pozzi, G., Grandinetti, P., & Martinotti, G. (2012). Aripiprazole, alcohol and substance abuse: A review. *European Review for Medical and Pharmacological Sciences, 16*(10), 1346–1354.

Christensen, R., Kristensen, P. K., Bartels, E. M., Bliddal, H., & Astrup, A. (2007). Efficacy and safety of the weight-loss drug rimonabant: A meta-analysis of randomised trials. *Lancet, 370*(9600), 1706–1713.

Cooper, Z. D., & Haney, M. (2010). Opioid antagonism enhances marijuana's effects in heavy marijuana smokers. *Psychopharmacology, 211*(2), 141–148.

Crippa, J. A., Chagas, M. H., Atakan, Z., Martin-Santos, R., Zuardi, A. W., & Hallak, J. E. (2012). Pharmacological interventions in the treatment of the acute effects of cannabis: A systematic review of literature. *Harm Reduction Journal, 9*, 7.

Cui, S. S., Bowen, R. C., Gu, G.-B., Hannesson, D. K., Yu, P. H., & Zhang, X. (2001). Prevention of cannabinoid withdrawal syndrome by lithium: Involvement of oxytocinergic neuronal activation. *Journal of Neuroscience, 21*(24), 9867–9876.

D'Souza, D. C., Braley, G., Blaise, R., Vendetti, M., Oliver, S., Pittman, B., et al. (2008). Effects of haloperidol on the behavioral, subjective, cognitive, motor, and neuroendocrine effects of Δ-9-tetrahydrocannabinol in humans. *Psychopharmacology, 198*(4), 587–603.

De Hert, M., Vancampfort, D., Correll, C. U., Mercken, V., Peuskins, J., Sweers, K., et al. (2011). Guidelines for screening and monitoring of cardiometabolic risk in schizophrenia: Systematic evaluation. *The British Journal of Psychiatry, 199*(2), 99–105.

De Witte, N. A., Crunelle, C. L., Sabbe, B., Moggi, F., & Dom, G. (2014). Treatment for outpatients with comorbid schizophrenia and substance use disorders: A review. *European Addiction Research, 20*, 105–114.

Drake, R. E., Altereman, A. L., & Rosenberg, S. R. (1993). Detection of substance use disorders in severely mentally ill patients. *The Journal of Nervous and Mental Disease, 24*, 589–608.

Drake, R. E., Mueser, K. T., Brunette, M. F., & McHugo, G. J. (2004). A review of treatments for people with severe mental illness and co-occurring substance use. *Psychiatric Rehabilitation Journal, 27*, 360–374.

Drake, R. E., O'Neal, E. L., & Wallach, M. A. (2008). A systematic review of psychosocial interventions for people with co-occurring severe mental and substance use disorders. *Journal of Substance Abuse Treatment, 34*, 123–138.

Edwards, J., Elkins, K., Hinton, M., Harrigan, S. M., Donovan, K., Athanasopoulos, O., et al. (2006). Randomized controlled trial of a cannabis-focused intervention for young people with first-episode psychosis. *Acta Psychiatrica Scandinavica, 114*(2), 109–117.

Grech, A., Van Os, J., Jones, P. B., Lewis, S. B., & Murray, R. M. (2005). Cannabis use and outcome of recent onset psychosis. *European Psychiatry, 20*(4), 349–353.

Green, A. I., Burgess, E. S., Dawson, R., Zimmet, S. V., & Strous, R. D. (2003). Alcohol and cannabis use in schizophrenia: Effects of clozapine vs. risperidone. *Schizophrenia Research*, *60*(1), 81–85.

Haney, M. (2007). Opioid antagonism of cannabinoid effects: Differences between marijuana smokers and nonmarijuana smokers. *Neuropsychopharmacology*, *32*(6), 1391–1403.

Haney, M., Ramesh, D., Glass, A., Pavlicova, M., Bedi, G., & Cooper, Z. D. (2015). Naltrexone maintenance decreases cannabis self-administration and subjective effects in daily cannabis smokers. *Neuropsychopharmacology*, *40*(11), 2489–2498.

Hashibe, M., Morgenstern, H., Cui, Y., Tashkent, D. P., Zhang, Z. F., Cozen, W., et al. (2006). Marijuana use and the risk of lung and upper aerodigestive tract cancers: Results of a population-based case-control study. *Cancer Epidemiology, Biomarkers & Prevention*, *15*(10), 1829–1834.

Hjorthoj, C. R., Fohlmann, A., Larsen, A. M., & Nordentoft, M. (2013). Specialized psychosocial treatment plus treatment as usual (TAU) versus TAU for patients with cannabis use disorder and psychosis: The CapOpus randomized trial. *Psychological Medicine*, *43*, 1499–1510.

Huestis, M. A., Boyd, S. J., Heishman, S. J., & Preston, K. L. (2007). Single and multiple doses of rimonabant antagonize acute effects of smoked cannabis in male cannabis users. *Psychopharmacology (Berlin)*, *194*(4), 505–515.

Koola, M. M., Boggs, D. L., Kelly, D. L., Liu, F., Linthicum, J. A., Turner, H. E., et al. (2013). Relief of cannabis withdrawal symptoms and cannabis quitting strategies in people with schizophrenia. *Psychiatry Research*, *209*(3), 273–278.

Kuepper, R., van Os, J., Lieb, R., Wittchen, H., Höfler, M., & Henqet, C. (2011). Continued cannabis use and risk of incidence and persistence of psychotic symptoms: 10 year follow-up cohort study. *BMJ*, *342*, d738. http://dx.doi.org/10.1136/bmj.d738.

Levin, F. R., Mariani, J. J., Brooks, D. J., Pavlicova, M., Cheng, W., & Nunes, E. (2011). Dronabinol for the treatment of cannabis dependence: A randomized, double-blind, placebo-controlled trial. *Drug and Alcohol Dependence*, *116*(1–3), 142–150.

Levin, F. R., McDowell, D., Evans, S. M., Nunes, E., Akerele, E., Donovan, S., et al. (2004). Pharmacotherapy for marijuana dependence: A double-blind, placebo-controlled pilot study of divalproex sodium. *The American Journal on Addictions*, *13*, 21–32.

Machielsen, M., Beduin, A. S., Dekker, N.Genetic Risk and Outcome of Psychosis (GROUP) investigators, Kahn, R. S., Linszen, D. H., et al. (2012). Differences in craving for cannabis between schizophrenia patients using risperidone, olanzapine or clozapine. *Journal of Psychopharmacology*, *26*(1), 189–195.

Machielsen, M. W., Veltman, D. J., van den Brink, W., & de Haan, L. (2014). The effect of clozapine and risperidone on attentional bias in patients with schizophrenia and a cannabis use disorder: An fMRI study. *Journal of Psychopharmacology*, *28*(7), 633–642.

Madigan, K., Brennan, D., Lawlor, E., Turner, N., Kinsella, A., O'Connor, J. J., et al. (2013). A multi-center, randomized controlled trial of a group psychological intervention for psychosis with comorbid cannabis dependence over the early course of illness. *Schizophrenia Research*, *143*, 138–142.

Mariani, J. J., Pavlicova, M., Mamczur, A. K., Bisaga, A., Nunes, E., & Levin, F. (2014). Open-label pilot study of quetiapine treatment for cannabis dependence. *The American Journal of Drug and Alcohol Abuse*, *40*(4), 280–284.

Marshall, K., Gowing, L., Ali, R., & Le Foll, B. (2014). Pharmacotherapies for cannabis dependence. *Cochrane Database of Systematic Reviews*, *12*, CD008940.

Martino, S., Carroll, K. M., Nich, C., & Rounsaville, B. J. (2006). A randomized controlled pilot study of motivational interviewing for patients with psychotic and drug use disorders. *Addiction*, *101*(10), 1479–1492.

McConnaughy, E. N., Prochaska, J. O., & Velicer, W. F. (1983). Stages of change in psychotherapy: Measurement and sample profiles. *Psychotherapy: Theory, Research and Practice*, *20*, 368–375.

McLoughlin, B. C., Pushpa-Rajah, J. A., Gillies, D., Rathbone, J., Variend, H., Kalakouti, E., et al. (2014). Cannabis and schizophrenia. *Cochrane Database of Systematic Reviews*, *10*, CD004837.

Mueser, K. T., Noordsy, D. L., Drake, R. E., & Fox, L. (2003). *Integrated treatment for dual disorders: A guide to effective practice*. New York: Guilford Press.

Mueser, K. T., Glynn, S. M., Cather, C., Xie, H., Fox Smith, L., Clark, R. E., et al. (2013). A randomized controlled trial of family intervention for co-occurring substance use and severe psychiatric disorders. *Schizophrenia Bulletin*, *39*, 658–672.

National Institute for Health and Care Excellence (NICE). (2015). *Psychosis with substance misuse in over 14s: assessment and management*. https://www.nice.org.uk/guidance/cg120.

Potvin, S., Stip, E., & Roy, J. Y. (2003). Clozapine, quetiapine, and olanzapine among addicted schizophrenic patients: Towards testable hypotheses. *International Clinical Psychopharmacology*, *18*(3), 121–132.

Potvin, S., Stip, E., & Roy, J. Y. (2004). The effect of quetiapine on cannabis use in 8 psychosis patients with drug dependency. *Canadian Journal of Psychiatry*, *49*(10), 711.

Rolland, B., Geoffroy, P. A., Jardri, R., & Cottencin, O. (2013). Aripiprazole for treating cannabis-induced psychotic symptoms in ultrahigh-risk individuals. *Clinical Neuropharmacology*, *36*, 98–99.

Rubio, G., Martínez, I., Recio, A., Ponce, G., López-Muñoz, F., Alamo, C., et al. (2006). Risperidone versus zuclopenthixol in the treatment of schizophrenia with substance abuse comorbidity: A long-term randomized, controlled, crossover study. *The European Journal of Psychiatry*, *20*(3), 133–146.

Schizophrenia Commission. (2012). *The abandoned illness*. https://www.rethink.org/media/514093/TSC_main_report_14_nov.pdf.

Schnell, T., Koethe, D., Krasnianski, A., et al. (2014). Ziprasidone versus clozapine in the treatment of dually diagnosed (DD) patients with schizophrenia and cannabis use disorders: A randomized study. *American Journal on Addictions*, *23*, 308–312.

Schulte, S. J., Meier, P. S., & Stirling, J. (2011). Dual diagnosis client's treatment satisfaction—A systematic review. *BMC Psychiatry*, *11*, 64. http://dx.doi.org/10.1186/1471-244X-11-64.

Schwarcz, G., Karajgi, B., & McCarthy, R. (2009). Synthetic delta-9-tetrahydrocannabinol (dronabinol) can improve the symptoms of schizophrenia. *Journal of Clinical Psychopharmacology*, *29*(3), 255–258.

Sevy, S., Robinson, D. G., Sunday, S., Napolitano, B., Miller, R., McCormack, J., et al. (2011). Olanzapine vs. risperidone in patients with first-episode schizophrenia and a lifetime history of cannabis use disorders: 16-Week clinical and substance use outcomes. *Psychiatry Research*, *188*(3), 310–314.

Shafa, R., Abdolmaleky, H. M., Yaqubi, S., Smith, C., & Ghaemi, S. N. (2009). COMT—Inhibitors may be a promising tool in treatment of marijuana addiction trials. *American Journal on Addictions*, *18*, 322.

Spencer, D., Castle, D., & Michie, P. T. (2002). Motivations that maintain substance use among individuals with psychotic disorders. *Schizophrenia Bulletin*, *28*(2), 233–247.

Tashkin, D. P. (2013). Effects of marijuana smoking on the lung. *Annals of the American Thoracic Society*, *10*, 239–247.

Thomas, G., Kloner, R. A., & Rezkalla, S. (2014). Adverse cardiovascular, cerebrovascular, and peripheral vascular effects of marijuana inhalation: What cardiologists need to know. *The American Journal of Cardiology*, *113*, 187–190.

van Nimwegen, L. J., de Haan, L., van Beveren, N. J. M., & Linszen, D. H. (2008). Effect of olanzapine and risperidone on subjective well-being and craving for cannabis inpatients with schizophrenia or related disorders: A double-blind randomized controlled trial. *Canadian Journal of Psychiatry*, *53*(6), 400–405.

Walsh, E., Buchanan, A., & Fahy, T. (2002). Violence and schizophrenia: Examining the evidence. *The British Journal of Psychiatry*, *180*, 490–495.

Winstock, A. R., Lea, T., & Copeland, J. (2009). Lithium carbonate in the management of cannabis withdrawal in humans: An open-label study. *Journal of Psychopharmacology*, *23*(1), 84–93.

Witt, K., van Dom, R., & Fazel, S. (2013). Risk factors for violence in psychosis: Systematic review and meta-regression analysis of 110 studies. *PLoS ONE*, *8*(2), e55942. http://dx.doi.org/10.1371/journal.pone.0055942.

Wynne, J., & Castle, D. (2012). Addressing cannabis use in people with psychosis. In D. Castle, R. M. Murray, & C. D. D'Souza (Eds.), *Marijuana and madness*. (2nd ed.)(pp. 225–233). Cambridge: Cambridge University Press.

Chapter 14

Cannabidiol as a Potential Novel Therapeutic Agent for Psychotic Disorders

Serena Deiana*, Erica Zamberletti†

**Boehringer Ingelheim Pharma GmbH & Co. KG, Biberach an der Riss, Germany †University of Insubria, Busto Arsizio, Italy*

INTRODUCTION

Psychotic disorders represent a group of severe and heterogeneous mental diseases characterized by a complex and still unclear pathophysiology. Although continuous antipsychotic drug treatment is often thought to be fundamental to controlling the symptoms of these disorders, currently available therapies have several limitations, leading to negative outcomes such as lack of adherence and subsequent risk of early stage relapse. The search for safer and more effective antipsychotic drugs is therefore a main objective of current research. In recent years, increasing evidence has suggested a potential use for cannabidiol (CBD), the major nonpsychotropic constituent of the plant *Cannabis sativa*, in the treatment of psychotic disorders, and schizophrenia in particular. This chapter will describe available data supporting the emerging potential antipsychotic properties of CBD as assessed by preclinical models, clinical data, and epidemiological studies. Brief overviews of cannabis, the endocannabinoid system, and schizophrenia are also provided.

PSYCHOSIS, CANNABIS, AND ENDOCANNABINOIDS

Psychotic conditions refer to a range of mental disorders causing abnormal perception and thinking, and a loss of ability to assess reality. Hallucinations and delusions are two main symptoms, consisting of false auditory, tactile, and/or visual perceptions, and of fixed, false beliefs, respectively. Cognitive dysfunction, bizarre behavior, abnormal social interactions, and diminished performance in daily life activities are also common features of psychotic ailments, often preventing normal socio-occupational life functioning.

The Complex Connection between Cannabis and Schizophrenia. http://dx.doi.org/10.1016/B978-0-12-804791-0.00014-8

The *Diagnostic and Statistical Manual of Mental Disorders, Fifth Edition* (DSM-V, 2013) delineates psychotic disorders as "schizophrenia spectrum and other psychotic disorders," including schizophrenia, schizotypal personality disorder, brief psychotic disorder, delusional disorder, schizoaffective disorder, schizophreniform disorder, catatonia, substance/medication-induced psychotic disorder, and psychosis due to another medical condition.

The National Institute of Mental Health (NIMH) has recently launched a new approach to the diagnosis of mental disorders, including psychosis, called the Research Domain Criteria, or RDoC. Symptoms are believed to result from alterations in the physiological functioning of brain circuits that subserve complex human behaviors and processes. The RDoC structures these neurobehavioral processes into five distinct domains (cognitive systems, negative valence systems, positive valence systems, systems for social processes, and arousal/regulatory systems), which include a variable number of specific constructs that can be studied and quantified using "Units of Analysis" (Genes, Molecules, Cells, Circuits, Physiology, Behavior, Self-Report, Paradigms). Table 1 depicts such structure for the social processes domain. First attempts to study psychotic symptoms within the RDoC framework have been aimed (Ford et al., 2014) at drawing a path toward a more biological- and quantitative-based analysis of symptoms. Such work may allow for the identification of RDoC domains and neuronal substrates that are relevant to explain the typical symptoms of psychosis. Ford et al. (2014) questioned matters such as whether the neural basis of hallucinated and external sounds overlap and whether memory-related auditory hallucinations are governed by the same circuits supporting normal memory retrieval. With the advent of the RDoC, clinically observed psychotic symptoms can be integrated with quantitative analyses gained from basic research in genetics, molecular science, cognitive neuroscience, physiology, and behavioral science.

With an incidence of 10–22 per 100,000 and a prevalence of 0.3%–0.7% (McGrath, Saha, Chant, & Welham, 2008), schizophrenia is the most prevalent of all psychotic disorders. Although the etiology and pathophysiology of this debilitating mental illness remains elusive, dysregulation of the dopaminergic (DA), glutamatergic, gamma-aminobutyric acid (GABA)-ergic, and serotoninergic (5-HT) neurotransmitter systems have been shown to be involved in the development of the typical "positive" (hallucinations, delusions, disordered thinking, and paranoia), "negative" (affective flattening, alogia, avolition, apathy, anhedonia, and asociality), and "cognitive" (working memory, executive function, and sustained attention deficits) symptoms associated with schizophrenia (Abi-Dargham & Guillin, 2007).

Continuous pharmacological treatment is thought by many to be fundamental to controlling the symptoms of the disorder, but despite evolving and extensive research, the discovery of additional effective and safe medications remains a challenge. Indeed, the available treatments are often effective for the positive but not cognitive or negative symptoms; moreover, metabolic and neurological adverse effects represent still unresolved problems (Tandon, Keshavan, & Nasrallah, 2008).

TABLE 1 Example of the NIMH RDoC Matrix Representing the 5th Domain (Social Processes)

		Domain: Social Processes							
		Units of Analysis							
Construct/Subconstruct		Genes	Molecules	Cells	Circuits	Physiology	Behavior	Self-Report	Paradigms
1. Affiliation and attachment									
2. Social Communication	2.1 Reception of facial communication								
	2.2 Production of facial communication								
	2.3 Reception of nonfacial communication								
	2.4 Production of nonfacial communication								
3. Perception and understanding of self	3.1 Agency								
	3.2 Self-knowledge								
4. Perception and understanding of others	4.1 Animacy perception								
	4.2 Action perception								
	4.3 Understanding mental states								

Note: Rows indicate behavioral constructs and subconstructs while columns represent units of analysis.
Adapted from https://www.nimh.nih.gov/research-priorities/rdoc/constructs/rdoc-matrix.shtml.

One promising novel pharmacological target in the context of schizophrenia is the endocannabinoid system (Manseau & Goff, 2015). This neuromodulatory system, which is described in-depth in Chapter 3 of this book, consists of at least two types of receptors (CB_1 and CB_2). CB_2 receptors are predominantly expressed in peripheral regions and glial cells, while CB_1 receptors are mainly found throughout the central nervous system, with the highest concentrations observed in the basal ganglia, cerebellum, hippocampus, and cortex. The main endogenous cannabinoid ligands are anandamide (AEA) and 2-arachidonoyl glycerol (2-AG). Acting as retrograde messengers, they are synthesized and released postsynaptically and bind to presynaptic receptors, thus regulating the release of both inhibitory and excitatory neurotransmitters. Endocannabinoid synthesis occurs mainly on demand following increased intracellular Ca^{2+} at postsynaptic sites in response to sustained synaptic activity. After their reuptake, AEA is hydrolyzed by the enzyme fatty acid amide hydrolase (FAAH), while 2-AG is primarily metabolized by monoacylglycerol (MAG) lipase (Mechoulam & Parker, 2013). The endocannabinoid system is thought to regulate many neurological functions such as mood, memory, and reward processing.

The role of the endocannabinoid system in the pathophysiology of schizophrenia is supported by the distribution of cannabinoid receptors in brain areas that are shown to be dysfunctional in the disorder, such as the prefrontal cortex, basal ganglia, hippocampus, and anterior cingulate cortex. Moreover, the dysregulation of cannabinoid receptors and endocannabinoid levels in animal models of psychosis, as well as in individuals with schizophrenia, strongly supports the involvement of the endocannabinoid system in the pathophysiology of this disorder (Schubart et al., 2014; Zamberletti, Rubino, & Parolaro, 2012).

Further supporting the relationship between the endocannabinoid system and schizophrenia, epidemiological studies consistently report higher cannabis use and misuse among individuals with schizophrenia (Green, Young, & Kavanagh, 2005). Prospective studies have demonstrated an increased risk for schizophrenia in subjects who frequently use cannabis (Kuepper et al., 2011). Cannabinoids can produce transient, schizophrenia-like positive, negative, and cognitive symptoms in healthy individuals (Bhattacharyya et al., 2009), and can aggravate symptoms in psychotic subjects (D'Souza, Sewell, & Ranganathan, 2009). However, because only a small proportion of the general population exposed to cannabinoids develop a psychotic illness, it has been suggested that cannabis consumption interacts with other factors to cause psychotic disorders in vulnerable individuals (Ferdinand et al., 2005).

Of particular note is that young subjects with genetic susceptibility to schizophrenia are particularly vulnerable to the effects of cannabis (Hollis et al., 2008). A 27-year longitudinal study (Zammit, Allebeck, Andreasson, Lundberg, & Lewis, 2002) documented that in adolescence, the risk of developing schizophrenia is associated in a dose-dependent fashion with the use of cannabis, such that individuals using cannabis on more than 50 occasions showed a nearly seven times higher probability of developing schizophrenia compared to individuals who had never used cannabis. It has also been shown that occasional cannabis smokers have a 1.6-times higher chance of hospitalization for a psychotic episode, whereas heavy users are at 6.2-times greater risk (Schubart, Boks, et al., 2011), hence further corroborating the link between cannabis use and the development of psychosis. While research supports the idea that cannabis is pro-psychotic, it should be noted that not all cannabis users develop schizophrenia or psychotic symptoms, and not all people who develop schizophrenia have used cannabis.

As an alternative explanation of the association between cannabis use and psychosis, the self-medication hypothesis suggests that the use of cannabis might relieve negative symptoms and dysphoric states or counteract the side effects of antipsychotic therapies (Dixon, Haas, Weiden, Sweeney, & Frances, 1990; Krystal, D'Souza, Madonick, & Petrakis, 1999). A possible explanation for this discrepancy could be that different components of cannabis have been shown to exert opposing actions on schizophrenia-related symptoms. While Δ^9-tetrahydrocannabinol (THC) is responsible for the major psychotropic effects associated with cannabis consumption, CBD (which is devoid of psychotropic activity) has been found to counteract some of the effects of THC. For instance, the first report that CBD interferes with the psychotomimetic actions of THC was published in 1974 (Karniol, Shirakawa, Kasinski, Pfeferman, & Carlini, 1974), providing the first indication that CBD may have potential as an antipsychotic agent. It has also been reported that CBD counteracts THC-induced depersonalization, anxiety, perceptual disturbance, and disconnected thoughts (Campos, Moreira, Gomes, Del Bel, & Guimaraes, 2012). As a result, CBD pharmacology is currently under intense investigation; a summary of CBD receptor targets is depicted in Table 2.

TABLE 2 Receptor Targets of Cannabidiol

Receptor	Concentration (μM)	EC50/ IC50 (μM)	Assay
5-HT$_{1A}$	8–32 ↓	n.s.	[3H]-8-OH-DPAT ligand binding
	16 ↓	n.s.	[35S]-GTPγS
	16 ↓	n.s.	Forskolin
5-HT$_{2A}$	8–32 ↑	n.s.	[3H]-Ketanserin
CB$_1$	↔	>30	GTPγS
CB$_2$	↔	>30	GTPγS
Glycine receptor α1 subunit	1–300 ↑	12.3	Patch clamp/current with glycine
	1–300 ↑	132.4	Patch clamp/current without glycine

Continued

TABLE 2 Receptor Targets of Cannabidiol—cont'd

Receptor	Concentration (μM)	EC50/ IC50 (μM)	Assay
Glycine receptor α1β subunit	1–300 ↑	18.1	Patch clamp/current with glycine
	1–300 ↑	144.3	Patch clamp/current without glycine
Glycine receptor α3 subunit	0.01–50.00 ↑	3[a]	Patch clamp/current without glycine
GPR18	10^{-4}–100 ↑	51.1	p44/42; MAPK activation
GPR55	10^{-3}–1 ↓	0.45	Rho and ERK1/2 activation
	10^{-3}–10^{-2} ↓	n.s.	Contraction
nAChRα-7	0.1–100.0 ↓	11.3	Patch clamp/current/acetylcholine
Opioid (δ)	0.1–100.0 ↓[a]	10.7[a]	[3H]-NTI binding assay
Opioid (μ)	0.1–100.0 ↓[a]	10[a]	[3H]-DAMGO binding assay
PPARγ	3 ↑	n.s.	mRNA RT-PCR/Western blot
	3 ↑	n.s.	mRNA RT-PCR/Western blot

Note: Further details can be found in Ibeas Bih et al. (2015).
[3H]-3-OH-DPAT, 7-(dipropylamino)-5,6,7,8-tetrahydronaphthalen-1-ol; *[3H]-DAMGO*, D-Ala2, N-MePhe4, Glyol; *[3H]-NTI*, naltrindole; *5-HT*, serotonin; CB_1, cannabinoid receptor type 1; CB_2, cannabinoid receptor type 2; *ERK*, extracellular regulated kinase; *GPR*, G protein-coupled receptor; *GTPγS*, guanosine 5′-O-[gamma-thio]triphosphate; *MAPK*, mitogen-activated protein kinase; *n.s.*, not specified; *nAChR*, nicotinic acetylcholine receptor; *PPAR*, peroxisome proliferator-activated receptor; *RT-PCR*, reverse transcription polymerase chain reaction; (↑), stimulation; (↓), inhibition; ↔, no change.
aEstimated from plots in cited paper.

PRECLINICAL DATA

Of central importance for in vivo studies is the fact that the CBD molecule readily crosses the blood-brain barrier in rodents (Deiana et al., 2012). The pharmacokinetic details of CBD (120 mg/kg) have been reported in mouse and rat, when administered both orally (p.o.) and intraperitoneally (i.p.). Mouse brain C_{Max} values are 1.3 μg/g (p.o.) and 6.9 μg/g (i.p.), while plasma C_{Max} values are 2.2 (p.o.) μg/ml and 14.3 μg/ml (i.p.). Rat brain C_{Max} has been reported as 8.6 μg/ml (p.o.) and 6.8 μg/ml (i.p.), while reported plasma C_{Max} values have been 2 μg/ml (p.o.) and 2.6 μg/ml (i.p.).

ANIMAL MODELS OF SCHIZOPHRENIA

Schizophrenia is a uniquely human disorder. As such, it is impossible to mimic the entire spectrum of symptoms and biological markers in animals. However, some behavioral endophenotypes can be reproduced in rodents with some degree of construct, face, and predictive validity for schizophrenia. The literature on animal models of the positive symptoms of schizophrenia has focused on two main categories of behavior: disruption of prepulse inhibition (PPI) and locomotor hyperactivity.

PPI is a neurological response consisting of a physiological decrease in the acoustic startle response to an intense acoustic stimulus when it is immediately preceded by a lower intensity, nonstartling stimulus (Swerdlow, Geyer, & Braff, 2001). Individuals with schizophrenia show no or milder reduction of this startle response indicating a sensorimotor gating deficit associated with the disease. In the context of schizophrenia research, PPI assessment of sensorimotor gating deficits certainly represents a robust translational behavioral assay for studying antipsychotic drugs. In addition to PPI, locomotor hyperactivity, either at baseline or induced by psychostimulant administration, has been widely used to investigate psychotic-like behaviors and to assess the efficacy and side effects of antipsychotic drugs in animal models, mainly based on its predictive validity (Swerdlow et al., 2001).

Concerning the negative signs of schizophrenia, social interaction tasks represent the behavioral construct most widely used to preclinically model this category of symptoms. Social interaction in rodents can be easily assessed in an open field arena or using a three-chamber apparatus, where a test rodent encounters a stranger rodent and the amount of time spent exploring the unfamiliar congener is measured. As measures of social interaction in rodents are closely related to measures of social interaction in humans, these tasks represent useful behavioral paradigms for investigating aspects of social withdrawal and social cognition that are relevant to schizophrenia (Gururajan, Taylor, & Malone, 2010).

Finally, with reference to cognitive deficits, a series of behavioral paradigms have been developed in order to assess specific components of learning and memory in rodents (Salgado & Sandner, 2013). Procedures that have been used for developing drugs for schizophrenia include object and social recognition tasks, the Morris water maze, operant delayed alternation tasks, conditioned avoidance behaviors, and attentional set-shifting tasks.

All of these endophenotypes can be reproduced in mice and rats via pharmacological, genetic, or neurodevelopmental manipulations. Pharmacological models are mainly based on the acute or chronic administration of DA-ergic or glutamatergic agents in an attempt to mimic the DA-ergic and glutamatergic dysregulation observed in patients with schizophrenia. The majority of genetic models are derived from human genetic studies in psychotic patients. Moreover, since human epidemiological studies provide evidence that exposure to adverse environmental insults during gestation or the perinatal period increases the risk of developing schizophrenia, other animal models have been developed based on environmental manipulations or drug administration during these sensitive periods, in order to produce irreversible changes in brain development. In particular, animal models of prenatal immune activation appear to be particularly interesting in the context of recent schizophrenia research. Indeed, maternal immune activation during gestation via systemic administration of bacterial or viral agents in rodents has been associated with the development of several schizophrenia-like behaviors in the adult offspring (Meyer, 2013).

All of these models have been used to evaluate the antipsychotic potential of CBD, and the following paragraphs will describe the currently available data (Table 3).

Dopamine-Based Models

Acute CBD administration in a dose range of 15–480 mg/kg i.p. was effective in reversing stereotyped behaviors induced by the nonselective dopamine agonist apomorphine in rats (Zuardi, Rodrigues, & Cunha, 1991). In addition, CBD at doses of 15–60 mg/kg i.p. was able to counteract the hyperlocomotion induced by the indirect dopamine agonist amphetamine in mice (Moreira & Guimaraes, 2005). In both studies, CBD treatment did not induce catalepsy even at the highest doses tested. However, these findings were not replicated in a subsequent study by Long et al. (2010), in which acute CBD administration (1, 50 mg/kg i.p.) failed to counteract the hyperlocomotion induced by the catecholaminergic stimulant dexamphetamine in mice. As opposed to acute administration, a reversal of dexamphetamine-induced hyperactivity was observed after chronic CBD administration at a dose of 50 mg/kg (Long et al., 2010). More recently, acute treatment with CBD (15–60 mg/kg i.p.) was shown to reverse the PPI deficits induced by amphetamine administration in mice (Pedrazzi, Issy, Gomes, Guimaraes, & Del-Bel, 2015). The authors suggested that CBD's antipsychotic effect could be mediated via its direct action in the nucleus accumbens, as CBD infusion (60 nmol) into this brain region was sufficient for the attenuation of amphetamine-induced PPI disruption.

Glutamate-Based Models

Acute CBD administration (15–60 mg/kg i.p.) attenuated the hyperlocomotion induced by the noncompetitive NMDA (N-methyl-d-aspartate) receptor antagonist ketamine in mice (Moreira & Guimaraes, 2005). Interestingly, acute CBD treatment at the dose of 5 mg/kg i.p. also reversed PPI deficits induced by the noncompetitive NMDA receptor antagonist MK-801 in a manner similar to the atypical antipsychotic clozapine, without affecting the PPI responses in control mice *per se* (Long et al., 2006). However, Gururajan, Taylor, and Malone (2011) did not replicate these findings. Indeed, acute CBD treatment in a dose range of 3–30 mg/kg i.p. failed to reverse MK-801-induced hyperlocomotion and disruption of PPI in rats and, when administered alone, it disrupted PPI and produced hyperactivity in control animals, thus pointing to a potential pro-psychotic effect of CBD. Interestingly, these authors also demonstrated that in the same

TABLE 3 Preclinical Studies Testing Cannabidiol Effects on Animal Models of Psychosis

Species	Model	CBD Dose	CBD Regime	Outcome	Reference
Dopamine-based models					
Rats	Apomorphine-induced stereotypies	15–480 mg/kg	Acute	Improvement; no catalepsy	Zuardi et al. (1991)
Mice	Amphetamine-induced hyperlocomotion	15–60 mg/kg	Acute	Improvement; no catalepsy	Moreira and Guimaraes (2005)
Mice	Dexamphetamine-induced hyperlocomotion	1–50 mg/kg	Acute	No effect	Long et al. (2010)
Mice	Dexamphetamine-induced disruption of PPI	50 mg/kg	Chronic	Improvement	
Mice	Amphetamine-induced disruption of PPI	15–60 mg/kg	Acute	Improvement	Pedrazzi et al. (2015)
Glutamate-based models					
Mice	Ketamine-induced hyperlocomotion	15–60 mg/kg	Acute	Improvement; no catalepsy	Moreira and Guimaraes (2005)
Mice	MK-801-induced disruption of PPI	1–15 mg/kg	Acute	Improvement	Long et al. (2006)
Rats	MK-801-induced hyperlocomotion and PPI deficits	3–30 mg/kg	Acute	No improvement; CBD *per se* produced hyperactivity and disrupted PPI	Gururajan et al. (2011)
Rats	MK-801-induced social withdrawal	3–30 mg/kg	Acute	Improvement	
Rats	MK-801-induced hyperlocomotion and social withdrawal	3 mg/kg	Acute	Improvement	Gururajan et al. (2012)
Mice	MK-801-induced disruption of PPI	30–60 mg/kg	Chronic	Improvement	Gomes et al. (2015)

Continued

TABLE 3 Preclinical Studies Testing Cannabidiol Effects on Animal Models of Psychosis—cont'd

Species	Model	CBD Dose	CBD Regime	Outcome	Reference
Rats	MK-801-induced impairment in social recognition memory	12–30 mg/kg	Acute	No improvement; CBD *per se* impaired social recognition memory	Deiana et al. (2015)
Genetic and neurodevelopmental models					
Mice	Hyperlocomotion in Nrg1 TM HET mice	1–50–100 mg/kg	Acute/chronic	No effect	Long et al. (2012)
Mice	Disruption of PPI in Nrg1 TM HET mice	100 mg/kg	Acute	Improvement	
Mice	Social withdrawal in Nrg1 TM HET mice	50–100 mg/kg	Chronic	Improvement	
Rats	Deficit in contextual fear conditioning in Spontaneously Hypertensive Rats (SHR)	1 mg/kg	Acute	Improvement	Levin et al. (2012)
Rats	Hyperlocomotion and social withdrawal in Spontaneously Hypertensive Rats (SHR)	15–30–60 mg/kg	Acute	No effect	Almeida et al. (2013)
Rats	Disruption of PPI in Spontaneously Hypertensive Rats (SHR)	30 mg/kg	Acute	Improvement	Levin et al. (2014)
Rats	Prenatal poly I:C-induced hyperlocomotion	1 mg/kg	Chronic	Improvement	Peres et al. (2016)
Rats	Prenatal poly I:C-induced social withdrawal	1 mg/kg	Chronic	No effect	

Note: Psychotic endophenotypes can be reproduced in rodents via pharmacological (DA-ergic or glutamatergic intervention), genetic, or neurodevelopmental manipulations. This table summarizes in vivo studies assessing cannabidiol (CBD) effects on animal models of schizophrenia and psychosis. *Nrg1 TM HET*, transmembrane domain neuregulin 1 heterozygous mutant; *PPI*, prepulse inhibition.

animal model, CBD was able to reverse MK-801-induced social withdrawal, providing for the first time experimental evidence for a potential beneficial effect of CBD to treat the negative symptoms of schizophrenia. More recently, the same group demonstrated that CBD, at lower doses than the ones tested in their previous study (1 and 3 mg/kg i.p.), effectively inhibited MK-801-induced hyperlocomotion and social withdrawal when tested in an experimental setup that allowed for the simultaneous assessment of social and locomotor behaviors (Gururajan, Taylor, & Malone, 2012).

Of particular note, a recent article by Deiana et al. investigated the ability of acute CBD administration to specifically recover the cognitive deficits in an animal model of schizophrenia in rats (Deiana et al., 2012). In this study, aripiprazole, but not CBD (12–30 mg/kg i.p.), reversed impairments in social recognition memory induced by MK-801 administration. Furthermore, CBD alone impaired social recognition memory. The authors speculated that the lack of efficacy in this animal model might be due to CBD agonism at the 5-HT_{1A} receptor (Russo, Burnett, Hall, & Parker, 2005), as full 5-HT_{1A} agonists are also reported to impair memory (Carli, Luschi, Garofalo, & Samanin, 1995). Whether this mechanism of action also underlies social cognition has not been explored in the present experimental setup.

Most of the above-mentioned data refer to acute CBD administration. To date, only two studies have investigated the potential antipsychotic properties of chronic CBD administration in a glutamate-based animal model (Gomes et al., 2014, 2015). In these reports, the animals received daily injections of saline or MK-801 for 28 days and CBD or vehicle treatment began on the 6th day after the start of MK-801 administration; behavioral testing was performed 24 hours after discontinuation of treatment. Chronic treatment with CBD in a range of 30–60 mg/kg i.p. was effective in attenuating the PPI impairment as well as the impairments in social interaction and recognition memory induced by chronic MK-801 administration in mice. This suggests that chronic, but not acute, CBD application may be necessary for treating schizophrenia-like deficits.

Genetic and Neurodevelopmental Models

In addition to pharmacological models, there are also a number of genetic models in which CBD has been tested. Long et al. investigated the effects of acute and chronic CBD administration (1, 50, 100 mg/kg i.p.) in the transmembrane domain neuregulin 1 heterozygous mutant (Nrg1 TM HET) mouse (Long et al., 2012), a model of a schizophrenia susceptibility gene that offers partial construct, predictive, and face validity (Falls, 2003). Neither acute nor chronic CBD treatments were able to reverse hyperlocomotion in Nrg1 TM HET mice, whereas acute CBD selectively increased both PPI and the startle response in Nrg1 TM HET mice at the highest dose tested. More interestingly, chronic CBD treatment (50 and 100 mg/kg i.p.) completely reversed the deficit in sociability present in Nrg1 TM HET mice.

The potential antipsychotic actions of CBD were also assessed in the spontaneously hypertensive rat (SHR) strain, an animal model used to study several aspects of schizophrenia, including hyperlocomotion, social withdrawal, and abnormalities in sensorimotor gating (Levin et al., 2011). In this model, acute CBD administration at doses of 15, 30, and 60 mg/kg i.p. did not ameliorate the hyperlocomotion and the deficit in social interaction (Almeida et al., 2013), but administration of the dose of 30 mg/kg completely reversed PPI deficits (Levin et al., 2014). Moreover, using a contextual fear conditioning paradigm to study the abnormalities in emotion processing presumed to be related to the negative symptoms of schizophrenia (Calzavara et al., 2009), the same group demonstrated that acute CBD treatment (1 mg/kg i.p.) mitigated the emotional processing impairment present in the SHR strain (Levin et al., 2012).

Interestingly, a very recent study performed in a neurodevelopmental animal model of schizophrenia based on prenatal treatment with the synthetic viral analogue, polyinosinic:polycytidylic acid (poly I:C), in rats investigated for the first time the potential use of chronic CBD treatment as a preventive treatment (Peres et al., 2016). The authors found that chronic CBD (1 mg/kg i.p.) administration at postnatal days 30–60 prevented hyperlocomotion when tested at postnatal day 90. Although preliminary, these findings suggest that chronic CBD treatment could partially prevent the development of psychotic-like symptoms in early adulthood.

HUMAN STUDIES

The effects of CBD on in vivo models seem to point to its potential as an antipsychotic agent, and an increasing number of human studies have followed encouraging in vivo investigations in animal models (Schubart et al., 2014; Zuardi et al., 2012). CBD tolerability has previously been studied, with no side effects or toxicity observed in humans. CBD showed a Lethal Dose (LD50) of 212 mg/kg in rhesus monkeys (Rosenkrantz, Fleischman, & Grant, 1981), and it has been reported as being well tolerated in humans up to doses of 1500 mg/day (Bergamaschi, Queiroz, Zuardi, & Crippa, 2011). Moreover, it was shown to be nonmutagenic and nonteratogenic (Dalterio, Steger, Mayfield, & Bartke, 1984; Matsuyama & Fu, 1981).

Epidemiological Studies in Cannabis Users

Remarkably, what appears to be fundamental to the ability of cannabis to produce psychosis is the THC-to-CBD ratio. In 1982, a study performed in 20 psychotic subjects with high urinary cannabinoid levels showed a protective effect of CBD on THC-induced psychosis (Rottanburg, Robins, Ben-Arie, Teggin, & Elk, 1982). The authors suggested that South African cannabis strains, known to have a higher THC content and to be virtually devoid of CBD, were associated with the higher incidence of cannabis-induced psychosis. A number of studies have been conducted to determine the role of the concentrations of THC and

CBD in *C. sativa* strains in psychosis (Table 4). When 28 first-episode psychosis subjects were compared to healthy cannabis users, it was found that first-episode psychosis subjects smoked cannabis with higher THC concentrations for longer periods and at a higher frequency as compared to healthy individuals (Di Forti et al., 2009). The authors suggested that these results are in line with the hypothesis that THC is the active ingredient in cannabis that increases the risk of psychosis. High-potency cannabis or "skunk" is characterized by high concentrations of THC (about 15%), and low concentrations of CBD (<0.1%). By contrast, low-potency cannabis or hashish usually contains roughly 5% THC and about 4% CBD (Hardwick & King, 2008). Daily use of high-potency cannabis has been associated with a much earlier onset (6 years earlier) of psychosis as compared to nonusers (Di Forti et al., 2014). In addition, a case-control analysis of first-episode psychosis conducted between 2005 and 2011 in 410 patients and 370 population controls reported that the risk of psychosis has increased three-fold in users of high-potency cannabis compared to nonsmokers. Intriguingly, the use of low-potency cannabis showed no increased risk of psychosis compared to nonuse (Di Forti et al., 2015).

TABLE 4 Epidemiological Studies Focusing on the Role of CBD-to-THC Ratio in Smoked Cannabis

Subjects	Measurement	Outcome	Reference
Cannabis users vs healthy controls	Hair sample analysis, hippocampal volume and gray matter concentrations measured with VBM	Low CBD-to-THC ratio associated with low right hippocampal volume and low bilateral hippocampal gray matter concentrations	Demirakca et al. (2011)
First-episode psychosis patients	Cannabis Experiences Questionnaire	First-episode psychosis patients used higher potency cannabis more frequently and for a longer period than healthy controls	Di Forti et al. (2009)
First-episode psychosis patients	Cannabis Experiences Questionnaire	Low CBD-to-THC ratio associated with lower age of psychosis onset	Di Forti et al. (2014)
First-episode psychosis patients	Cannabis Experiences Questionnaire	Low CBD content increased the risk of psychosis compared to nonusers. High CBD content did not increase the risk of psychosis	Di Forti et al. (2015)

Continued

TABLE 4 Epidemiological Studies Focusing on the Role of CBD/THC Ratio in Smoked Cannabis—cont'd

Subjects	Measurement	Outcome	Reference
Cannabis users vs controls	Neuropsychological tests. Hair sample analysis, N-acetylaspartate/ total creatine ratio measured with proton magnetic resonance spectroscopy	CBD hair concentration was positively correlated with NAA/tCr ratio in putamen/ globus pallidus and with improved performance on the D_2 test for attention and concentration, but impaired performance on the Wisconsin Card Sorting Task	Hermann et al. (2007)
Cannabis users	Hair sample analysis, Oxford Liverpool Inventory of Life Experiences (OLIFE), Peter's Delusion Inventory (PDI)	THC associated with higher hallucinations and delusions, compared to CBD+THC and no cannabinoids	Morgan and Curran (2008)
Cannabis users	Prose recall, source memory, Psychotomimetic States Inventory (PSI) after smoking the usual cannabis	High CBD content was associated with absence of memory impairment. No effects of CBD on psychotic symptoms	Morgan et al. (2010)
Cannabis users	Hair sample analysis, prose recall, source memory, Schizotypal Personality Questionnaire (SPQ), Brief Psychiatric Rating Scale (BPRS)	High CBD content was associated with lower psychosis-like symptoms (only in recreational users with high THC). Better recognition memory in individuals with CBD in hair	Morgan et al. (2012)
Cannabis users	Community Assessment of Psychic Experiences (CAPE)	Low CBD content was associated with self-reported positive symptoms, but not negative symptoms or depression	Schubart, Sommer, et al., 2011

Note: CBD, Cannabidiol; *THC*, Δ9-tetrahydrocannabinol; *VBM*, voxel based morphometry.

Along these same lines of evidence, a study conducted by Morgan and Curran investigating cannabinoid content in hair samples from 140 cannabis users revealed that individuals with only THC in their hair presented higher levels of hallucinations and delusions than those whose hair revealed both THC and CBD or those with no hair cannabinoid content (Morgan & Curran, 2008). In a later study, the same authors demonstrated that subjects who smoked cannabis low in CBD showed impaired prose recall, while participants smoking cannabis high in CBD showed no memory impairment (Morgan, Schafer, Freeman, & Curran, 2010). In 2012, Morgan and colleagues conducted another study in 120 cannabis smokers, of which 66 were daily users and 54 were occasional users. Subjects were divided into groups depending on the presence or absence of CBD and on high versus low levels of THC in their hair. Psychosis-like symptoms, depression, anxiety, and memory were measured. Subjects whose hair contained CBD showed less psychotic-like symptoms when compared to subjects without CBD in their hair. Notably, this phenomenon was observed only in recreational users with high THC hair content. Moreover, higher THC levels in hair were associated with increased depression and anxiety, and with impaired prose recall and source memory, while recognition memory was improved in individuals with CBD present in their hair. This finding further corroborated the notion that CBD mitigates the cannabis-induced psychotic-like effects in recreational users and that high THC levels are detrimental for memory and psychological health (Morgan et al., 2012).

In a study including 1877 subjects regularly smoking their usual cannabis strain, a significant inverse relationship between CBD content in the cannabis itself and self-reported positive symptoms—but not negative symptoms or depression—was found (Schubart, Sommer, et al., 2011). Using magnetic resonance spectroscopy (H1-MRS), a study conducted in cannabis users and controls found a positive correlation between CBD concentration in hair and N-acetylaspartate (NAA)/total creatine (tCr) ratios (a marker of neuronal integrity) in the globus pallidus and putamen. Performance on a test of attention and concentration was also positively correlated with hair CBD levels (Hermann et al., 2007). A further correlation between the CBD content of cannabis and biomarkers relevant to schizophrenia was found in a study conducted by Demirakca et al. (2011) in cannabis users. The authors demonstrated an inverse correlation of the THC-to-CBD ratio with right hippocampal volume and showed that CBD levels positively correlated with hippocampal gray matter volume, a hallmark biomarker that has been consistently implicated in schizophrenia, thus indicating neurotoxic effects of THC and neuroprotective effects of CBD (Demirakca et al., 2011).

These studies converge on the conclusion that high CBD content in smoked cannabis correlates with a decreased risk of developing psychosis, fewer delusions and hallucinations, better cognitive abilities, and a delay in the onset of psychosis when compared to cannabis with low CBD content; by contrast high THC concentration does much the opposite.

Studies in Healthy Volunteers

Some evidence in healthy volunteers suggests the capacity of CBD to antagonize THC-induced psychotic effects as well as cognitive impairment (Table 5). Indeed, THC administered to healthy volunteers induces cognitive and

TABLE 5 Studies in Healthy Volunteers Administered CBD and THC

Measurement	THC Dose/ Route	CBD Dose/ Route	CBD Regime	Outcome	Reference
fMRI Paired associate learning task and objective/ subjective ratings of psychotic symptoms	10 mg (p.o.)	600 mg (p.o.)	Alone	No effect of CBD on cognition and psychological measurements	Bhattacharyya et al. (2009)
PANSS	1.25 mg (i.v.)	5 mg (i.v.)	Combination	CBD reduced THC-induced psychotic symptoms	Bhattacharyya et al. (2010)
FMRI visual Oddball detection paradigm	10 mg (p.o.)	600 mg (p.o.)	Oral	CBD augmented left caudate and hippocampal activation and attenuated right prefrontal activation	Bhattacharyya et al. (2012)
fMRI Go/no-go task	10 mg (p.o.)	600 mg (p.o.)	Alone	CBD deactivated the left temporal cortex and insula	Borgwardt et al. (2008)
Subjective euphoria, psychomotor function	THC 25 μg/kg (inhalation)	150 μg/kg (inhalation)	Alone and combined (simultaneous or prior to THC)	CBD reduced THC-induced euphoria when given simultaneously. No effects on THC-induced action by CBD pretreatment (30 min prior THC)	Dalton et al. (1976)
PANSS; HVLT-R; SSPS	1.5 mg (i.v.)	600 mg (p.o.)	Combination	CBD reduced THC-induced paranoia, memory impairment and positive symptoms	Englund et al. (2013)
Regional brain activation, electrodermal activity, fMRI gender discrimination task, objective/subjective ratings of anxiety.	10 mg (p.o.)	600 mg (p.o.)	Alone	CBD reduced amygdala L, anterior/posterior cingulate cortex activity, which correlated with skin conductance response. Trend for a reduction in anxiety following CBD	Fusar-Poli et al. (2009)
EEG MMN amplitude	10 mg (p.o.)	5.4 mg (p.o.)	Combination	CBD + THC, but not THC alone enhanced MMN amplitudes at central EEG electrodes	Juckel et al. (2007)

Pulse rate, psychological reactions, time production task	30 mg (p.o.)	15, 30, 60 mg (p.o.)	Alone and combined	CBD reduced THC-induced psychological effects, anxiety and impaired performance	Karniol et al. (1974)
Emotional processing task; BDI; STAI; SPQ; STWT	8 mg (inhalation)	16 mg (inhalation)	Alone and combined	CBD alone improved emotional facial affect recognition; THC was detrimental to the recognition of ambiguous faces. THC + CBD produced no impairment	Hindocha et al. (2015)
Subjective reports, cognitive task performance, EEG and ERP	1.8% vs 3.6% (inhalation)	0.2% vs 1.0% (inhalation)	Combined	No effects of CBD	Ilan et al. (2005)[a]
Binocular depth inversion	1 mg (p.o.)	200 mg (p.o.)	Alone and combined (simultaneous or prior to THC)	CBD improved THC-induced impairment in binocular depth inversion	Leweke et al. (2000)
Finger tapping task for psychomotor performance	10 mg (p.o.)	5.4 mg (p.o.)	Combination	CBD + THC, but not THC alone reduced right-hand tapping frequencies	Roser et al. (2009)
EEG P300 during choice reaction task	10 mg (p.o.)	5.4 mg (p.o.)	Combination	CBD had no effects on THC-induced P300 reduction	Roser et al. (2008)
fMRI visual and auditory processing	10 mg (p.o.)	600 mg (p.o.)	Alone	During auditory processing CBD was associated with activation in right temporal cortex. During visual processing CBD increased activation in the right occipital lobe	Winton-Brown et al. (2011)
Anxiety and physiological subjective effects	0.5 mg/kg (p.o.)	1 mg/kg (p.o.)	Alone and combined (simultaneous or prior to THC)	CBD reduced THC-induced anxiety and subjective effects	Zuardi et al. (1982)

Note: *BDI*, Beck Depression Inventory; *CBD*, cannabidiol; *EEG*, electroencephalography; *ERP*, event-related potential; *fMRI*, functional magnetic resonance imaging; *HVLT-R*, Hopkins Verbal Learning Task-revised; *MMN*, mismatch negativity; *PANNS*, Positive and Negative Syndrome Scale; *SPQ*, Schizotypal Personality Questionnaire; *SSPS*, State Social Paranoia Scale; *STAI*, Spielberger State Trait Anxiety Inventory; *STWT*, Spot the Word Test; *THC*, Δ9-tetrahydrocannabinol; *p.o.*, per os (oral); *i.v.*, intravenous.
aStudy performed in cannabis users.

psychological impairments, which are mitigated when THC is administered in combination with CBD (Karniol et al., 1974). Furthermore, subjects inhaling CBD (150 μg/kg) and THC (25 μg/kg) showed attenuated THC-induced euphoria (Dalton, Martz, Lemberger, Rodda, & Forney, 1976). Along these same lines of evidence, oral administration of 1 mg/kg CBD antagonized the anxiety and psychotic-like effects induced by 0.5 mg/kg THC (Zuardi, Shirakawa, Finkelfarb, & Karniol, 1982), and decreased the impairment in binocular depth inversion (Leweke, Schneider, Radwan, Schmidt, & Emrich, 2000) (a model of impaired perception during psychotic states) induced by nabilone, a synthetic THC derivative marketed under the name of Cesamet as a medication for nausea and vomiting caused by cancer chemotherapy. Moreover, episodic memory impairment (scored with the Hopkins Verbal Learning Task-revised) and paranoia induced by intravenous THC (1.5 mg) were antagonized by 600 mg oral CBD pretreatment (Englund et al., 2013). *C. sativa* extract containing both THC and CBD enhanced auditory evoked mismatch negativity (MMN) amplitude (Juckel, Roser, Nadulski, Stadelmann, & Gallinat, 2007), which is considerably lower in patients with schizophrenia (Light et al., 2012). By contrast, THC alone had no effect on MMN. Of note, MMN is a brain response to an odd stimulus within a sequence of stimuli; it reflects the ability to make comparisons between consecutive and similar stimuli and provides an electrophysiological hallmark of sensory learning and perceptual accuracy. Several fMRI studies conducted in healthy subjects have explored the effects of CBD on brain function, suggesting opposing effects of CBD and THC in cognitive domains typically impaired in schizophrenia, such as working memory, salience and emotional processing, and response inhibition. THC and CBD were shown to evoke differential brain activation of areas affected in schizophrenia such as the amygdala, hippocampus, striatum, prefrontal area, and auditory cortex (Bhattacharyya et al., 2009, 2010, 2012; Borgwardt et al., 2008; Fusar-Poli et al., 2009; Winton-Brown et al., 2011).

Altered codification of emotional expression has been reported in psychiatric conditions such as depression, anxiety, and schizophrenia (Phillips, Drevets, Rauch, & Lane, 2003), as well as in cannabis users (Hindocha et al., 2014; Platt, Kamboj, Morgan, & Curran, 2010), suggesting an involvement of the endocannabinoid system in emotional processing. This notion led Hindocha et al. (2015) to perform a study in 48 volunteers, aimed at investigating the effects of THC (8 mg) and CBD (16 mg), administered either alone or in combination, on facial emotion recognition. Subjects were selected for high and low frequency of cannabis use and for schizotypy. THC alone and combined THC + CBD increased feelings of being "stoned," and CBD was reported not to influence these feelings. Remarkably, CBD improved facial affect recognition as compared to placebo. By contrast, THC alone was detrimental for the recognition of ambiguous faces, whereas combined THC + CBD administration produced no impairment. Hence, CBD was not only beneficial for emotional processing, but it also reversed THC-induced deficits. Interestingly, the frequency of cannabis use was not found to play a significant role in the findings. Conversely, a study performed in 23 healthy cannabis users randomly allocated to a low (1.8%) or a high (3.6%) THC group, and low (0.2%) or high (1%) levels of CBD, showed that different levels of CBD did not counteract THC-induced detrimental effects on working memory and episodic memory, electroencephalography (EEG), or event-related potentials (ERPs) (Ilan, Gevins, Coleman, ElSohly, & de Wit, 2005). In the same manner, a prospective, crossover study performed in 20 healthy volunteers showed that CBD did not reverse THC-induced P300 amplitude reduction recorded at central, midline, frontal, and parietal electrodes during a choice reaction task (Roser et al., 2008). P300 is a cognitive ERP component that reflects attention and active working memory, and which is reduced in patients with schizophrenia. The authors could not determine whether the lack of an effect of CBD was due to neurotransmitter systems influencing the P300 not being affected by CBD, or the CBD doses being too low to influence this neurophysiological endophenotype.

Studies Among Individuals With Schizophrenia

Several clinical trials are currently investigating the effects of CBD in patients with psychosis, particularly schizophrenia (Table 6); however, the majority are currently recruiting or have not yet reported results. The first study reporting the effects of CBD on patients with schizophrenia was conducted in 1995 by Zuardi, Morais, Guimaraes, and Mechoulam (1995). They described a single case study of a 19-year-old patient with schizophrenia, who was administered CBD 1500 mg/day orally for four weeks and reported a subsequent improvement of symptomatology as measured with the Brief Psychiatric Rating Scale (BPRS). Later, the same author reported the cases of three treatment-resistant schizophrenia patients treated with escalating doses of CBD for four weeks (maximum 1280 mg/day orally). However, only one patient showed mild improvement on the BPRS after CBD monotherapy (Zuardi et al., 2006). The authors hypothesized that the treatment resistance in these patients and the low initial CBD doses might explain the negative findings. An open-label pilot study investigated the efficacy, tolerability, and safety of CBD on Parkinson's disease patients with psychotic symptoms. Six patients received CBD in flexible oral doses (starting dose: 150 mg/day) for four weeks in addition to their standard therapy. Psychotic symptoms evaluated by the BPRS and the Parkinson Psychosis Questionnaire showed a significant decrease with CBD treatment. Motor function and cognition were not affected and no adverse effects were observed. The authors suggested that CBD may be effective, safe, and well tolerated for the treatment of psychosis in the context of Parkinson's disease (Zuardi et al., 2009).

Selective attention was tested in 28 patients with schizophrenia using the Stroop Color Word Test. Subjects underwent two experimental sessions: the first drug-free, and the second after receiving a single oral administration of

TABLE 6 Studies in Individuals With Psychosis Administered Cannabidiol

Subjects	Measurement	CBD Dose/ Regime	Outcome	Reference
28 patients with schizophrenia	Stroop Color Word Test, BPRS, PANSS	Single doses of 300 or 600 mg p.o.	Patients treated with placebo and CBD 300 mg performed better than those who received CBD 600 mg	Hallak et al. (2010)
21 acute paranoid patients with schizophrenia, 21 controls. Phase II clinical trial of cannabidiol vs amisulpride	BPRS, PANSS	Up to 800 mg/day p.o. for 4 weeks	CBD was as effective as amisulpride in terms of improvement of symptomatology; CBD displayed superior side effect profile	Leweke et al. (2012)
One single case. 19-year-old woman with psychosis	BPRS	Up to 1500 mg/day p.o. for 26 days	Improvement of symptomatology, no side effects	Zuardi et al. (1995)
Three patients with treatment-resistant schizophrenia	BPRS	Up to 1280 mg/day p.o. for 4 weeks	Mild improvement of symptomatology of 1 patient, no side effects	Zuardi et al. (2006)
Six PD patients who had psychosis for at least 3 months	BPRS, PPQ	Up to 600 mg/day p.o. for 4 weeks	Improvement of symptomatology, no side effects	Zuardi et al. (2009)

Note: *BPRS*, Brief Psychiatric Rating Scale; *PD*, Parkinson's Disease; *PANNS*, Positive and Negative Syndrome Scale; *PPQ*, Parkinson Psychosis Questionnaire.

either 300mg or 600mg of CBD, or placebo. Comparison of the first to the second session showed improved performance in all three treatment groups, with subjects receiving placebo and CBD 300mg performing better than those administered CBD 600mg (Hallak et al., 2010). This led to the conclusion that CBD was not beneficial for performance in the Stroop Color Word Test in patients with schizophrenia. Leweke et al. conducted a subsequent clinical trial in which 20 schizophrenia patients received CBD (oral, 800mg/day for 4 weeks) and 19 received amisulpride, a D_2 and D_3 receptor antagonist. Both groups of patients showed improvement in the Positive and Negative Syndrome Scale (PANSS) and BPRS and, remarkably, the CBD group showed reduced extrapyramidal symptoms, weight gain, and prolactin levels. CBD administration also enhanced serum AEA levels; this increase in AEA led the authors to suggest that CBD may exert its antipsychotic properties by a moderate FAAH inhibition (Leweke et al., 2012).

In 2015, a phase II proof-of-concept clinical trial supported by GW Pharmaceuticals (UK) investigated the effect of six weeks of treatment with CBD (1000mg/day p.o.) as an add-on to an antipsychotic medication in 88 patients affected by schizophrenia or related disorders (ClinicalTrials.gov, 2015). Data are not reported in a peer-reviewed article, or in the United States (US) National Institutes of Health website, ClinicalTrials.gov, but the results are currently outlined in the website of GW Pharmaceuticals and have been recently presented at the 5th Biennial Schizophrenia International Research Society Conference (McGuire, Cubala, Vasile, Morrison, & Wright, 2016). It was reported that CBD was consistently superior to placebo with regard to psychopathology and showed no relevant undesired effects (GW-Pharmaceuticals, 2016). Positive and cognitive domains were improved, while negative symptoms were only marginally ameliorated as compared to placebo, but did reach statistical significance in patients treated with CBD in combination with an antipsychotic drug.

Currently available data obtained from epidemiological and clinical studies conducted in patients with schizophrenia support a potential role for CBD in the treatment of schizophrenia, with a desirable side effect profile. Of note, however, a recent systematic review suggested caution in interpreting these results, as clear-cut evidence of the antipsychotic effects of CBD is still absent due to the size and length of the studies to date (Pushpa-Rajah et al., 2015).

REGULATION OF MEDICAL MARIJUANA AND CANNABIS-DERIVED MEDICINES

The current regulation of the medical use of cannabis is mixed across countries, not always univocal, and certainly evolving. Medical use of cannabis is legalized in Canada, Austria, Belgium, Israel, Chile, Colombia, the Czech Republic, Finland, the Netherlands, Spain, and the United Kingdom (UK), and since 2015, cannabis has been legalized for medical use in 23 states and Washington D.C. in the US.

The American Food and Drug Administration (FDA) has neither recognized nor approved cannabis as a medication. Together with heroin, lysergic acid diethylamide (LSD), 3,4-methylenedioxymethamphetamine (ecstasy), methaqualone, and peyote, cannabis is classified by the Drug Enforcement Agency (DEA) as a Schedule I substance of the Controlled Substances Act (CSA). This qualifies cannabis as a drug having high potential for abuse, with no currently accepted medical use, and a lack of accepted safety for use under medical supervision. Two drugs containing synthetic cannabinoids strictly related to THC, namely Marinol (THC, dronabinol) and Cesamet (nabilone), received FDA approval and are currently controlled in Schedule III and II of the CSA.

In Canada, the Marijuana for Medical Purposes Regulations was enacted in 2013, facilitating the production of cannabis for medical uses and assuring quality and safety. Medical cannabis regulation in Europe is undergoing a wave of reforms, moving toward legalization. According to a draft law, Germany currently plans to establish a state cannabis agency to regulate its cultivation by releasing cultivation licenses, and to streamline rules for its distribution to treat seriously ill patients. In particular cases, the cost of cannabis can be covered by health insurance (Bundesministerium für Gesundheit, 2016). Italy has recently launched a new law proposing to decriminalize cultivation of cannabis for medical and research purposes (Law proposal C. 3864/2016). In the UK, the pharmaceutical company GW Pharmaceuticals received an exclusive license to cultivate cannabis for medicinal use in 2003, and in 2013, the THC/CBD-based nasal spray, Sativex, was rescheduled from Schedule I to IV (like diazepam and codeine) of the UK Drugs Act, hence allowing it to be prescribed by doctors (GW-Pharmaceuticals, 2013). This distinguished the THC/CBD-based drug from herbal cannabis, which remains in Schedule I of the Act.

REGULATION OF CBD

The legal status of CBD varies among countries and often remains a matter of legal interpretation. In Canada, CBD is a Schedule II drug in the Controlled Drugs and Substances Act (S.C. 1996, c. 19). In Australia, it is classified as a Schedule IV drug, and hence is a Prescription Only Medicine. Understanding the regulation of CBD use in the US at the federal level is not straightforward. The CSA does not specifically list CBD as a controlled substance in any Schedules. On the other hand, the DEA has stated that CBD is a cannabis derivative, and therefore it is a Schedule I drug and has its own DEA Code Number, 7372 (DEA, 2016). The FDA considers CBD to be a drug, and not a dietary supplement (FDA, 2016); hence any company raising medical claims about their CBD-based products must demonstrate FDA compliance.

In Europe, CBD is listed in the EU Cosmetic Ingredient Database (CosIng) with four functional claims as antioxidant, antiseborrheic, skin conditioning, and skin protecting (European Commission website, 2016a). The Cosmetic Ingredient Database "may also list ingredients known to be used in medicinal products. If, due to such ingredients, a product restores, corrects, or modifies physiological functions by exerting a pharmacological, immunological, or metabolic action, the product shall be qualified as a medicinal product. However, products that, while having an effect on the human body, do not significantly affect the metabolism and thus do not strictly modify the way in which it functions, may be qualified as cosmetic products" (European Commission website, 2016b). In the UK, CBD is marketed as an oral-mucosal spray formulation combined with THC (Sativex), and it is available by prescription for treating spasticity in multiple sclerosis. In 2015, CBD reached the European market as a nutritional supplement. It can be legally purchased online from the Satipharm website, a Swiss nutraceutical manufacturing subsidiary of an Australian firm, MMJ PhytoTech. The tablet contains a consistent 95% CBD, extracted from a high-CBD-content strain of the cannabis plant.

CONCLUSIONS AND NEEDED RESEARCH

Collectively, the available preclinical and clinical data suggest that CBD may represent a promising compound for the treatment of the symptoms of schizophrenia. Its potential ability to relieve the negative symptoms of the disease appears particularly interesting and certainly deserves further investigation, as current antipsychotic therapies effectively reduce positive but not negative or cognitive deficits. Of note, CBD's antipsychotic action is associated with reduced extrapyramidal side effects, thus supporting its remarkably safe profile in animal studies as well as in humans. However, the safety profile of CBD has been largely explored with acute administration experiments, and little preclinical or clinical research has focused on the safety and side effect profile of chronic CBD administration. Thus, given the chronic nature of schizophrenia and the need for continuous antipsychotic therapy, the response to CBD in patients with schizophrenia as well as the safety profile of chronic CBD treatment requires further exploration.

The molecular mechanisms underlying CBD's potential antipsychotic effects are currently under intense investigation and are still a matter of debate. One such mechanism could be its direct action on the endocannabinoid system. Indeed, CBD's ability to inhibit AEA degradation, thereby increasing AEA levels, has been associated with decreased psychotic symptoms in patients (Leweke et al., 2012), although more recent computational analysis and ligand displacement assays do not seem to confirm the ability of CBD to inhibit the AEA catabolic enzyme FAAH in humans (Elmes et al., 2015). Elmes et al. (2015) not only showed that human FAAH is not inhibited by CBD, but they also demonstrated that CBD inhibits AEA cellular uptake by competing for the fatty acid binding proteins (FABPs) that mediate the transport of AEA to FAAH. Hence, CBD-induced AEA increase may be explained by its ability to compete for binding to FABPs, thereby reducing the rate of AEA breakdown by FAAH. Increased levels of AEA in prodromal or first-episode psychosis (Volk & Lewis, 2016) have been suggested to represent an attempt to compensate for dysregulated brain DA transmission (Curran et al., 2016). However, it remains unclear whether AEA alterations are a compensatory mechanism or part of the disease itself. The fact that AEA is negatively correlated with psychotic symptoms suggests that the ability of CBD to enhance the levels of this endocannabinoid may be a plausible antipsychotic mechanism.

The weak partial antagonist action of CBD at CB_1 receptors (Thomas et al., 2007) has been suggested to contribute to its putative antipsychotic effects (Roser & Haussleiter, 2012). However, CBD has a low affinity at this receptor and presents an antagonistic profile against CB_1 receptor agonists, hence indicating functional CB_1 receptor signaling antagonism (Pertwee, 2008; Thomas et al., 2007). Indeed more recent findings suggest that there might be no direct interaction between CBD and the CB_1 receptor orthosteric site, but rather it may behave as a noncompetitive negative allosteric modulator of the CB_1 receptor (Laprairie, Bagher, Kelly, & Denovan-Wright, 2015). Given the weak activity at the CB_1 receptor, CBD antagonism at this receptor has been suggested to not be the main mechanism underlying its antipsychotic effects. Other mechanisms that could account for CBD's antipsychotic effects are its ability to activate 5-HT_{1A} receptors (Russo et al., 2005), an effect shared by the atypical antipsychotic aripiprazole (Sumiyoshi, 2012), as well as its antiinflammatory and neuroprotective properties (Gomes et al., 2015). In addition to these mechanisms, CBD's pharmacological actions also include inhibition of adenosine uptake, inverse CB_2 receptor agonism, G protein-coupled receptor 55 (GPR55) antagonism, agonism at peroxisome proliferator-activated receptor gamma (PPARγ), and activation of transient receptor potential vanilloid 1 (TRPV1) channels; but no studies thus far have investigated the possible involvement of these mechanisms in CBD's antipsychotic effects (see Table 2).

In conclusion, CBD represents an attractive compound for the possible treatment of schizophrenia, although assessing a chronic CBD administration regimen, human bioavailability, toxicity, increased sample sizes, and multiple doses should be objectives of further investigation to refine CBD's profile as a convincing and desirable antipsychotic agent.

Key Chapter Points

- Cannabis use can induce psychosis in normal individuals, exacerbate psychotic symptoms in individuals with schizophrenia, and facilitate the onset of schizophrenia in vulnerable individuals.
- The psychosis-inducing effects of cannabis are closely related to its THC-to-CBD ratio: high levels of THC are correlated with psychosis, whereas high CBD levels are linked to antipsychotic effects.
- Data from animal models provide evidence for an antipsychotic potential of CBD for the positive and possibly negative symptoms of schizophrenia. As opposed to current antipsychotics, its administration is associated with few side effects.

- Human investigations suggest CBD may be a promising candidate to treat psychosis; however, further clinical studies with larger samples and chronic CBD exposure are needed to warrant more solid conclusions on the efficacy, safety, and tolerability of CBD.
- The regulation of CBD is dependent on the laws of individual countries. European countries currently classify CBD as a dietary supplement, while in the US, its use, cultivation, and marketing are regulated as a Schedule I drug.

REFERENCES

Abi-Dargham, A., & Guillin, O. (2007). Integrating the neurobiology of schizophrenia. Preface. *International Review of Neurobiology*, *78*, xiii–xvi.

Almeida, V., Levin, R., Peres, F. F., Niigaki, S. T., Calzavara, M. B., Zuardi, A. W., et al. (2013). Cannabidiol exhibits anxiolytic but not antipsychotic property evaluated in the social interaction test. *Progress in Neuro-Psychopharmacology & Biological Psychiatry*, *41*, 30–35.

Bergamaschi, M. M., Queiroz, R. H., Zuardi, A. W., & Crippa, J. A. (2011). Safety and side effects of cannabidiol, a *Cannabis sativa* constituent. *Current Drug Safety*, *6*, 237–249.

Bhattacharyya, S., Crippa, J. A., Allen, P., Martin-Santos, R., Borgwardt, S., Fusar-Poli, P., et al. (2012). Induction of psychosis by Delta9-tetrahydrocannabinol reflects modulation of prefrontal and striatal function during attentional salience processing. *Archives of General Psychiatry*, *69*, 27–36.

Bhattacharyya, S., Fusar-Poli, P., Borgwardt, S., Martin-Santos, R., Nosarti, C., O'Carroll, C., et al. (2009). Modulation of mediotemporal and ventrostriatal function in humans by Delta9-tetrahydrocannabinol: A neural basis for the effects of *Cannabis sativa* on learning and psychosis. *Archives of General Psychiatry*, *66*, 442–451.

Bhattacharyya, S., Morrison, P. D., Fusar-Poli, P., Martin-Santos, R., Borgwardt, S., Winton-Brown, T., et al. (2010). Opposite effects of delta-9-tetrahydrocannabinol and cannabidiol on human brain function and psychopathology. *Neuropsychopharmacology*, *35*, 764–774.

Borgwardt, S. J., Allen, P., Bhattacharyya, S., Fusar-Poli, P., Crippa, J. A., Seal, M. L., et al. (2008). Neural basis of Delta-9-tetrahydrocannabinol and cannabidiol: Effects during response inhibition. *Biological Psychiatry*, *64*, 966–973.

Bundesministerium für Gesundheit (2016). Cannabis als Medizin. http://www.bmg.bund.de/ministerium/meldungen/2016/cannabisarzneimittel-kabinett.html (Accessed June 2016).

Calzavara, M. B., Medrano, W. A., Levin, R., Kameda, S. R., Andersen, M. L., Tufik, S., et al. (2009). Neuroleptic drugs revert the contextual fear conditioning deficit presented by spontaneously hypertensive rats: A potential animal model of emotional context processing in schizophrenia? *Schizophrenia Bulletin*, *35*, 748–759.

Campos, A. C., Moreira, F. A., Gomes, F. V., Del Bel, E. A., & Guimaraes, F. S. (2012). Multiple mechanisms involved in the large-spectrum therapeutic potential of cannabidiol in psychiatric disorders. *Philosophical Transactions of the Royal Society of London Series B, Biological Sciences*, *367*, 3364–3378.

Carli, M., Luschi, R., Garofalo, P., & Samanin, R. (1995). 8-OH-DPAT impairs spatial but not visual learning in a water maze by stimulating 5-HT1A receptors in the hippocampus. *Behavioural Brain Research*, *67*, 67–74.

ClinicalTrials.gov (2015). A Study of GWP42003 as Adjunctive Therapy in the First Line Treatment of Schizophrenia or Related Psychotic Disorder. In, p ClinicalTrials.gov Identifier: NCT02006628.

Curran, H. V., Freeman, T. P., Mokrysz, C., Lewis, D. A., Morgan, C. J., & Parsons, L. H. (2016). Keep off the grass? Cannabis, cognition and addiction. *Nature Reviews Neuroscience*, *17*, 293–306.

Dalterio, S., Steger, R., Mayfield, D., & Bartke, A. (1984). Early cannabinoid exposure influences neuroendocrine and reproductive functions in mice: II. Postnatal effects. *Pharmacology, Biochemistry, and Behavior*, *20*, 115–123.

Dalton, W. S., Martz, R., Lemberger, L., Rodda, B. E., & Forney, R. B. (1976). Influence of cannabidiol on delta-9-tetrahydrocannabinol effects. *Clinical Pharmacology and Therapeutics*, *19*, 300–309.

Drug Enforcement Administration DEA (2016). DEA Form 225—New Application for Registration. pp SCHEDULE 1 NARCOTIC & NON-NARCOTIC. http://www.deadiversion.usdoj.gov/drugreg/reg_apps/225/225_instruct.htm (Accessed June 2016).

Deiana, S., Watanabe, A., Yamasaki, Y., Amada, N., Arthur, M., Fleming, S., et al. (2012). Plasma and brain pharmacokinetic profile of cannabidiol (CBD), cannabidivarine (CBDV), Delta(9)-tetrahydrocannabivarin (THCV) and cannabigerol (CBG) in rats and mice following oral and intraperitoneal administration and CBD action on obsessive-compulsive behaviour. *Psychopharmacology*, *219*, 859–873.

Deiana, S., Watanabe, A., Yamasaki, Y., Amada, N., Kikuchi, T., Stott, C., et al. (2015). MK-801-induced deficits in social recognition in rats: reversal by aripiprazole, but not olanzapine, risperidone, or cannabidiol. *Behavioral Pharmacology*, *26*, 748–765.

Demirakca, T., Sartorius, A., Ende, G., Meyer, N., Welzel, H., Skopp, G., et al. (2011). Diminished gray matter in the hippocampus of cannabis users: Possible protective effects of cannabidiol. *Drug and Alcohol Dependence*, *114*, 242–245.

Di Forti, M., Morgan, C., Dazzan, P., Pariante, C., Mondelli, V., Marques, T. R., et al. (2009). High-potency cannabis and the risk of psychosis. *The British Journal of Psychiatry*, *195*, 488–491.

Di Forti, M., et al. (2014). Daily use, especially of high-potency cannabis, drives the earlier onset of psychosis in cannabis users. *Schizophrenia Bulletin*, *40*, 1509–1517.

Di Forti, M., et al. (2015). Proportion of patients in south London with first-episode psychosis attributable to use of high potency cannabis: A case-control study. *The Lancet Psychiatry*, *2*, 233–238.

Dixon, L., Haas, G., Weiden, P., Sweeney, J., & Frances, A. (1990). Acute effects of drug abuse in schizophrenic patients: Clinical observations and patients' self-reports. *Schizophrenia Bulletin*, *16*, 69–79.

D'Souza, D. C., Sewell, R. A., & Ranganathan, M. (2009). Cannabis and psychosis/schizophrenia: Human studies. *European Archives of Psychiatry and Clinical Neuroscience*, *259*, 413–431.

Elmes, M. W., Kaczocha, M., Berger, W. T., Leung, K., Ralph, B. P., Wang, L., et al. (2015). Fatty acid-binding proteins (FABPs) are intracellular carriers for Delta9-tetrahydrocannabinol (THC) and cannabidiol (CBD). *The Journal of Biological Chemistry*, *290*, 8711–8721.

Englund, A., Morrison, P. D., Nottage, J., Hague, D., Kane, F., Bonaccorso, S., et al. (2013). Cannabidiol inhibits THC-elicited paranoid symptoms and hippocampal-dependent memory impairment. *Journal of Psychopharmacology*, *27*, 19–27.

European Commission website (2016a). European Commission-Growth-Sectors-Cosmetics-CosIng. Ingredient: Cannabidiol. http://ec.europa.eu/growth/tools-databases/cosing/index.cfm?fuseaction=search.details_v2&id=93486 (Accessed June 2016).

European Commission website (2016b). European Commission-Growth-Sectors-Cosmetics-CosIng. Cosmetic ingredient database. Important notice. http://ec.europa.eu/growth/sectors/cosmetics/cosing/index_en.htm (Accessed June 2016).

Falls, D. L. (2003). Neuregulins: Functions, forms, and signaling strategies. *Experimental Cell Research*, *284*, 14–30.

Ferdinand, R. F., Sondeijker, F., van der Ende, J., Selten, J. P., Huizink, A., & Verhulst, F. C. (2005). Cannabis use predicts future psychotic symptoms, and vice versa. *Addiction (Abingdon, England)*, *100*, 612–618.

Food and Drug Administration (2016). FDA and Marijuana: Questions and Answers. http://www.fda.gov/NewsEvents/PublicHealthFocus/ucm421168.html (Accessed June 2016).

Ford, J. M., Morris, S. E., Hoffman, R. E., Sommer, I., Waters, F., McCarthy-Jones, S., et al. (2014). Studying hallucinations within the NIMH RDoC framework. *Schizophrenia Bulletin*, *40*(Suppl. 4), S295–S304.

Fusar-Poli, P., Crippa, J. A., Bhattacharyya, S., Borgwardt, S. J., Allen, P., Martin-Santos, R., et al. (2009). Distinct effects of {delta}9-tetrahydrocannabinol and cannabidiol on neural activation during emotional processing. *Archives of General Psychiatry*, *66*, 95–105.

Gomes, F. V., Issy, A. C., Ferreira, F. R., Viveros, M. P., Del Bel, E. A., & Guimaraes, F. S. (2014). Cannabidiol attenuates sensorimotor gating disruption and molecular changes induced by chronic antagonism of NMDA receptors in mice. *International Journal of Neuropsychopharmacology*, *18*(5), http://dx.doi.org/10.1093/ijnp/pyu041.

Gomes, F. V., Llorente, R., Del Bel, E. A., Viveros, M. P., Lopez-Gallardo, M., & Guimaraes, F. S. (2015). Decreased glial reactivity could be involved in the antipsychotic-like effect of cannabidiol. *Schizophrenia Research*, *164*, 155–163.

Green, B., Young, R., & Kavanagh, D. (2005). Cannabis use and misuse prevalence among people with psychosis. *The British Journal of Psychiatry*, *187*, 306–313.

Gururajan, A., Taylor, D. A., & Malone, D. T. (2010). Effect of testing conditions on the propsychotic action of MK-801 on prepulse inhibition, social behaviour and locomotor activity. *Physiology & Behavior*, *99*, 131–138.

Gururajan, A., Taylor, D. A., & Malone, D. T. (2011). Effect of cannabidiol in a MK-801-rodent model of aspects of schizophrenia. *Behavioural Brain Research*, *222*, 299–308.

Gururajan, A., Taylor, D. A., & Malone, D. T. (2012). Cannabidiol and clozapine reverse MK-801-induced deficits in social interaction and hyperactivity in Sprague-Dawley rats. *Journal of Psychopharmacology*, *26*, 1317–1332.

GW-Pharmaceuticals (2013). GW Pharmaceuticals cannabinoid-medicine Sativex moved to Schedule 4 of UK Drugs Act. http://www.gwpharm.com/gw%20pharmaceuticals%20cannabinoid-medicine%20sativex%20moved%20to%20schedule%204%20of%20uk%20drugs%20act.aspx (Accessed June 2016).

GW-Pharmaceuticals (2016). GW Pharmaceuticals Announces Positive Proof of Concept Data in Schizophrenia. http://www.gwpharm.com/GW%20Pharmaceuticals%20Announces%20Positive%20Proof%20of%20Concept%20Data%20in%20Schizophrenia.aspx (Accessed June 2016).

Hallak, J. E., Machado-de-Sousa, J. P., Crippa, J. A., Sanches, R. F., Trzesniak, C., Chaves, C., et al. (2010). Performance of schizophrenic patients in the Stroop Color Word Test and electrodermal responsiveness after acute administration of cannabidiol (CBD). *Revista Brasileira de Psiquiatria (Sao Paulo, Brazil: 1999)*, *32*, 56–61.

Hardwick, S., & King, L. A. (2008). *Home Office cannabis potency study 2008*. London: Home Office Scientific Development Branch.

Hermann, D., Sartorius, A., Welzel, H., Walter, S., Skopp, G., Ende, G., et al. (2007). Dorsolateral prefrontal cortex N-acetylaspartate/total creatine (NAA/tCr) loss in male recreational cannabis users. *Biological Psychiatry*, *61*, 1281–1289.

Hindocha, C., Freeman, T. P., Schafer, G., Gardener, C., Das, R. K., Morgan, C. J., et al. (2015). Acute effects of delta-9-tetrahydrocannabinol, cannabidiol and their combination on facial emotion recognition: A randomised, double-blind, placebo-controlled study in cannabis users. *European Neuropsychopharmacology : The Journal of the European College of Neuropsychopharmacology*, *25*, 325–334.

Hindocha, C., Wollenberg, O., Carter Leno, V., Alvarez, B. O., Curran, H. V., & Freeman, T. P. (2014). Emotional processing deficits in chronic cannabis use: A replication and extension. *Journal of Psychopharmacology*, *28*, 466–471.

Hollis, C., Groom, M. J., Das, D., Calton, T., Bates, A. T., Andrews, H. K., et al. (2008). Different psychological effects of cannabis use in adolescents at genetic high risk for schizophrenia and with attention deficit/hyperactivity disorder (ADHD). *Schizophrenia Research*, *105*, 216–223.

Ibeas Bih, C., Chen, T., Nunn, A. V., Bazelot, M., Dallas, M., & Whalley, B. J. (2015). Molecular targets of cannabidiol in neurological disorders. *Neurotherapeutics: The Journal of the American Society for Experimental NeuroTherapeutics*, *12*, 699–730.

Ilan, A. B., Gevins, A., Coleman, M., ElSohly, M. A., & de Wit, H. (2005). Neurophysiological and subjective profile of marijuana with varying concentrations of cannabinoids. *Behavioural Pharmacology*, *16*, 487–496.

Juckel, G., Roser, P., Nadulski, T., Stadelmann, A. M., & Gallinat, J. (2007). Acute effects of Delta9-tetrahydrocannabinol and standardized cannabis extract on the auditory evoked mismatch negativity. *Schizophrenia Research*, *97*, 109–117.

Karniol, I. G., Shirakawa, I., Kasinski, N., Pfeferman, A., & Carlini, E. A. (1974). Cannabidiol interferes with the effects of delta 9-tetrahydrocannabinol in man. *European Journal of Pharmacology*, *28*, 172–177.

Krystal, J. H., D'Souza, D. C., Madonick, S., & Petrakis, I. L. (1999). Toward a rational pharmacotherapy of comorbid substance abuse in schizophrenic patients. *Schizophrenia Research*, *35*(Suppl.), S35–S49.

Kuepper, R., van Os, J., Lieb, R., Wittchen, H. U., Hofler, M., & Henquet, C. (2011). Continued cannabis use and risk of incidence and persistence of psychotic symptoms: 10 year follow-up cohort study. *BMJ (Clinical Research Edition)*, *342*, d738.

Laprairie, R. B., Bagher, A. M., Kelly, M. E., & Denovan-Wright, E. M. (2015). Cannabidiol is a negative allosteric modulator of the cannabinoid CB1 receptor. *British Journal of Pharmacology*, *172*, 4790–4805.

Levin, R., Almeida, V., Peres, F. F., Calzavara, M. B., da Silva, N. D., Suiama, M. A., et al. (2012). Antipsychotic profile of cannabidiol and rimonabant in an animal model of emotional context processing in schizophrenia. *Current Pharmaceutical Design*, *18*, 4960–4965.

Levin, R., Calzavara, M. B., Santos, C. M., Medrano, W. A., Niigaki, S. T., & Abilio, V. C. (2011). Spontaneously Hypertensive Rats (SHR) present deficits in prepulse inhibition of startle specifically reverted by clozapine. *Progress in Neuro-Psychopharmacology & Biological Psychiatry*, *35*, 1748–1752.

Levin, R., Peres, F. F., Almeida, V., Calzavara, M. B., Zuardi, A. W., Hallak, J. E., et al. (2014). Effects of cannabinoid drugs on the deficit of prepulse inhibition of startle in an animal model of schizophrenia: The SHR strain. *Frontiers in Pharmacology*, *5*, 10.

Leweke, F. M., Piomelli, D., Pahlisch, F., Muhl, D., Gerth, C. W., Hoyer, C., et al. (2012). Cannabidiol enhances anandamide signaling and alleviates psychotic symptoms of schizophrenia. *Translational Psychiatry*, *2*, e94.

Leweke, F. M., Schneider, U., Radwan, M., Schmidt, E., & Emrich, H. M. (2000). Different effects of nabilone and cannabidiol on binocular depth inversion in Man. *Pharmacology, Biochemistry and Behavior*, *66*, 175–181.

Light, G. A., Swerdlow, N. R., Rissling, A. J., Radant, A., Sugar, C. A., Sprock, J., et al. (2012). Characterization of neurophysiologic and neurocognitive biomarkers for use in genomic and clinical outcome studies of schizophrenia. *PLoS ONE*, *7*(7), e39434.

Long, L. E., Chesworth, R., Huang, X. F., McGregor, I. S., Arnold, J. C., & Karl, T. (2010). A behavioural comparison of acute and chronic delta9-tetrahydrocannabinol and cannabidiol in C57BL/6JArc mice. *The International Journal of Neuropsychopharmacology*, *13*, 861–876.

Long, L. E., Chesworth, R., Huang, X. F., Wong, A., Spiro, A., McGregor, I. S., et al. (2012). Distinct neurobehavioural effects of cannabidiol in transmembrane domain neuregulin 1 mutant mice. *PLoS ONE*, *7*, e34129.

Long, L. E., Malone, D. T., & Taylor, D. A. (2006). Cannabidiol reverses MK-801-induced disruption of prepulse inhibition in mice. *Neuropsychopharmacology*, *31*, 795–803.

Manseau, M. W., & Goff, D. C. (2015). Cannabinoids and schizophrenia: Risks and therapeutic potential. *Neurotherapeutics*, *12*, 816–824.

Matsuyama, S. S., & Fu, T. K. (1981). In vivo cytogenetic effects of cannabinoids. *Journal of Clinical Psychopharmacology*, *1*, 135–140.

McGrath, J., Saha, S., Chant, D., & Welham, J. (2008). Schizophrenia: A concise overview of incidence, prevalence, and mortality. *Epidemiologic Reviews*, *30*, 67–76.

McGuire, P. R., Cubala, W. J., Vasile, D., Morrison, P., & Wright, S. (2016). A double-blind, randomised, placebo-controlled, parallel group trial of cannabidiol as adjunctive therapy in the first line treatment of schizophrenia or related psychotic disorder. In *5th Biennial Schizophrenia International Research Society Conference*.Firenze: NPJ Schizophrenia.

Mechoulam, R., & Parker, L. A. (2013). The endocannabinoid system and the brain. *Annual Review of Psychology*, *64*, 21–47.

Meyer, U. (2013). Developmental neuroinflammation and schizophrenia. *Progress in Neuro-Psychopharmacology & Biological Psychiatry*, *42*, 20–34.

Moreira, F. A., & Guimaraes, F. S. (2005). Cannabidiol inhibits the hyperlocomotion induced by psychotomimetic drugs in mice. *European Journal of Pharmacology*, *512*, 199–205.

Morgan, C. J., & Curran, H. V. (2008). Effects of cannabidiol on schizophrenia-like symptoms in people who use cannabis. *The British Journal of Psychiatry*, *192*, 306–307.

Morgan, C. J., Gardener, C., Schafer, G., Swan, S., Demarchi, C., Freeman, T. P., et al. (2012). Subchronic impact of cannabinoids in street cannabis on cognition, psychotic-like symptoms and psychological well-being. *Psychological Medicine*, *42*, 391–400.

Morgan, C. J., Schafer, G., Freeman, T. P., & Curran, H. V. (2010). Impact of cannabidiol on the acute memory and psychotomimetic effects of smoked cannabis: Naturalistic study: Naturalistic study [corrected]. *The British Journal of Psychiatry*, *197*, 285–290.

Pedrazzi, J. F., Issy, A. C., Gomes, F. V., Guimaraes, F. S., & Del-Bel, E. A. (2015). Cannabidiol effects in the prepulse inhibition disruption induced by amphetamine. *Psychopharmacology*, *232*, 3057–3065.

Peres, F. F., Diana, M. C., Suiama, M. A., Justi, V., Almeida, V., Bressan, R. A., et al. (2016). Peripubertal treatment with cannabidiol prevents the emergence of psychosis in an animal model of schizophrenia. *Schizophrenia Research*, *172*, 220–221.

Pertwee, R. G. (2008). The diverse CB1 and CB2 receptor pharmacology of three plant cannabinoids: Delta9-tetrahydrocannabinol, cannabidiol and delta9-tetrahydrocannabivarin. *British Journal of Pharmacology*, *153*, 199–215.

Phillips, M. L., Drevets, W. C., Rauch, S. L., & Lane, R. (2003). Neurobiology of emotion perception II: Implications for major psychiatric disorders. *Biological Psychiatry*, *54*, 515–528.

Platt, B., Kamboj, S., Morgan, C. J., & Curran, H. V. (2010). Processing dynamic facial affect in frequent cannabis-users: Evidence of deficits in the speed of identifying emotional expressions. *Drug and Alcohol Dependence*, *112*, 27–32.

Pushpa-Rajah, J. A., McLoughlin, B. C., Gillies, D., Rathbone, J., Variend, H., Kalakouti, E., et al. (2015). Cannabis and schizophrenia. *Schizophrenia Bulletin*, *41*, 336–337.

Rosenkrantz, H., Fleischman, R. W., & Grant, R. J. (1981). Toxicity of short-term administration of cannabinoids to rhesus monkeys. *Toxicology and Applied Pharmacology*, *58*, 118–131.

Roser, P., & Haussleiter, I. S. (2012). Antipsychotic-like effects of cannabidiol and rimonabant: Systematic review of animal and human studies. *Current Pharmaceutical Design, 18*, 5141–5155.

Roser, P., Gallinat, J., Weinberg, G., Juckel, G., Gorynia, I., & Stadelmann, A. M. (2009). Psychomotor performance in relation to acute oral administration of Δ^9-tetrahydrocannabinol and standardized cannabis extract in healthy human subjects. *European Archives of Psychiatry and Clinical Neuroscience, 259*(5), 284–292.

Roser, P., Juckel, G., Rentzsch, J., Nadulski, T., Gallinat, J., & Stadelmann, A. M. (2008). Effects of acute oral Delta9-tetrahydrocannabinol and standardized cannabis extract on the auditory P300 event-related potential in healthy volunteers. *European Neuropsychopharmacology: The Journal of the European College of Neuropsychopharmacology, 18*, 569–577.

Rottanburg, D., Robins, A. H., Ben-Arie, O., Teggin, A., & Elk, R. (1982). Cannabis-associated psychosis with hypomanic features. *Lancet (London, England), 2*, 1364–1366.

Russo, E. B., Burnett, A., Hall, B., & Parker, K. K. (2005). Agonistic properties of cannabidiol at 5-HT1a receptors. *Neurochemical Research, 30*, 1037–1043.

Salgado, J. V., & Sandner, G. (2013). A critical overview of animal models of psychiatric disorders: Challenges and perspectives. *Revista Brasileira de Psiquiatria (Sao Paulo, Brazil : 1999), 35*(Suppl. 2), S77–S81.

Schubart, C. D., Boks, M. P., Breetvelt, E. J., van Gastel, W. A., Groenwold, R. H., Ophoff, R. A., et al. (2011a). Association between cannabis and psychiatric hospitalization. *Acta Psychiatrica Scandinavica, 123*, 368–375.

Schubart, C. D., Sommer, I. E., Fusar-Poli, P., de Witte, L., Kahn, R. S., & Boks, M. P. (2014). Cannabidiol as a potential treatment for psychosis. *European Neuropsychopharmacology: The Journal of the European College of Neuropsychopharmacology, 24*, 51–64.

Schubart, C. D., Sommer, I. E., van Gastel, W. A., Goetgebuer, R. L., Kahn, R. S., & Boks, M. P. (2011b). Cannabis with high cannabidiol content is associated with fewer psychotic experiences. *Schizophrenia Research, 130*, 216–221.

Sumiyoshi, T. (2012). Serotonin 1A receptors in the action of antipsychotic drugs: Comment on 'Measurement of the serotonin 1A receptor availability in patients with schizophrenia during treatment with the antipsychotic medication ziprasidone' by Frankle et al. 2011;25(6):734–743. *Journal of Psychopharmacology, 26*, 1283–1284.

Swerdlow, N. R., Geyer, M. A., & Braff, D. L. (2001). Neural circuit regulation of prepulse inhibition of startle in the rat: Current knowledge and future challenges. *Psychopharmacology, 156*, 194–215.

Tandon, R., Keshavan, M. S., & Nasrallah, H. A. (2008). Schizophrenia, "Just the Facts": What we know in 2008 part 1: Overview. *Schizophrenia Research, 100*, 4–19.

Thomas, A., Baillie, G. L., Phillips, A. M., Razdan, R. K., Ross, R. A., & Pertwee, R. G. (2007). Cannabidiol displays unexpectedly high potency as an antagonist of CB1 and CB2 receptor agonists in vitro. *British Journal of Pharmacology, 150*, 613–623.

Volk, D. W., & Lewis, D. A. (2016). The role of endocannabinoid signaling in cortical inhibitory neuron dysfunction in schizophrenia. *Biological Psychiatry, 79*, 595–603.

Winton-Brown, T. T., Allen, P., Bhattacharyya, S., Borgwardt, S. J., Fusar-Poli, P., Crippa, J. A., et al. (2011). Modulation of auditory and visual processing by delta-9-tetrahydrocannabinol and cannabidiol: An FMRI study. *Neuropsychopharmacology, 36*, 1340–1348.

Zamberletti, E., Rubino, T., & Parolaro, D. (2012). The endocannabinoid system and schizophrenia: Integration of evidence. *Current Pharmaceutical Design, 18*, 4980–4990.

Zammit, S., Allebeck, P., Andreasson, S., Lundberg, I., & Lewis, G. (2002). Self reported cannabis use as a risk factor for schizophrenia in Swedish conscripts of 1969: Historical cohort study. *BMJ (Clinical Research Edition)*, *325*, 1199.

Zuardi, A. W., Crippa, J. A., Hallak, J. E., Bhattacharyya, S., Atakan, Z., Martin-Santos, R., et al. (2012). A critical review of the antipsychotic effects of cannabidiol: 30 years of a translational investigation. *Current Pharmaceutical Design*, *18*, 5131–5140.

Zuardi, A. W., Crippa, J. A., Hallak, J. E., Pinto, J. P., Chagas, M. H., Rodrigues, G. G., et al. (2009). Cannabidiol for the treatment of psychosis in Parkinson's disease. *Journal of Psychopharmacology*, *23*, 979–983.

Zuardi, A. W., Hallak, J. E., Dursun, S. M., Morais, S. L., Sanches, R. F., Musty, R. E., et al. (2006). Cannabidiol monotherapy for treatment-resistant schizophrenia. *Journal of Psychopharmacology*, *20*, 683–686.

Zuardi, A. W., Morais, S. L., Guimaraes, F. S., & Mechoulam, R. (1995). Antipsychotic effect of cannabidiol. *The Journal of Clinical Psychiatry*, *56*, 485–486.

Zuardi, A. W., Rodrigues, J. A., & Cunha, J. M. (1991). Effects of cannabidiol in animal models predictive of antipsychotic activity. *Psychopharmacology*, *104*, 260–264.

Zuardi, A. W., Shirakawa, I., Finkelfarb, E., & Karniol, I. G. (1982). Action of cannabidiol on the anxiety and other effects produced by delta 9-THC in normal subjects. *Psychopharmacology*, *76*, 245–250.

FURTHER READING

Glass, M., Dragunow, M., & Faull, R. L. (1997). Cannabinoid receptors in the human brain: A detailed anatomical and quantitative autoradiographic study in the fetal, neonatal and adult human brain. *Neuroscience*, *77*, 299–318.

Leweke, F. M., Giuffrida, A., Koethe, D., Schreiber, D., Nolden, B. M., Kranaster, L., et al. (2007). Anandamide levels in cerebrospinal fluid of first-episode schizophrenic patients: Impact of cannabis use. *Schizophrenia Research*, *94*, 29–36.

Index

Note: Page numbers followed by *f* indicate figures and *t* indicate tables.

D

T

U

W

Printed in the United States
By Bookmasters